中兽医术语

TERMINOLOGIES OF CHINESE VETERINARY MEDICINE

辛盛鹏　史万玉　主编
中国兽医协会　组编

中国农业科学技术出版社

图书在版编目（CIP）数据

中兽医术语 / 辛盛鹏，史万玉主编. -- 北京：中国农业科学技术出版社，2024.10. -- ISBN 978-7-5116-7141-7

Ⅰ. S853-61

中国国家版本馆 CIP 数据核字第 2024SX1310 号

责任编辑　张　羽
责任校对　王　彦
责任印制　姜义伟　王思文

出 版 者	中国农业科学技术出版社
	北京市中关村南大街 12 号　邮编：100081
电　　话	（010）82109705（编辑室）（010）82106624（发行部）
	（010）82109709（读者服务部）
网　　址	https://castp.caas.cn
经 销 者	各地新华书店
印 刷 者	北京地大彩印有限公司
开　　本	185 mm×260 mm　1/16
印　　张	21.875
字　　数	420 千字
版　　次	2024 年 10 月第 1 版　2024 年 10 月第 1 次印刷
定　　价	198.00 元

◆ 版权所有·侵权必究 ◆

《中兽医术语》
编委会

主　编： 辛盛鹏　史万玉

副主编： 刘秀丽　齐思锦　赵兴华　刘　祥　王　霄

编　者：（按姓氏笔画排序）

王　霄	王帅玉	王贵波	王洪宇	王晓丹	车　花
牛红日	仇正英	史万玉	冯亚楠	朱怡萱	任庆贤
刘　祥	刘秀丽	刘家国	刘朝阳	刘翠艳	齐思锦
孙书旺	杜西翠	李　杨	李云波	李定刚	李建喜
杨　倩	谷鹏飞	辛盛鹏	辛蕊华	张　妍	张　铁
张　璐	张翠霞	苑方重	范　开	范迎赛	林珈好
孟佳成	赵千惠	赵兴华	郝智慧	胡宇声	宣秋希
宫新城	啜亚南	崔宇擎	韩春杨	韩振兴	管　朔

主　审： 史万玉

序

在中华文明的浩瀚长河中，中兽医学作为中华医学宝库中的璀璨明珠，古老而神秘，独特且多彩，跨越千年，历久弥新。它不仅承载着诊治动物疾病的智慧，更蕴含着人与自然和谐共生的深邃哲理，为我们的人生观和价值观提供了宝贵的启示。长久以来，我们都听说过"望闻问切""阴阳五行""气血津液""八纲辨证"等中医理论术语，却知其然不知其所以然，甚至将中医、中兽医与"玄学"混为一谈。

作为一名对中兽医学怀有浓厚兴趣的兽医工作者，当我拿到这本《中兽医术语》时，心中涌动的不仅是期待，更有对中兽医学未来的无限憧憬。术语的专业解释是指在特定学科领域用来表示概念的称谓的集合，我的理解是，在特定领域对一些特定事物的统一称呼和共识。术语是一个学科的基础，它可以提供明确、精准的定义和沟通方式，保证学科内的知识传递与交流的准确性和效率，通过规范化的语言符号，帮助从业者精确表述复杂的概念、理论和过程，避免歧义和误解。

《中兽医术语》基于中国兽医协会团体标准，汇集了中兽医基础理论和中兽医临床诊疗各方面共九章内容，名词术语共2 400余条。书中的每一条术语、每一个解释，都是对中兽医药深厚文化的尊重和传承。统一的中兽医术语，使中兽医学概念更加清晰，阐述更加标准，应用更加规范和统一。对于中兽医的教学、实践和发展而言，标准的术语是必不可少的工具。在教学过程中，准确的术语能够帮助学生清晰地理解中兽医的理论和概念，建立正确的思维模式，提高教学质量；在临床诊疗中，规范的术语能够确保兽医之间的沟通顺畅，避免因术语理解的差异而导致的误诊、误治，提高诊疗的准确性和有效性；在发展过程中，通过明确术语的含义和用法，学科的基础理论和概念得以稳固，为后续的研究和标准化发展提供坚实的基础，才能共同推动中兽医学和动物医学的发展；在国际交流中，统一的术语成为了连接中国与世界的桥梁，让全球中兽医爱好者与从业者更精准地共享中国智慧与经验。中国是中兽医理论和实践的发祥地，标准化的中兽医也将吸引更多的中兽医爱好者来中国学习交流，传播中国文化。规范的术语体系，将对中兽医学理论与技术的传承、弘扬、推广做出巨大贡献。展望未来，中国的标准将成为世界的标准。

《中兽医术语》凝聚着众多中兽医专家学者的智慧与心血，是为中兽医学发展献上的一份厚礼。它像一盏明灯，指引着中兽医学科研究与发展前进的道路；像一座桥梁，连接着中兽医学与其他学科的有效交流与合作；像一把钥匙，打开渴望深入中兽医领域的读者通往那深奥的传统知识宝库的大门。希望大家喜欢它！

先睹为快，欣然序之。

国务院参事、农业农村部原副部长：

2024年10月

前 言

中兽医学与中医学一脉相承，是祖国医学遗产的重要组成部分。数千年来的中兽医临床实践和理论探索形成了独特的理论体系和丰富的诊治手段，有效地指导着动物疾病防治临床实践。在科学技术突飞猛进、国际学术交流日益频繁的当代，中兽医学越来越受到世界兽医同行的青睐。为了更好地继承和发展中兽医学，满足当前中兽医诊疗、教学、科研、管理、出版及国内外学术交流的需要，有必要遵循中兽医学理论体系，建立统一、科学的中兽医术语标准。

为使中兽医学相关术语形成规范、统一的标准，推动中兽医学的传承与发展，中国兽医协会组织中兽医学研究领域相关单位和知名专家制定了中兽医术语的九项团体标准（T/CVMA 139.1～T/CVMA 139.9）。为更好地在兽医行业内推广和应用相关中兽医术语，根据有关知识应用、出版要求及国内外学术交流的需求，在这九项团体标准基础上，整理形成了本书。本书主要内容包括中兽医学总论与阴阳五行、脏腑与气血津液精、经络、病因病机、治则、疾病、证候、治法、针灸9个部分，规定了总论与阴阳五行、脏腑和气血津液精、经络、病因病机、治未病与治则部分等中兽医基础理论以及疾病、证候、治法、针灸等临床诊疗部分的常用术语及其定义，共计2 461条，并按中兽医认识动物疾病和疾病发生发展的规律进行分类，基本术语和疾病的确定以中兽医为主，在符合中兽医学理论体系和临床实践的前提下，收录了部分经改进、新创和分化的病名和术语，以反映中兽医学不断发展的成熟内容。本书中部分基础理论术语引用了《中兽医基础理论》（GB/T 42953—2023）中的一些术语定义，并结合业务实践对部分引用术语进行了适当修改，对相关术语的具体引用来源和修改情况都进行了注明。同时本书列有部分术语的同义词，在术语应用和临床诊疗实践中，应尽量使用正名，避免使用同义词，以避免因名称不同造成的理解偏差和使用错误。

随着中兽医学理论、诊疗技术的不断发展，以及畜牧兽医行业对中兽医的需求增加，我们将对中兽医术语的内容进行增补和修订，同时鉴于中兽医学理论渊博，辨证施治的动物品种较多，且实践性强，需要我们不断地去拓展和发掘，坚持不懈地开展标准化、规范化工作，使中兽医术语内容不断充实和完善。

由于中兽医学发展历史漫长，不同地域的兽医在应用和理解这些术语的过程中多少有一些差异，因此，在术语标准制定过程中虽然所有术语均有文献出处，但由于编撰时间较紧，参考资料有限，书中难免有疏漏和不足之处，敬请广大同行和读者予以指正并提出宝贵意见，以便在今后修订时加以改正和完善。

编　者
2024 年 9 月

目 录

第一章　基础理论总论与阴阳五行 … 1
1.1 总论 … 1
1.1.1 中兽医学 Chinese veterinary medicine（CVM）… 1
1.2 阴阳类 … 2
1.2.1 阴阳学说 yin-yang theory … 2
1.3 五行类 … 4
1.3.1 五行学说 five-element theory … 4
1.3.2 五行 five elements … 4
1.3.3 五时 five periods of year … 4
1.3.4 五气 five qi … 5
1.3.5 五化 five changes … 5
1.3.6 五色 five colors … 5
1.3.7 五味 five flavors … 5
1.3.8 五音 five notes … 5
1.3.9 五官 five sense organs … 5
1.3.10 五方（宫）five directions … 5
1.3.11 五行相生 five elements mutually generate … 5
1.3.12 五行相克 five elements mutually restrain … 6
1.3.13 五行相乘 five elements mutually over-restrain one another … 6
1.3.14 五行相侮 five elements mutually counter-restrain one another … 7
1.3.15 五行生克 generation and restriction among five elements … 7
1.3.16 五行乘侮 over-restriction and counter-restriction among five elements … 7
1.3.17 五行制化 restriction and generation among five elements … 7
1.3.18 五行胜复 alternate preponderance among five elements … 7
1.3.19 亢害承制 harmful hyperactivity checked for harmony … 7

第二章　脏腑与气血津液精 … 8
2.1 脏腑类 … 8
2.1.1 藏象 visceral manifestation … 8
2.1.2 脏腑 zang-fu … 8
2.1.3 脏腑学说 theory of visceral state … 8
2.1.4 五脏 five zang organs … 8
2.1.5 六腑 six fu organs … 16

  2.1.6 小肠 small intestine ………………………………………………… 18
  2.1.7 大肠 large intestine ………………………………………………… 19
  2.1.8 膀胱 bladder ……………………………………………………… 19
  2.1.9 三焦 san jiao ……………………………………………………… 19
  2.1.10 奇恒之腑 extraordinary fu organs …………………………… 20
 2.2 气血津液精类 …………………………………………………………… 21
  2.2.1 气 qi ……………………………………………………………… 21
  2.2.2 血 blood …………………………………………………………… 22
  2.2.3 津液 body fluids ………………………………………………… 22
  2.2.4 精 essence ………………………………………………………… 23
  2.2.5 气为血帅 qi is the commander of the blood …………………… 23
  2.2.6 血为气母 blood is the mother of qi …………………………… 23
  2.2.7 气能生津 qi engenders body fluids …………………………… 23
  2.2.8 气能行津 qi circulates body fluids …………………………… 24
  2.2.9 气能摄津 qi controls body fluids ……………………………… 24
  2.2.10 津能载气 body fluids carries qi ……………………………… 24
  2.2.11 津血同源 body fluids and blood share the same source …… 24
  2.2.12 精血同源 essence and blood share the same source ………… 24
  2.2.13 血汗同源 blood and sweat share the same source …………… 24

第三章　经　络 …………………………………………………………… 25
 3.1 常用经络术语 …………………………………………………………… 25
  3.1.1 经络学说 meridian theory ……………………………………… 25
  3.1.2 经气 qi of meridians ……………………………………………… 25
  3.1.3 经隧 meridian passage …………………………………………… 25
  3.1.4 十二经脉 twelve meridians ……………………………………… 25
  3.1.5 十二经别 twelve divergent meridians ………………………… 27
  3.1.6 十二经筋 twelve muscle regions ……………………………… 27
  3.1.7 十二皮部 twelve cutaneous regions …………………………… 27
  3.1.8 奇经八脉 eight extraordinary meridians ……………………… 27
  3.1.9 十五络脉 fifteen collaterals …………………………………… 28
  3.1.10 孙络 tertiary collaterals ……………………………………… 28
  3.1.11 浮络 floating collaterals ……………………………………… 28
  3.1.12 阳脉 yang meridian …………………………………………… 28
  3.1.13 阴脉 yin meridian ……………………………………………… 28
  3.1.14 十二经脉流注次序 the order of perfusion in the twelve meridians ……… 29

第四章　病因病机 ………………………………………………………… 30
 4.1 病因类 …………………………………………………………………… 30
  4.1.1 病因学说 etiological theory …………………………………… 30

4.1.2	六淫 six pathogenic factors	30
4.1.3	疫疠 epidemic pestilence	33
4.1.4	内生五邪 five endogenous pathogenic factors	33
4.1.5	内伤七情 seven emotions	33
4.1.6	饲喂失宜 improper feeding	34
4.1.7	劳逸失度 work-rest imbalance	35
4.1.8	可致病的病理产物 pathological product that can cause disease	35
4.1.9	外伤 external injury	36
4.1.10	寄生虫 parasite	36
4.1.11	胎毒 fetal toxin	36
4.1.12	虫兽伤 injury by animal and insect	36
4.1.13	水土不服 non-acclimatization	36
4.1.14	禀赋不足 constitutional insufficiency	37

4.2 病机类 ··· 37

4.2.1	病机学说 pathogenesis theory	37
4.2.2	八纲病证病机 pathogenesis of eight principal syndromes	38
4.2.3	气血津液病证病机 pathogenesis of qi, blood and body fluid syndromes	44
4.2.4	脏腑病证病机 pathogenesis of zang-fu viscera syndromes	47
4.2.5	六经病证病机 pathogenesis of six meridians syndromes	55
4.2.6	卫气营血病证病机 pathogenesis of defense qi nutrient blood syndromes	57
4.2.7	三焦病证病机 pathogenesis of triple energizer syndromes	58

第五章 治　则 ··· 59

5.1 治未病类 ··· 59

5.1.1	治未病 preventive treatment	59
5.1.2	四季药 seasonal formula	59
5.1.3	针刺六脉血 bleeding at the six points	59

5.2 治则类 ··· 59

5.2.1	治则 treatment principles	59
5.2.2	正治 routine treatment	59
5.2.3	反治 paradoxical treatment	60
5.2.4	标本缓急 symptom, root-cause, non-urgency and urgency	60
5.2.5	扶正祛邪 reinforce healthy qi to eliminate pathogenic factors	61
5.2.6	同病异治 different treatments for the same disease	62
5.2.7	异病同治 same treatment for different diseases	62
5.2.8	三因制宜 treatment in accordance with three factors	62
5.2.9	调整阴阳 regulate and balance yin and yang	62
5.2.10	调和气血 regulate and harmonize qi and blood	63

第六章 疾病 ··· **64**

6.1 心系病 ··· **64**

- 6.1.1 口舌生疮 mouth and tongue ulcers ·· 64
- 6.1.2 血证 bleeding or hemorrhagic syndrome ·· 64
- 6.1.3 汗证 sweating syndrome ·· 64
- 6.1.4 木舌 swollen and rigid tongue ·· 64
- 6.1.5 心虚 heart deficiency ·· 64
- 6.1.6 心冷吐水 hydroptysis spleen and heart deficiency ···························· 64
- 6.1.7 心热风邪 apoplexy ·· 65
- 6.1.8 中暑 heat stroke ·· 65
- 6.1.9 痫病 epilepsy ·· 65
- 6.1.10 心黄 xin huang ·· 65
- 6.1.11 心痛 cardialgia ·· 65
- 6.1.12 脑黄 meningo encephalitis ·· 65
- 6.1.13 中风 stroke ··· 65

6.2 肺系病 ··· **66**

- 6.2.1 感冒 cold ·· 66
- 6.2.2 咳嗽 cough ·· 66
- 6.2.3 喘证 asthma ··· 67
- 6.2.4 流鼻 nasal discharge ··· 67
- 6.2.5 肺痈 pulmonary abscess ··· 67
- 6.2.6 胸痛 chest pain ·· 68
- 6.2.7 肺风毛燥 pruritus and alopecia ··· 68
- 6.2.8 肺败 deterioration of the lung ·· 68
- 6.2.9 鼻血 epistaxis ·· 68
- 6.2.10 肺寒吐沫 salivation due to pulmonary cold evil ······························· 68
- 6.2.11 肺把胸膊痛 pain on chest and foreleg ·· 69
- 6.2.12 肺黄 fei huang ··· 69
- 6.2.13 胸水 hydrothorax ·· 69
- 6.2.14 异物呛肺 aspiration or deglutition pneumonia ································ 69
- 6.2.15 锁喉风 locked throat ··· 69

6.3 脾系病 ··· **69**

- 6.3.1 呕吐 vomiting ··· 69
- 6.3.2 草噎 oesophagus obstruction ··· 69
- 6.3.3 料伤 indigestion due to improper feeding ······································· 70
- 6.3.4 水伤 indigestion due to improper watering ····································· 70
- 6.3.5 脾虚慢草 inappetence due to spleen deficiency ································ 70
- 6.3.6 百叶干 impaction of omasum ··· 70

6.3.7 异物伤胃 traumatic reticuloperitonitis ··· 70
6.3.8 脾虚不磨 indigestion due to spleen deficiency ··· 70
6.3.9 脾虚浮肿 edema due to spleen deficiency ··· 70
6.3.10 起卧症 equine colic ··· 70
6.3.11 结症 impaction of intestines ··· 70
6.3.12 肠积沙 intestinal sabulous ··· 71
6.3.13 痢疾 dysentery ··· 71
6.3.14 肠黄 enteritis ··· 71
6.3.15 肠痈 appendicitis ··· 71
6.3.16 肠风下血 intestinal wind bleeding ··· 71
6.3.17 马肠臌气 equine intestinal tympany ··· 72
6.3.18 牛羊气胀 cattle and sheep bloating ··· 72
6.3.19 泄泻 diarrhea ··· 72
6.3.20 便秘 constipation ··· 72
6.3.21 便血 bloody stool ··· 73
6.3.22 宿水停脐 ascites ··· 73
6.3.23 宿草不转 impaction of rumen ··· 73
6.3.24 胃寒 stomach cold ··· 73
6.3.25 胃热 stomach heat ··· 74
6.3.26 大肚结 gastric dilatation ··· 74
6.3.27 冷痛 spasmodic colic ··· 74
6.3.28 肠入阴 scrotal hernia ··· 74
6.3.29 肠绞痛 torsion incarceration and intussusception ··· 74
6.3.30 盘肠结 obstruction of colon ··· 74
6.3.31 翻胃吐草 osteomalacia ··· 74
6.3.32 肠嵌闭 incarceration or strangulation ··· 74
6.3.33 肚胀 abdominal fullness ··· 75
6.3.34 脾虚带下 over discharge in vagina due to spleen defcience ··· 75
6.3.35 困水膈痰 phlegm retention of diaphragm ··· 75
6.4 肝系病 ··· 75
6.4.1 黄疸 jaundice ··· 75
6.4.2 胆胀 gallbladder enlargement ··· 75
6.4.3 肝热传眼 acute conjunctivitis due to liver heat ··· 75
6.4.4 云翳遮睛 nebula over the eyes ··· 75
6.4.5 肝经风热 wind heat accumulated in the liver channel ··· 76
6.4.6 肝胀 liver enlargement ··· 76
6.4.7 肝胆风 hepatobiliary wind ··· 76
6.4.8 肝黄 liver huang ··· 76
6.4.9 月盲 moon blindness ··· 76

6.4.10 夜盲 nyctalopia ……76
6.5 肾系病 ……76
　　6.5.1 肾痛 nephralgia ……76
　　6.5.2 肾寒 kidney cold ……76
　　6.5.3 肾虚腿肿 edema of posterior limbs due to kidney asthenia ……76
　　6.5.4 肾虚带下 leucorrhea due to kidney asthenia ……77
　　6.5.5 肾虚骨痿 bone impotence due to kidney asthenia ……77
　　6.5.6 肾厥 renal syncope ……77
　　6.5.7 淋证 stranguria syndrome ……77
　　6.5.8 尿浊 turbid urine ……77
　　6.5.9 胞转 torsion of the urinary bladder ……78
　　6.5.10 癃闭 obstructive dysuria ……78
　　6.5.11 尿不禁 urinary incontinence ……78
　　6.5.12 尿崩 diabetes insipidus ……78
　　6.5.13 肾火症 yellow urine due to renal meridian fever ……78
　　6.5.14 风水 wind edema ……78
　　6.5.15 皮水 puffiness by wind cold wet and hot ……78
　　6.5.16 石水 proteinuric edema ……78
　　6.5.17 肾水 chronic edema ……78
　　6.5.18 正水 progressive edema ……78
　　6.5.19 外肾黄 swelling of the scrotum ……79
　　6.5.20 内肾黄 lumbar swelling ……79
　　6.5.21 阳痿 impotence ……79
　　6.5.22 垂缕不收 penial prolapse ……79
　　6.5.23 胞黄 dysuria due to hot and wet ……79
　　6.5.24 胞虚 stranguria due to deficiency of qi and blood ……80
　　6.5.25 尿闭 anuria ……80
　　6.5.26 尿血 hematuria ……80
　　6.5.27 肾虚腿肿 swollen of leg due to deficiency of kidney ……80
6.6 外伤及疮疡病 ……80
　　6.6.1 创伤 wounds ……80
　　6.6.2 豁鼻 laceration of the muzzle ……80
　　6.6.3 水火烫伤 scalds and burns ……81
　　6.6.4 闪伤 sudden sprain ……81
　　6.6.5 闪挫 sprain and contusion ……81
　　6.6.6 瘘管 fistula of inner and outer orifice ……81
　　6.6.7 窦道 fistula of outer orifice ……81
　　6.6.8 骨折 fracture ……81
　　6.6.9 角折 horn fracture ……81

6.6.10 疮黄疔毒 furuncle or pyogenic infection on body surface ·················· 81
6.6.11 黄 stasis and swelling ·················· 81
6.6.12 锁口黄 stomatitis due to hot and poison ·················· 82
6.6.13 鼻黄 nasal swelling due to hot evil ·················· 82
6.6.14 颊黄 cheek swelling due to hot evil ·················· 82
6.6.15 耳黄 ear swelling due to hot and poison ·················· 82
6.6.16 腮黄 gill swelling due to hot evil ·················· 82
6.6.17 背黄 back inflammation due to hot evil ·················· 82
6.6.18 胸黄 chest swelling due to hot evil ·················· 82
6.6.19 肚底黄 abdominal swelling ·················· 82
6.6.20 疖 furuncle ·················· 82
6.6.21 疮 sores ·················· 82
6.6.22 褥疮 bedsore ·················· 83
6.6.23 疔疮 nail like boil ·················· 83
6.6.24 黑疔 black nail like boil ·················· 83
6.6.25 筋疔 nail like boil of exposed membrane ·················· 83
6.6.26 气疔 nail like boil of foamy pus ·················· 83
6.6.27 水疔 nail like boil of exudate ·················· 83
6.6.28 血疔 nail like boil of pus and blood ·················· 83
6.6.29 丹毒 erysipelas ·················· 83
6.6.30 痈 acute suppurative disease ·················· 83
6.6.31 疽 cellulitis or phlegmon ·················· 84
6.6.32 有头疽 cellulitis in skin and muscle ·················· 84
6.6.33 无头疽 cellulitis in joints and bones ·················· 84
6.6.34 项痈 carbuncle of neck ·················· 84
6.6.35 肩痈 carbuncle of shoulder ·················· 84
6.6.36 脑颡黄 nasal discharge of pus and blood ·················· 84
6.6.37 无名肿毒 inflammation due to wind cold or hot evil ·················· 84
6.6.38 流注 carbuncle of pus ·················· 84
6.6.39 流痰 phthisis of bones and joints ·················· 84
6.6.40 瘰疬 scrofula ·················· 85
6.6.41 漏瘘 fistula ·················· 85

6.7 肢蹄病 ·················· 85
6.7.1 肩膊痛 pain in shoulder and up arm ·················· 85
6.7.2 膊尖痛 pains of the shoulder joint ·················· 85
6.7.3 传经痛 gout ·················· 85
6.7.4 脱臼 dislocation ·················· 85
6.7.5 脱膊 shoulder dislocation ·················· 85
6.7.6 抢风痛 paralysis of the radial nerve ·················· 86

6.7.7 四肢神经麻痹症 quadriplegia ... 86
6.7.8 夹气痛 sprain of triceps ... 86
6.7.9 乘重痛 contusion of the elbow joint ... 86
6.7.10 攒筋痛 flexor tendinitis ... 86
6.7.11 掌骨痛 metacarpal pains ... 86
6.7.12 缠腕痛 contusion of the fetlock joint ... 86
6.7.13 肾冷拖腰 lumbago pains due to wind cold and wet evil ... 86
6.7.14 腰胯痛 pains of the lumbus and hip ... 87
6.7.15 胯瓦痛 sprain of the hip joint ... 87
6.7.16 掠草痛 pain of the stifle joint ... 87
6.7.17 冷拖竿 string halt ... 87
6.7.18 合子骨肿痛 bone spavin ... 87
6.7.19 蹄伤 injuries to the hoof ... 87
6.7.20 筋断 tendon rupture ... 88
6.7.21 风蹄 sandcrack ... 88
6.7.22 滚蹄 contraction of flexor tendons ... 88
6.7.23 毛边漏 quittor ... 88
6.7.24 蹄头痛 pain in the head of the hoof ... 88
6.7.25 五攒痛 laminitis ... 88
6.7.26 败血凝蹄 chronic laminitis due to stasis ... 88
6.7.27 腐蹄病 foot rot ... 89

6.8 皮肤病 ... 89
6.8.1 遍身黄 urticaria ... 89
6.8.2 肺风毛燥 pruritus and alopecia ... 89
6.8.3 湿毒 eczema due to wet poison ... 89
6.8.4 热气疮 herpetic dermatosis due to hot evil ... 89
6.8.5 皮肤瘙痒症 pruritus ... 89
6.8.6 脱毛症 alopecia ... 89
6.8.7 疣 warts ... 90
6.8.8 臊疣 warts of perineum ... 90
6.8.9 黄水疮 sores of pus ... 90
6.8.10 圆癣 ringworm ... 90
6.8.11 阴癣 tinea of pudendum ... 90
6.8.12 疥疮 scabies ... 90
6.8.13 湿疮 sores of wet ... 90
6.8.14 风土疮 sores due to unacclimatization ... 90
6.8.15 顽湿结聚 sores due to wet evil ... 90
6.8.16 面游风 facial wandering wind ... 91
6.8.17 药毒 drug allergy ... 91

6.8.18 晒疮 sunburn … 91
6.8.19 恶虫叮咬伤 bite wound of insect … 91
6.8.20 茧唇 callus like disease of the lips … 91
6.8.21 紫癜风 purpura … 91
6.8.22 火赤疮 vesicular disease due to hot evil … 91
6.8.23 松皮癣 psoriasis … 91
6.8.24 流皮漏 spreading skin ulcer … 91
6.8.25 石疽 hard skin nodule … 92

6.9 胎产病 … 92
6.9.1 胎动不安 excessive fetal moovement … 92
6.9.2 流产 abortion … 92
6.9.3 产后恶露不尽 prolonged lochia … 92
6.9.4 胎衣不下 retention of placenta … 92
6.9.5 产后腹痛 postpartum abdominal pain … 93
6.9.6 难产 dystocia … 93
6.9.7 阴道脱出 prolapse of vagina … 93
6.9.8 乳痈 mastitis … 93
6.9.9 产后发热 postpartum fever … 93
6.9.10 缺乳 hypogalactia … 93
6.9.11 胎风 postpartum paralysis … 93
6.9.12 胎气 pregnant edema … 94
6.9.13 流产与死胎 early death of the embryo and abortion … 94

6.10 瘟病 … 94
6.10.1 温病 warm disease … 94
6.10.2 时疫 seasonal pestilence … 94
6.10.3 时行感冒 influenza … 94
6.10.4 破伤风 tetanus … 94
6.10.5 三喉症 three kinds of laryngo-pharyngeal diseases … 94
6.10.6 牛红眼病 bovine acute conjunctivitis … 95
6.10.7 兔流涎病 infectious vesicular stomatitis of rabbits … 95

6.11 其他病症 … 95
6.11.1 寒结 cold constipation … 95
6.11.2 水肿 edema … 95
6.11.3 虚劳病 consumptive diseases … 95
6.11.4 项脊悸 neck rheumatism … 95
6.11.5 疝 hernia … 95
6.11.6 直肠脱出 retum prolapse … 96
6.11.7 幼畜惊风 eclampsia of young stock … 96
6.11.8 新驹奶泻 diarrhoea in foal … 96

6.11.9 幼畜胎粪不下 constipation of young stock ········· 97
6.11.10 幼驹尿血 hematuria of foal ········· 97
6.11.11 跳肷 spasm of diaphragm ········· 97
6.11.12 风瘫 paralysis due to wind evil ········· 97
6.11.13 痹证 arthralgia syndrome ········· 97
6.11.14 痿证 flaccidity syndrome ········· 98
6.11.15 骨眼 protrusion of the swollen third eyelid ········· 98
6.11.16 内障眼 glaucoma ········· 98
6.11.17 外障眼 external oculopathy ········· 98
6.11.18 口㖞 facial paralysis ········· 98

第七章 证 候 ········· 99

7.1 证 ········· 99
7.1.1 证 ········· 99

7.2 基本虚证类 ········· 99
7.2.1 气虚证 qi deficiency pattern ········· 99
7.2.2 气陷证 qi sinking pattern ········· 99
7.2.3 气脱证 qi desertion pattern ········· 99
7.2.4 血虚证 blood deficiency pattern ········· 99
7.2.5 阴虚证 yin deficiency pattern ········· 100
7.2.6 亡阴证 yin fluids exhausting pattern ········· 100
7.2.7 阳虚证 yang deficiency pattern ········· 100
7.2.8 亡阳证 yang qi exhausting pattern ········· 101
7.2.9 虚阳浮越证 deficiency yang floating upward pattern ········· 101
7.2.10 气血两虚证 deficiency of both qi and blood pattern ········· 101
7.2.11 气阴两虚证 both qi and yin fluids deficiency pattern ········· 101
7.2.12 阴血亏虚证 deficiency of yin and blood pattern ········· 101
7.2.13 阴阳两虚证 deficiency of yin and yang pattern ········· 101
7.2.14 津液亏虚证 body fluids deficiency pattern ········· 102
7.2.15 津气亏虚证 both body fluids and qi deficiency pattern ········· 102
7.2.16 精气亏虚证 deficiency of vital essence pattern ········· 102
7.2.17 精血亏虚证 both vital essence and blood deficiency pattern ········· 102
7.2.18 卫虚证 defensive qi deficiency pattern ········· 103
7.2.19 营虚证 deficiencies of ying and qi pattern ········· 103

7.3 基本实证类 ········· 103
7.3.1 外风证 external wind evil pattern ········· 103
7.3.2 寒凝证 cold coagulation pattern ········· 104
7.3.3 暑热证 summer heat pattern ········· 105
7.3.4 湿阻证 dampness blocking qi pattern ········· 106

7.3.5 外燥证 external dryness pattern ……………………………………… 106
7.3.6 火热炽盛证 severe heat pattern ……………………………………… 107
7.3.7 痰证 sputum pattern ……………………………………………………… 108
7.3.8 水饮内停证 excessive fluid collecting internally pattern ………… 110
7.3.9 水停证 water dampness stagnation pattern ………………………… 110
7.3.10 气滞证 qi stagnation pattern ………………………………………… 110
7.3.11 气逆证 qi upwards reverse pattern ………………………………… 111
7.3.12 气闭证 qi blocked pattern …………………………………………… 111
7.3.13 血瘀证 blood stasis pattern ………………………………………… 111
7.3.14 邪毒炽盛证 severe evil toxin pattern ……………………………… 112
7.3.15 食积证 dyspepsia pattern …………………………………………… 114
7.3.16 虫积证 parasite clumping pattern …………………………………… 114
7.3.17 石阻证 lithiasis blocking pattern …………………………………… 114
7.3.18 真实假虚证 excess pattern with pseudo deficiency pattern …… 114

7.4 虚实夹杂证类 …………………………………………………………… 114
7.4.1 气虚挟实证 qi deficiency mixed excess pattern …………………… 114
7.4.2 血虚挟实证 blood deficiency mixed excess pattern ……………… 115
7.4.3 阴虚挟实证 yin deficiency mixed excess pattern ………………… 116
7.4.4 阳虚挟实证 yang deficiency mixed excess pattern ……………… 117
7.4.5 津亏热结证 body fluids loss and heat accumulating pattern …… 118
7.4.6 正虚邪恋证 vital qi deficiency causing evil detaining pattern … 118
7.4.7 风热血燥证 wind heat causing blood dryness pattern …………… 120
7.4.8 暑伤津气证 summer heat injuring body fluids and qi pattern … 120
7.4.9 邪热伤阴证 evil heat injuring yin pattern ………………………… 120
7.4.10 痰热阴虚证 sputum heat causing yin deficiency pattern ……… 121
7.4.11 血瘀风燥证 blood stasis causing wind dryness pattern ………… 121
7.4.12 实中挟虚证 excess pattern carrying deficiency pattern ………… 121
7.4.13 邪陷正脱证 evil toxin attacking while healthy qi exhausting pattern …… 121
7.4.14 内闭外脱证 excess evil blocking and healthy qi exhausting pattern … 121

7.5 心系证类 ………………………………………………………………… 121
7.5.1 心气虚证 heart qi deficiency pattern ……………………………… 121
7.5.2 心气虚血瘀证 heart qi deficiency causing blood stasis pattern … 121
7.5.3 心气血两虚证 both qi and blood deficiency in heart pattern …… 121
7.5.4 心气阴两虚证 both qi and yin deficiency in heart pattern ……… 122
7.5.5 心阳虚证 yang deficiency of heart pattern ………………………… 122
7.5.6 心阳虚血瘀证 heart yang deficiency causing blood stasis pattern … 122
7.5.7 心血虚证 heart blood deficiency pattern …………………………… 122
7.5.8 心阴血虚证 heart yin and blood deficiency pattern ……………… 122
7.5.9 心阴虚火旺证 deficiency of heart yin induces fire hyperactivity pattern … 122

7.5.10 心阴虚血瘀证 heart yin deficiency causing blood stasis pattern ············ 122
7.5.11 心阴阳两虚证 heart yin and yang deficiency pattern ················ 123
7.5.12 心血瘀阻证 heart blood stasis pattern ································ 123
7.5.13 痰阻心脉证 sputum blocking heart channel pattern ················· 123
7.5.14 寒滞心脉证 cold evil stagnated in heart pattern ····················· 123
7.5.15 心脉气滞证 heart qi stagnation pattern ································ 123
7.5.16 饮停心包证 excessive body fluids stagnated in pericardium pattern ······ 123
7.5.17 心火炽盛证 heart fire blazing pattern ··································· 123
7.5.18 心火上炎证 upward flaming of heart fire pattern ····················· 124
7.5.19 心热阴虚证 heart heat and yin deficiency pattern ···················· 124
7.5.20 热闭心包证 heat blocking pericardium pattern ························ 124
7.5.21 热扰心神证 heat disturbing mind pattern ··············· 124
7.5.22 热入心营证 heat attacking heart ying pattern ························ 124
7.5.23 血热扰神证 blood heat disturbing mind pattern ······················· 124
7.5.24 暑热闭神证 summer heat blocking mind pattern ······················· 124
7.5.25 痰火扰神证 sputum heat disturbing mind pattern ····················· 124
7.5.26 痰迷心窍证 sputum blocked heart mind pattern ······················· 125
7.5.27 风痰闭神证 wind sputum blocked mind pattern ······················· 125
7.5.28 浊毒闭神证 toxin blocked heart apertures pattern····················· 125
7.5.29 气闭神厥证 qi stagnation causing mind syncope pattern ············· 125
7.5.30 心虚神怯证 heart deficiency and mind timidity pattern ·············· 125
7.5.31 囊虫侵脑证 cysticercus attacking brain pattern·························· 125
7.5.32 惊恐伤神证 scare caused mental disorder pattern ···················· 125
7.5.33 心神不宁证 nervous pattern ··· 125
7.6 肺系证类 ··· 126
　7.6.1 肺气虚证 lung qi deficiency pattern ·· 126
　7.6.2 肺气阴两虚证 both qi and yin deficiency in lung pattern ············ 126
　7.6.3 肺阳虚证 lung yang deficiency pattern ··································· 126
　7.6.4 肺阴虚证 lung yin deficiency pattern ····································· 126
　7.6.5 肺卫气虚证 lung protective qi deficiency pattern ······················ 126
　7.6.6 阴虚肺燥证 yin deficiency and lung dryness pattern ·················· 126
　7.6.7 肺热炽盛证 lung heat over vigorous pattern ··························· 126
　7.6.8 肺热阴虚证 lung heat and yin deficiency pattern ······················ 127
　7.6.9 肺热移肠证 lung heat caused intestinal dysfunction pattern ········· 127
　7.6.10 风热犯肺证 wind heat attacking the lung pattern ···················· 127
　7.6.11 风热闭肺证 wind heat blocking lung pattern ·························· 127
　7.6.12 肺经风热证 wind heat in lung channel pattern ······················· 127
　7.6.13 肺经郁火证 lung channel fire stagnation pattern ····················· 127
　7.6.14 肺热血瘀证 lung heat and blood stasis pattern ······················· 127

7.6.15 暑伤肺络证 summer-heat injured lung channel pattern ·················· 127
7.6.16 痰热壅肺证 sputum dampness accumulating in the lung pattern ············ 127
7.6.17 痰热闭肺证 sputum heat blocking lung pattern ························ 128
7.6.18 痰浊阻肺证 sputum dampness accumulating in the lung pattern ············ 128
7.6.19 痰瘀阻肺证 sputum stasis blocking lung pattern ······················· 128
7.6.20 风寒袭肺证 wind cold evil attacking lung pattern ······················ 128
7.6.21 寒饮停肺证 cold fluids accumulating in lung pattern ···················· 128
7.6.22 肺郁水停证 lung qi stagnation and body fluids stagnation pattern ·········· 128
7.6.23 肺热饮停证 lung heat and body fluids stagnation pattern ················· 128
7.6.24 寒痰阻肺证 cold sputum blocking lung pattern ························ 128
7.6.25 表寒肺热证 exterior cold and lung heat pattern ······················· 128
7.6.26 燥邪犯肺证 dryness evil attacking lung pattern ······················· 129
7.6.27 凉燥袭肺证 cool dryness attacking lung pattern ······················· 129
7.6.28 温燥袭肺证 warm dryness attacking lung pattern ······················ 129
7.6.29 燥痰结肺证 dryness sputum evil blocking lung pattern ··················· 129
7.6.30 肺燥郁热证 lung dryness and heat stagnation pattern ···················· 129
7.6.31 肺燥肠热证 lung dryness and intestine heat pattern ···················· 129
7.6.32 肺燥肠闭证 lung dryness and intestine blocking pattern ·················· 129
7.6.33 瘀阻肺络证 blood stasis blocking lung channel pattern ·················· 129
7.6.34 热毒闭肺证 heat toxin blocking lung pattern ·························· 130
7.6.35 虫毒犯肺证 parasite toxin attacking lung pattern ······················ 130
7.6.36 气郁伤肺证 qi stagnation caused lung injuring pattern ·················· 130

7.7 脾系证类 ··· 130
 7.7.1 脾气虚证 spleen qi deficiency pattern ································ 130
 7.7.2 脾气下陷证 sinking of spleen qi pattern ······························· 130
 7.7.3 脾气郁结证 spleen qi stagnation pattern ······························ 130
 7.7.4 脾不统血证 spleen failing to control the blood pattern ··················· 130
 7.7.5 脾阳虚证 spleen yang deficiency pattern ······························ 130
 7.7.6 脾阴虚证 spleen yin deficiency pattern ······························· 131
 7.7.7 脾虚营亏证 spleen ying deficiency pattern ··························· 131
 7.7.8 脾虚血亏证 spleen deficiency and blood depletion pattern ················ 131
 7.7.9 脾虚血燥证 spleen deficiency caused blood dryness pattern ··············· 131
 7.7.10 脾虚气滞证 spleen deficiency caused qi stagnation pattern ················ 131
 7.7.11 脾虚水泛证 spleen deficiency caused water retention pattern ············· 131
 7.7.12 脾阳虚水泛证 spleen yang deficiency caused water retention pattern ········ 131
 7.7.13 脾虚湿困证 spleen deficiency with dampness pattern ·················· 131
 7.7.14 脾虚湿热证 spleen deficiency and dampness heat stagnation pattern ········ 131
 7.7.15 脾虚痰湿证 spleen deficiency and sputum dampness stagnation pattern ····· 132
 7.7.16 脾虚食积证 spleen deficiency caused food stagnating pattern ············· 132

7.7.17 脾虚虫积证 spleen deficiency and parasite stagnation pattern ………… 132
7.7.18 湿热蕴脾证 damp heat affecting the spleen pattern …………………… 132
7.7.19 脾经热毒证 heat toxin in spleen channel pattern ……………………… 132
7.7.20 寒湿困脾证 cold dampness affecting the spleen pattern ……………… 132
7.7.21 思伤脾气证 worry injuring spleen qi pattern……………………………… 132
7.7.22 胃气虚证 stomach qi deficiency pattern ………………………………… 132
7.7.23 胃气阴两虚证 both stomach qi and yin deficiency pattern …………… 133
7.7.24 胃气虚血瘀证 stomach qi deficiency and blood stasis pattern ……… 133
7.7.25 胃气上逆证 ascending of stomach qi pattern………………………… 133
7.7.26 胃阳虚证 stomach yang deficiency pattern …………………………… 133
7.7.27 胃气滞血瘀证 stomach qi stagnation and blood stasis pattern ……… 134
7.7.28 胃阴虚证 stomach yin deficiency pattern ……………………………… 134
7.7.29 胃阴虚气滞证 stomach yin deficiency and qi stagnation pattern …… 134
7.7.30 胃阴虚血瘀证 stomach yin deficiency and blood stasis pattern …… 134
7.7.31 胃火证 stomach heat pattern …………………………………………… 134
7.7.32 胃热气滞证 stomach heat and qi stagnation pattern ………………… 134
7.7.33 胃热津伤证 stomach heat injuring body fluids pattern ……………… 134
7.7.34 胃热阴虚证 stomach heat and yin deficiency pattern ………………… 134
7.7.35 胃燥津伤证 stomach dryness injuring body fluids pattern …………… 134
7.7.36 寒邪犯胃证 cold evil attacking stomach pattern ……………………… 135
7.7.37 寒饮停胃证 cold water accumulating in stomach pattern …………… 135
7.7.38 瘀阻胃络证 blood stasis blocking stomach channel pattern ………… 135
7.7.39 寒滞肠道证 cold stagnated intestine pattern ………………………… 135
7.7.40 肠道湿热证 intestine dampness heat pattern ………………………… 135
7.7.41 肠道实热证 intestine excess heat pattern …………………………… 135
7.7.42 大肠热结证 heat tangling in the large intestine pattern ……………… 136
7.7.43 热毒蕴肠证 heat toxin accumulating in intestine pattern …………… 136
7.7.44 肠道津亏证 intestine dryness and fluids exhausting pattern ………… 136
7.7.45 血虚肠燥证 blood deficiency and intestine dryness pattern ………… 136
7.7.46 阴虚肠燥证 yin deficiency and intestine dryness pattern …………… 136
7.7.47 血热肠燥证 blood heat and intestine dryness pattern ………………… 136
7.7.48 湿阻肠道证 dampness blocking intestine pattern …………………… 136
7.7.49 肠道寒湿证 intestine cold dampness pattern ………………………… 136
7.7.50 肠道气滞证 qi stagnation in intestine pattern ………………………… 136
7.7.51 虫积肠道证 parasite accumulating in intestine pattern ……………… 137
7.7.52 虫扰魄门证 parasite gathering at anus pattern ……………………… 137
7.7.53 风伤肠络证 wind injuring intestine channel pattern ………………… 137
7.7.54 肛门热毒证 heat toxin at anus pattern ………………………………… 137
7.7.55 肛门湿热证 anus dampness heat pattern …………………………… 137

7.7.56 气血瘀滞肛门证 qi and blood stagnating at anus pattern ·················· 137
7.7.57 脾胃气虚证 spleen and stomach qi deficiency pattern ·················· 137
7.7.58 脾胃阴虚证 yin deficiency of the spleen and stomach pattern ·················· 137
7.7.59 脾胃阳虚证 spleen and stomach yang deficiency pattern ·················· 137
7.7.60 脾胃阳虚气滞证 spleen and stomach yang deficiency causing qi stagnation pattern ·················· 138
7.7.61 脾胃气阴两虚证 deficiency of qi and yin in spleen and stomach pattern ··· 138
7.7.62 脾胃实热证 spleen and stomach excess heat pattern ·················· 138
7.7.63 脾胃湿热证 damp heat in the spleen and stomach pattern ·················· 138
7.7.64 湿困脾胃证 dampness blocking spleen and stomach pattern ·················· 138
7.7.65 胃热脾虚证 stomach heat and spleen deficiency pattern ·················· 138
7.7.66 脾胃不和证 spleen and stomach qi disharmonyp pattern ·················· 138
7.7.67 胃肠湿热证 stomach and intestine dampness heat pattern ·················· 138
7.7.68 胃肠实热证 stomach and intestine excess heat pattern ·················· 139
7.7.69 食滞胃肠证 stomach and intestine food stagnation pattern ·················· 139
7.7.70 食滞胃热证 food stagnation in stomach and intestine pattern ·················· 139
7.7.71 胃肠气滞证 stomach and intestine qi stagnation pattern ·················· 139
7.7.72 瘀滞胃肠证 blood stasis blocking stomach and intestine pattern ·················· 139
7.7.73 寒滞胃肠证 cold evil attacking stomach and intestine pattern ·················· 139
7.7.74 痰湿中阻证 sputum dampness blocking zhongjiao pattern ·················· 139
7.7.75 肠道瘀滞证 intestinal stasis pattern ·················· 139

7.8 肝系证类·················· 140
 7.8.1 肝阴虚证 liver yin pattern ·················· 140
 7.8.2 肝血虚证 liver blood pattern ·················· 140
 7.8.3 肝气虚证 deficiency of liver qi pattern ·················· 140
 7.8.4 肝阳虚证 deficiency of liver yang pattern ·················· 140
 7.8.5 肝阳上亢证 hyperactivity of the liver yang pattern·················· 140
 7.8.6 肝阳暴亢证 sudden hyperactivity of the liver yang pattern·················· 140
 7.8.7 肝阴虚阳亢证 deficiency of liver yin and hyperactivity of liver yang pattern ·················· 140
 7.8.8 肝郁证 stagnation of liver qi pattern ·················· 140
 7.8.9 肝郁血虚证 stagnation of liver qi and blood deficiency pattern ·················· 141
 7.8.10 肝郁血瘀证 stagnation of liver qi and blood stasis pattern ·················· 141
 7.8.11 肝郁阴虚证 stagnation of liver qi and yin deficiency pattern ·················· 141
 7.8.12 肝瘀化热证 liver stasis transforming heat pattern·················· 141
 7.8.13 肝瘀证 blood stasis pattern·················· 141
 7.8.14 肝虚血瘀证 liver deficiency and blood stasis pattern ·················· 141
 7.8.15 肝阴虚血瘀证 liver yin deficiency and blood stasis pattern ·················· 141
 7.8.16 肝气虚血瘀证 liver qi deficiency and blood stasis pattern ·················· 141

7.8.17 肝郁化火证 liver depression transforming into fire pattern ············· 142
7.8.18 肝郁血热证 liver qi stagnation and blood heat pattern ············· 142
7.8.19 肝郁痰火证 liver qi stagnation and sputum heat pattern ············· 142
7.8.20 肝瘀痰结证 liver qi stagnant and sputum stasis pattern ············· 142
7.8.21 肝火炽盛证 liver fire blazing pattern ············· 142
7.8.22 肝火上炎证 upward flaming of liver heat pattern ············· 142
7.8.23 肝经火旺证 fire hyperactivity in liver channel pattern ············· 142
7.8.24 肝经风热证 wind heat affecting the liver channel pattern ············· 142
7.8.25 肝经湿热证 damp heat affecting the liver channel pattern ············· 142
7.8.26 肝郁湿热证 stagnation of liver qi and dampness heat pattern ············· 143
7.8.27 热毒淤肝证 heat toxicity and liver stasis pattern ············· 143
7.8.28 肝热气滞证 liver heat and qi stasis pattern ············· 143
7.8.29 肝热血瘀证 liver heat and blood stasis pattern ············· 143
7.8.30 肝热阴虚证 liver heat and yin deficiency pattern ············· 143
7.8.31 寒滞肝脉证 cold stagnating in the liver vessel pattern ············· 143
7.8.32 肝风内动证 internal stirring of liver wind pattern ············· 143
7.8.33 胆气虚证 gallbladder qi deficiency pattern ············· 144
7.8.34 胆郁痰扰证 depressed gallbladder with harassing sputum pattern ············· 144
7.8.35 胆热痰扰证 bile heat and sputum disturbance pattern ············· 144
7.8.36 虫扰胆膈证 parasite disturbing biliary and diaphragmatic pattern ············· 144
7.8.37 肝胆湿热证 liver gallbladder dampness heat pattern ············· 144
7.8.38 肝胆火旺证 fire hyperactivity in liver and gallbladder pattern ············· 144
7.8.39 胆经郁热证 stagnant heat of gallbladder channel pattern ············· 144
7.8.40 肝胆寒湿证 dampness and cold in liver and gallbladder pattern ············· 145
7.8.41 肝胆瘀滞证 liver gallbladder stasis pattern ············· 145
7.8.42 肝胆气滞证 liver gallbladder qi stagnation pattern ············· 145

7.9 肾系证类 ············· 145
  7.9.1 肾气虚证 kidney qi deficiency pattern ············· 145
  7.9.2 肾气不固证 kidney qi insecurity pattern ············· 145
  7.9.3 肾虚水泛证 water retention due to kidney deficiency pattern ············· 145
  7.9.4 肾阳虚证 kidney yang deficiency pattern ············· 145
  7.9.5 肾阴虚证 kidney yin deficiency pattern ············· 146
  7.9.6 肾阴虚火旺 kidney yin deficiency and fire hyperactivity pattern ············· 146
  7.9.7 肾精亏虚证 kidney essence depletion pattern ············· 146
  7.9.8 肾虚髓亏证 kidney deficiency and marrow depletion pattern ············· 146
  7.9.9 肾阴阳两虚证 consumptive disease with deficiency of both kidney yin and kidney yang pattern ············· 146
  7.9.10 肾虚寒湿证 cold dampness due to kidney deficiency pattern ············· 146
  7.9.11 湿热蕴肾证 damp heat brewing in the kidney pattern ············· 147
  7.9.12 脓毒蕴肾证 septic kidney pattern ············· 147

7.9.13 膀胱湿热证 damp heat in the urinary bladder pattern	147
7.9.14 膀胱蕴热证 bladder heat amassment pattern	147
7.9.15 膀胱蓄水证 bladder water amassment pattern	147
7.9.16 膀胱蓄血证 bladder blood amassment pattern	147
7.9.17 膀胱虚寒证 bladder cold deficiency pattern	147
7.9.18 寒凝胞宫证 coagulated cold in womb pattern	148
7.9.19 痰凝胞宫证 coagulated sputum in womb pattern	148
7.9.20 瘀阻胞宫证 static blood blocking in womb pattern	148
7.9.21 胞宫虚寒证 deficient cold in uterus pattern	148
7.9.22 胞宫湿热证 dampness heat in uterus pattern	148
7.9.23 胞宫血热证 uterine blood heat pattern	148
7.9.24 冲任失调证 disharmony of the thoroughfare and controlling vessels pattern	148
7.9.25 冲任不固证 insecurity of the thoroughfare and controlling vessels pattern	148
7.9.26 冲任瘀阻证 thoroughfare and controlling vessel stasis obstruction pattern	149
7.9.27 湿热阻滞精室证 dampness heat blocks essence chambe pattern	149
7.9.28 痰湿阻滞精室证 sputum dampness blocking essence chamber pattern	149
7.9.29 瘀血阻滞精室证 blood stasis and sperm blocking essence chamber pattern	149
7.9.30 瘀浊阻滞精室证 stasis turbidity blocking essence chambe pattern	149
7.9.31 惊恐伤肾证 fright and fear damage the kidney pattern	149
7.10 脏腑兼证类	149
7.10.1 心肾阴虚证 heart and kidney yin deficiency pattern	149
7.10.2 心肾不交证 disharmony between the heart and kidney pattern	149
7.10.3 心肾阳虚证 yang deficiency of the heart and kidney pattern	150
7.10.4 水气凌心证 water retention affecting the heart pattern	150
7.10.5 心肾气虚证 heart kidney qi deficiency pattern	150
7.10.6 心肾气阴两虚证 both heart and kidney deficiency of qi and yin pattern	150
7.10.7 心肾阴阳两虚证 both heart and kidney deficiency of yin and yang pattern	150
7.10.8 心肺气虚证 qi deficiency of the heart and lung pattern	150
7.10.9 心肺阴虚证 heart and lung yin deficiency pattern	150
7.10.10 心肺气阴两虚证 both heart and lung deficiency of yin and qi pattern	150
7.10.11 心肺阴虚血瘀证 heart and lung yin deficiency ang blood stasis pattern	150
7.10.12 心肺阳虚证 heart and lung yang deficiency pattern	151
7.10.13 心肺热盛证 heart and lung exuberant heat pattern	151
7.10.14 心肾火热证 heart and kidney fiery pattern	151
7.10.15 心脾两虚证 deficiency of the heart and spleen pattern	151
7.10.16 心脾气虚证 heart and spleen qi deficiency pattern	151
7.10.17 心脾阳虚证 heart and spleen yang deficiency pattern	151

7.10.18 心脾气血两虚证 heart and spleen qi deficiency pattern ………… 151
7.10.19 心脾积热证 heart and spleen heat accumulation pattern ………… 151
7.10.20 心肝火旺证 fire hyperactivity of the heart and liver pattern ………… 151
7.10.21 心肝血瘀证 heart and liver blood stasis pattern ………… 152
7.10.22 心肝血虚证 blood deficiency of the heart and liver pattern ………… 152
7.10.23 心肝阴虚证 heart liver yin deficiency pattern ………… 152
7.10.24 心肝血虚挟瘀证 heart and liver blood deficiency and stasis pattern ………… 152
7.10.25 心肝气血两虚证 heart and liver deficiency of qi and blood pattern …… 152
7.10.26 心肝气虚血瘀证 heart and liver both qi deficiency and blood stasis pattern ………… 152
7.10.27 肝肾亏虚证 deficiency of kidney essence pattern ………… 152
7.10.28 肝肾阴虚证 liver and kidney yin deficiency pattern ………… 152
7.10.29 肝肾阴虚阳亢证 liver and kidney yin deficiency and hyperactive yang pattern ………… 153
7.10.30 肝脾两虚证 liver spleen deficiency pattern ………… 153
7.10.31 肝脾气血两虚证 liver and spleen deficiency of qi and blood pattern ………… 153
7.10.32 肝脾气阴两虚证 liver and spleen deficiency of qi and yin pattern ……… 153
7.10.33 肝郁脾虚证 liver stagnant and spleen deficiency pattern ………… 153
7.10.34 土壅侮木证 effulgent liver and weak spleen pattern ………… 153
7.10.35 肝脾湿热证 liver and spleen dampness heat pattern ………… 153
7.10.36 肝脾气滞证 liver and spleen qi stagnation pattern ………… 153
7.10.37 肝脾血瘀证 liver and spleen blood stasis pattern ………… 153
7.10.38 肝热脾虚证 liver heat and spleen deficiency pattern ………… 154
7.10.39 肝胃不和证 liver and stomach disharmony pattern ………… 154
7.10.40 肝胃热盛证 liver and stomach heat filled pattern ………… 154
7.10.41 肝火犯胃证 liver fire attacking the stomach pattern ………… 154
7.10.42 肝胃气滞证 liver and stomach qi stagnation pattern ………… 154
7.10.43 肝胃气滞血瘀证 liver and stomach qi stagnation and blood stasis pattern … 154
7.10.44 肝胃气虚血瘀证 liver and stomach qi deficiency and blood stasis pattern … 154
7.10.45 肝胃气滞阴虚证 liver and stomach qi stagnation and yin deficiency pattern ………… 154
7.10.46 肝胃阴虚证 yin deficiency of liver and stomach pattern ………… 155
7.10.47 肝胃阴虚血瘀证 liver and stomach yin deficiency and blood stasis pattern ………… 155
7.10.48 肝胃虚寒证 liver and stomach cold deficiency pattern ………… 155
7.10.49 肝火犯肺证 liver fire attacking the lung pattern ………… 155
7.10.50 肝肺风热证 liver lung wind heat pattern ………… 155
7.10.51 肝肺热盛证 liver and lung heat flourishing pattern ………… 155
7.10.52 肝郁肾虚证 liver stagnant and kidney deficiency pattern ………… 155
7.10.53 脾肺两虚证 deficiency of both spleen and lung pattern ………… 155

7.10.54 脾肺气阴两虚证 spleen and lung both qi and yin deficiency pattern …… 155
7.10.55 脾肾阳虚证 yang deficiency of the spleen and kidney pattern ………… 155
7.10.56 脾肾气虚证 spleen and kidney qi deficiency pattern ……………… 156
7.10.57 脾肾两虚证 deficiency of both spleen and kidney pattern ………… 156
7.10.58 肺肾阴虚证 yin deficiency of the lung and kidney pattern ………… 156
7.10.59 肺肾气虚证 qi deficiency of the lung and kidney pattern ………… 156
7.10.60 肺肾阳虚证 yang deficiency of lung and kidney pattern……………… 156
7.10.61 肺胃风热证 lung and stomach wind heat pattern ………………… 156
7.10.62 肺胃火热证 intense lung stomach fire pattern……………………… 156
7.10.63 肺胃阴虚证 yin deficiency of lung and stomach pattern …………… 157
7.10.64 毒陷心肝证 toxin attacking heart and liver pattern ………………… 157
7.10.65 肺虚肠脱证 lung deficiency and intestinal withdrawal pattern ………… 157
7.10.66 脾虚肠脱证 spleen deficiency and intestinal withdrawal pattern ……… 157
7.10.67 肾虚肠脱证 kidney deficiency and intestinal withdrawal pattern ……… 157
7.10.68 肝肠气滞证 liver intestinal qi stagnation pattern…………………… 157

7.11 卫表肌肤证类 ……………………………………………………………… 157
 7.11.1 邪袭卫表证 evil attack of wei-defence surface pattern……………… 157
 7.11.2 风袭表疏证 wind evil attacking exterior pattern ………………… 157
 7.11.3 风寒束表证 wind cold fettering the exterior pattern ……………… 158
 7.11.4 风热犯表证 wind heat attacking exterior pattern ………………… 158
 7.11.5 风湿袭表证 wind dampness attacking exterior pattern ……………… 158
 7.11.6 暑湿袭表证 exterior attacked by summer heat dampness pattern ……… 158
 7.11.7 外燥袭表证 external dryness attacking the exterior pattern ………… 158
 7.11.8 风毒犯表证 wind and poison attacking the exterior pattern ………… 158
 7.11.9 温毒袭表证 wind poison attacking the exterior pattern……………… 158
 7.11.10 风湿蕴肤证 warm and toxic attacking skin pattern ……………… 158
 7.11.11 风毒蕴肤证 wind poison attacking skin pattern ………………… 158
 7.11.12 湿毒蕴结肌肤证 dampness and poison accumulate to the skin pattern … 159
 7.11.13 热毒蕴结肌肤证 heat toxicity in the skin pattern ………………… 159
 7.11.14 虫毒蕴肤证 parasite poison smolder at skin pattern ……………… 159
 7.11.15 寒湿蕴肤证 cold dampness smolder at skin pattern ……………… 159
 7.11.16 湿热组结肌肤证 damp heat smolder at skin pattern ……………… 159
 7.11.17 湿痰蕴结肌肤证 dampness sputum smolder at skin pattern ………… 159
 7.11.18 风热郁滞肌肤证 wind heat stagnation skin pattern………………… 159
 7.11.19 虫毒侵袭肌肤证 parasite poison attacking skin pattern ……………… 159
 7.11.20 风水证 wind and water retention with each other pattern …………… 160
 7.11.21 表闭水停证 skin is closed and the water stasis pattern ……………… 160
 7.11.22 瘀滞肌肤证 skin stasis pattern ………………………………… 160
 7.11.23 寒凝血涩肌肤证 cold coagulation astringent skin pattern ………… 160
 7.11.24 肌肤失养证 skin dysplasia pattern ……………………………… 160

7.12 头面官窍证类 ·· 160
7.12.1 实邪犯头证 excess evil attacking the head pattern ·· 160
7.12.2 实邪犯目证 excess evil attacking the eyes pattern ·· 161
7.12.3 气轮证 qi ring pattern ··· 162
7.12.4 血轮证 blood ring pattern ··· 162
7.12.5 肉轮证 flesh ring pattern ·· 163
7.12.6 风轮证 wind ring pattern ··· 163
7.12.7 水轮证 water ring pattern ··· 164
7.12.8 耳窍证 ear pattern ··· 165
7.12.9 鼻窍证 nasal pattern ··· 166
7.12.10 咽喉证 pharyngeal pattern ··· 168
7.12.11 齿龈证 gum pattern ·· 169
7.12.12 口唇证 mouth and lips pattern ·· 170
7.12.13 邪犯清窍证 evil crime clearing the orifice pattern ··· 171
7.13 经脉筋骨证类 ·· 171
7.13.1 风中经络证 channel hit by wind pattern ··· 171
7.13.2 经气不利证 adverse qi pattern ·· 173
7.14 六经病证 ·· 174
7.14.1 太阳病证 taiyang disease pattern ··· 174
7.14.2 太阳腑证 taiyang-fu organ pattern ·· 175
7.14.3 阳明病证 yangming disease pattern ·· 175
7.14.4 阳明经证 yangming channel pattern ··· 175
7.14.5 阳明腑证 yangming-fu organ pattern ·· 175
7.14.6 少阳病证 shaoyang disease pattern ··· 175
7.14.7 太阴病证 taiyin disease pattern ·· 175
7.14.8 少阴病证 shaoyin disease pattern ··· 175
7.14.9 厥阴病证 jueyin disease pattern ··· 176
7.15 其他证类 ·· 176
7.15.1 邪扰胸膈证 chest diaphragm evil disturbance pattern ······································· 176
7.15.2 湿热弥漫三焦证 diffusive dampness heat in sanjiao pattern ······························· 177
7.15.3 上焦湿热证 damp heat in the upper jiao pattern ··· 177
7.15.4 中焦湿热证 damp heat in middle jiao pattern ·· 177
7.15.5 下焦湿热证 damp heat in lower jiao pattern ·· 178
7.15.6 邪犯少腹证 hypoabdominal of evil offenders pattern ······································· 178
7.15.7 邪入少阳证 evil entering shaoyang pattern ·· 179
7.15.8 胎毒蕴热证 accumulated heat due to fetal toxicity pattern ·································· 179
7.15.9 营卫不和证 patterns of disharmony between ying and wei-defence pattern
··· 179
7.15.10 表寒里热证 patterns of exterior cold and interior heat pattern ···························· 179
7.15.11 表热里寒证 superficies heat and interior cold pattern ····································· 179

7.15.12 表里俱寒证 cold in both superficies and interior pattern ……… 179
7.15.13 表里俱热证 heat in both exterior and interior pattern ……… 179
7.15.14 上盛下虚证 upper body exuberance and lower body deficience pattern … 179
7.15.15 上热下寒证 upper heat and lower cold pattern ……… 179
7.15.16 上寒下热证 upper cold and lower heat pattern ……… 179

第八章 治 法 ……… **180**

8.1 内治法 ……… 180
8.1.1 汗法 ……… 180
8.1.2 吐法 ……… 181
8.1.3 下法 ……… 181
8.1.4 和法 ……… 182
8.1.5 温法 ……… 183
8.1.6 清法 ……… 184
8.1.7 消法 ……… 188
8.1.8 补法 ……… 192
8.1.9 八法并用 ……… 197
8.1.10 其他内治法 ……… 198

8.2 外治法 ……… 211
8.2.1 解毒散痈 removing toxin for eliminating carbuncles ……… 211
8.2.2 活血解毒 promoting blood circulation and detoxication ……… 211
8.2.3 解毒护阴 removing toxin for protecting yin ……… 211
8.2.4 解毒消肿 removing toxin and reducing swelling ……… 211
8.2.5 清热排脓 clearing heat and evacuating pus ……… 211
8.2.6 去腐生肌 eliminating slough and promoting granulation ……… 211
8.2.7 消痈散疖 resolving carbuncle and expulsing boil ……… 211
8.2.8 燥湿敛疮 eliminating dampness and astringing sores ……… 211
8.2.9 清解余毒 expelling retained toxin ……… 212
8.2.10 敛疮止痛 healing sore and relieving pain ……… 212
8.2.11 明目 improving vision ……… 212
8.2.12 退翳明目 removing nebula and improving vision ……… 212
8.2.13 祛风明目 dispelling wind and improving vision ……… 212
8.2.14 清热明目 clearing heat and improving vision ……… 212
8.2.15 补肾明目 tonifying the kidney and improving vision ……… 212
8.2.16 滋肝明目 enriching the liver and improving vision ……… 212
8.2.17 养血明目 nourishing the blood and improving vision ……… 212
8.2.18 通耳 unblocking the ears ……… 213
8.2.19 疏风宣耳 dispersing wind and diffusing the ears ……… 213
8.2.20 解毒利耳 removing toxin and disinhibiting the ears ……… 213
8.2.21 补血养耳 tonifying the blood and nourishing the ears ……… 213

8.2.22 益气通耳 replenishing qi and unblocking the ears ……………… 213
8.2.23 滋阴濡耳 enriching yin to moisten the ears ………………… 213
8.2.24 滋肝肾濡耳 enriching the liver and kidney to moisten the ears …… 213
8.2.25 通鼻 unblocking the nose ………………………………………… 213
8.2.26 疏风通鼻 dispelling wind and unblock the nose ……………… 213
8.2.27 散寒通鼻 dissipating cold and unblocking the nose ………… 213
8.2.28 清燥润鼻 clearing dryness and moistening the nose ………… 214
8.2.29 祛瘀通鼻 dispelling stasis and unblocking the nose ………… 214
8.2.30 芳香通鼻 unblocking the nose with aroma …………………… 214
8.2.31 利咽 disinhibiting the pharynx ……………………………… 214
8.2.32 疏风利咽 dispersing wind and disinhibiting the pharynx …… 214
8.2.33 散寒利咽 dissipating cold and disinhibiting the pharynx …… 214
8.2.34 祛痰利咽 dispelling phlegm and disinhibiting the pharynx … 214
8.2.35 理气利咽 regulating qi and disinhibiting the pharynx ……… 214
8.2.36 生津利咽 producing fluid and disinhibiting the pharynx …… 214
8.2.37 固齿 strengthening the teeth ………………………………… 215
8.2.38 补肾固齿 tonifying the kidney to strengthen the teeth ……… 215
8.2.39 益气固齿 replenishing qi to strengthen the teeth ……………… 215
8.2.40 补血健齿 tonifying the blood and fortifying the teeth ……… 215
8.2.41 滋阴润齿 enriching yin to moisten the teeth …………………… 215
8.2.42 温阳健齿 warming yang to fortify the teeth …………………… 215
8.2.43 降火固齿 downbearing fire to strengthen the teeth …………… 215
8.2.44 解毒利龈 removing toxin and disinhibiting the gums ………… 215
8.2.45 药熨疗法 medicated ironing therapy ………………………… 215
8.2.46 热敷疗法 hot compress therapy ……………………………… 215
8.2.47 敷贴疗法 paste application therapy …………………………… 216
8.2.48 膏药疗法 plaster application therapy ………………………… 216
8.2.49 药膏疗法 ointment application therapy ……………………… 216
8.2.50 湿敷疗法 wet compress therapy ……………………………… 216
8.2.51 熏洗疗法 fumigating and washing therapy ………………… 216
8.2.52 冲洗疗法 flushing and washing therapy …………………… 216
8.2.53 浸洗疗法 steeping and washing therapy …………………… 216
8.2.54 腐蚀疗法 corrosion therapy …………………………………… 216
8.2.55 切开疗法 incision therapy ……………………………………… 216
8.2.56 引流疗法 drainage therapy …………………………………… 217
8.2.57 放血疗法 blood-letting therapy ……………………………… 217
8.2.58 火烙疗法 fire cauterizing therapy …………………………… 217
8.2.59 刮痧疗法 scraping therapy …………………………………… 217
8.2.60 点眼疗法 eye drop therapy …………………………………… 217
8.2.61 吹耳疗法 ear-insufflation therapy …………………………… 217

8.2.62 滴耳疗法 ear drop therapy ·········· 217
8.2.63 洗耳疗法 ear washing therapy ·········· 217
8.2.64 塞鼻疗法 nose insertion therapy ·········· 217
8.2.65 吹鼻疗法 nose insufflation therapy ·········· 218
8.2.66 滴鼻疗法 medicated liquid nose drops therapy ·········· 218
8.2.67 取嚏疗法 catching therapy ·········· 218
8.2.68 喷雾疗法 aerosol therapy ·········· 218
8.2.69 灌肠疗法 coloclysis therapy ·········· 218
8.2.70 包扎固定疗法 bandage-fixing therapy ·········· 218
8.2.71 夹板固定疗法 splint-fixing therapy ·········· 218
8.2.72 火烧战船 burning warships ·········· 218

8.3 针灸疗法 ·········· 219
8.3.1 火针疗法 fire needling therapy ·········· 219
8.3.2 七星针疗法 seven-star needling therapy ·········· 219
8.3.3 三棱针疗法 three-edged needling therapy ·········· 219
8.3.4 点刺疗法 pricking therapy ·········· 219
8.3.5 隔姜灸疗法 moxibustion on ginger therapy ·········· 219
8.3.6 隔蒜灸疗法 moxibustion on garlic therapy ·········· 219
8.3.7 艾灸疗法 moxibustion therapy ·········· 219
8.3.8 拔罐疗法 cupping therapy ·········· 219
8.3.9 电针疗法 electro-acupuncture therapy ·········· 220
8.3.10 激光针疗法 laser-acupuncture therapy ·········· 220
8.3.11 水针疗法 hydro-acupuncture therapy ·········· 220
8.3.12 穴位埋线疗法 catgut embedding therapy ·········· 220
8.3.13 针刺麻醉法 acupuncture anesthesia ·········· 220
8.3.14 耳针疗法 auriculo-acupuncture therapy ·········· 220

8.4 推拿疗法 ·········· 220
8.4.1 推拿疗法 massage therapy ·········· 220

8.5 饮食疗法 ·········· 220
8.5.1 食疗 dietotherapy ·········· 220
8.5.2 药酒疗法 medicated liquor therapy ·········· 221

8.6 杂疗法 ·········· 221
8.6.1 热蜡疗法 hot wax therapy ·········· 221
8.6.2 药浴疗法 medicinal bath therapy ·········· 221
8.6.3 烟熏疗法 fumigation therapy ·········· 221
8.6.4 夹板固定疗法 splint-fixing therapy ·········· 221

第九章 针 灸 ·········· 222

9.1 基础术语 ·········· 222
9.1.1 针灸 acupuncture and moxibustion ·········· 222

9.1.2 针灸学 science of acupuncture and moxibustion ……………… 222
9.1.3 兽医针灸学 science of veterinary acupuncture and moxibustion ………… 222
9.1.4 经络学 subject of meridian and collateral ……………… 222
9.1.5 腧穴学 subject of acupuncture points ……………… 222
9.1.6 刺法灸法学 subject of acupuncture and moxibustion technique ……… 222
9.1.7 针灸治疗学 subject of acupuncture and moxibustion therapy ……… 222
9.1.8 针刺疗法 acupuncture therapy ……………… 222
9.1.9 针刺手法 needling technique ……………… 222
9.1.10 灸法 moxibustion ……………… 223
9.1.11 针灸刺激量 stimulating quantity of acupuncture and moxibustion ……… 223
9.1.12 针灸刺激强度 stimulus intensity of acupuncture and moxibustion ……… 223
9.1.13 得气 obtaining qi ……………… 223
9.1.14 调气 regulating qi ……………… 223
9.1.15 候气 awaiting qi ……………… 223
9.1.16 疏通经络 unblocking the meridian and collateral ……… 224
9.1.17 针灸治则 ……………… 224
9.1.18 选穴法 point selection ……………… 225
9.1.19 配穴法 points combination ……………… 226
9.2 针术 ……………… 227
9.2.1 九针 nine kinds of classical needles ……………… 227
9.2.2 兽用传统针具 ……………… 228
9.2.3 兽用现代针具 ……………… 230
9.2.4 押手法 ……………… 230
9.2.5 持针法 ……………… 230
9.2.6 进针法 ……………… 231
9.2.7 运针法 ……………… 232
9.2.8 留针法 ……………… 233
9.2.9 出针法 ……………… 233
9.2.10 针刺异常情况 ……………… 234
9.2.11 针刺补泻法 ……………… 234
9.2.12 练针法 ……………… 235
9.2.13 常用针术 ……………… 236
9.3 灸术 ……………… 237
9.3.1 灸具 ……………… 237
9.3.2 常用灸法 ……………… 238
9.3.3 其他针灸疗法 ……………… 240

参 考 文 献 ……………… 244
汉语拼音索引 ……………… 245
英文对应词索引 ……………… 279

第一章 基础理论总论与阴阳五行

1.1 总论

1.1.1 中兽医学 Chinese veterinary medicine（CVM）

研究和应用中国兽医学的理、法、方、药及针灸技术，防治动物病证的一门综合性学科。

1.1.1.1 中兽医基础理论 basic theory of Chinese veterinary medicine

研究和阐明中兽医学的基本概念、基本知识、基本原理和基本规律的知识体系。主要包括阴阳学说、五行学说、脏腑学说、气血津液精学说、经络学说、病因病机等内容。

1.1.1.1.1 整体观念 holistic concept

动物体本身各组成部分之间，在结构上不可分割，在生理功能上相互协调，在病理变化上相互影响，是一个有机的整体；同时，动物与环境之间紧密相关，也构成一个有机整体。

1.1.1.1.2 辨证论治 treatment based on pattern identification

中兽医学认识疾病、确定防治措施的基本过程。根据四诊所获取的病情资料进行分析综合，以判断疾病的性质，进而确定治则和治法。

1.1.1.1.3 证候 pattern clinical manifestation

证的外候。疾病过程中一定阶段的机体脏腑经络的综合反应状态，表现为临床可观察到的症状与体征。

1.1.1.1.4 理法方药 theory, principle, prescription and medicinal

根据中兽医学理论，明确临床证候，分析病因病机，确定治则治法，进行组方遣药。

1.1.1.1.5 八纲辨证 pattern differentiation by the eight principles

综合分析四诊获得的临床资料，将病证按照阴阳表里寒热虚实八纲归纳为八类证型，以辨明疾病的病位深浅、病性寒热以及邪正盛衰的辨证方法。

1.1.1.1.6 脏腑辨证 pattern differentiation of zang-fu organs

根据脏腑的生理功能和病理变化，对疾病的证候进行分析归纳，推断发病的脏腑以及病因、病机、病性和正邪盛衰等情况。

1.1.1.1.7 气血津液辨证 pattern differentiation of qi, blood and body fluids

应用有关气血津液的理论，分析临床上所见的一系列证候，以辨别气、血、津液病理变化的一种辨证方法。

1.1.1.1.8 六经辨证 pattern differentiation of six meridians

以阴阳为总纲，根据机体抗病力的强弱、病因的属性、病势的进退缓急等因素，将外感病演变过程中所表现的各种证候，归纳为太阳、阳明、少阳、太阴、少阴、厥阴六个阶段，并以这六个阶段出现的不同症状和体征作为辨证论治的根据。

1.1.1.1.9 卫气营血辨证 pattern differentiation of wei-defence, qi, ying nutrients and blood

将外感温病发生发展过程中所出现的证候概括为卫分、气分、营分、血分四类病证，以辨明病位深浅、病情轻重以及病势趋向的辨证方法。

1.1.1.1.10 三焦辨证 pattern differentiation of triple warmer

将温病发生发展过程中所出现的证候按上、中、下三焦归纳为三类证型，以辨明病位深浅、病情轻重和病势趋向的辨证方法。

1.2 阴阳类

1.2.1 阴阳学说 yin-yang theory

以阴和阳的相对属性及其对立、互根、消长、转化变化来说明动物体的组织结构、生理功能和疾病的发生、发展和变化，指导疾病的诊断和防治，是中国古代朴素的对立统一理论。

1.2.1.1 阴阳 yin-yang

代表相互联系又相互对立的两个不同事物，或同一事物内部相互对立的两个不同方面。

注：在中兽医学中用于阐述与动物体生理病理相关的病因、症状、治疗、用药等内容。

1.2.1.1.1 阴 yin

与阳相对，具有向下的、静止的、有形的、寒凉的、向内的、晦暗的、减退的、抑制的、虚弱的或属于物质方面等属性的事物及其运动。

1.2.1.1.2 阳 yang

与阴相对，具有向上的、运动的、无形的、温热的、向外的、明亮的、亢进的、兴奋的、强壮的或属于功能方面等属性的事物及其运动。

1.2.1.1.3 阴气 yin qi

与阳气相对。气之属于阴者，具有凝聚、滋润、抑制等作用。

1.2.1.1.4 阳气 yang qi

与阴气相对。气之属于阳者，具有温煦、推动、兴奋等作用。

1.2.1.1.5 阳化气 yang forming qi

阳主化气的功能，与"阴成形"相对。阳主动而散，散则无形而化气。

1.2.1.1.6 阴成形 yin forming substance

阴主成形的功能，与"阳化气"相对。阴主静而凝，聚则有形而成形。

1.2.1.2 阴阳交感 interaction of yin and yang
　　阴阳二气在运动中相互感应而交合的过程，是宇宙万物生长与发展变化之究极本原。

1.2.1.2.1 阴阳对立 opposition between yin and yang
　　阴阳双方存在着相互排斥、相互斗争和相互制约的关系。

1.2.1.2.2 阴阳互根 yin and yang reciprocally root
　　阴阳双方具有相互依存、互为根本的关系，即阴或阳的任何一方，都不能脱离另一方而单独存在，并且双方也存在着相互资生、相互促进的关系。

1.2.1.2.3 阴阳消长 dynamic waxing and waning between yin and yang
　　对立互根的阴阳双方始终处于不断运动变化中，此消彼长，力求维系动态平衡的关系。

1.2.1.2.4 阴阳转化 yin and yang mutually transform
　　对立的阴阳双方在一定条件下，可向其属性相反的方面转化，即阴可以转化为阳，阳可以转化为阴。

1.2.1.2.4.1 重阴必阳 extreme yin transforming into yang
　　阴重复积累至极必转化为阳。
　　同义词：阴极反阳

1.2.1.2.4.2 重阳必阴 extreme yang transforming into yin
　　阳重复积累至极必转化为阴。
　　同义词：阳极反阴

1.2.1.3 阴平阳秘 sound yin and firm yang
　　阴气平顺，阳气固守，相互为用，动态平衡。

1.2.1.4 阴阳平衡 equilibrium between yin and yang
　　阴阳之间和谐、均衡、相对稳定。
　　同义词：阴阳匀平

1.2.1.5 阴阳自和 natural harmony of yin and yang
　　机体通过自身调节，使阴阳恢复平衡。

1.2.1.6 阴阳离决 yin-yang separation
　　阴阳之间不能维系相互依存关系而分离决裂的一种状态。

1.2.1.6.1 孤阳不生 solitary yang failing to rise
　　无阴则阳无以生。

1.2.1.6.2 独阴不长 solitary yin failing to increase
　　无阳则阴无以长。

1.2.1.7 阴中之阴 yin within yin
　　阴阳之中复有阴阳，阴中再分阴阳之属于阴者。

1.2.1.8 阴中之阳 yang within yin
阴阳之中复有阴阳,阴中再分阴阳之属于阳者。

1.2.1.9 阳中之阴 yin within yang
阴阳之中复有阴阳,阳中再分阴阳之属于阴者。

1.2.1.10 阳中之阳 yang within yang
阴阳之中复有阴阳,阳中再分阴阳之属于阳者。

1.2.1.11 阳生阴长 growth of yang and generation of yin
阴阳相互为用,阳气生化正常,阴气才能不断滋长。

1.2.1.12 阳杀阴藏 decline of yang and safekeeping of yin
阳气肃杀收束,阴气封蛰闭藏。

1.2.1.13 生杀之本 origin of life
阴阳,阴阳为万物生长毁灭的根本。生,指生长;杀,指消亡;本,指根本。

1.3 五行类

1.3.1 五行学说 five-element theory
以木、火、土、金、水五种物质的特性及其"相生"和"相克"规律来认识世界、解释世界和探求宇宙规律的一种世界观和方法论。

注:中兽医学借以说明动物体的组织结构、生理功能、病理变化并指导临床实践,是中国古代朴素的哲学理论。

1.3.2 五行 five elements
木、火、土、金、水五种物质及其运动、变化的规律和相互关系。

1.3.2.1 五行归类 classification of five elements
按五行的特征以"取类比象"或"推演络绎"的方法,将自然界的事物和现象、动物体脏腑及其生理、病理现象分别归属于木、火、土、金、水五行之中。

1.3.3 五时 five periods of year
春、夏、长夏、秋、冬五季。

同义词:五季

1.3.3.1 长夏 long summer
农历六月。

1.3.4 五气　five qi

风、暑、湿、燥、寒五种气候。

1.3.5 五化　five changes

生、长、化、收、藏五个生化阶段。

1.3.6 五色　five colors

青、赤、黄、白、黑五种颜色。

1.3.7 五味　five flavors

酸、苦、甘、辛、咸五种滋味。

1.3.8 五音　five notes

角、徵、宫、商、羽五个音阶。

1.3.9 五官　five sense organs

动物的目、舌、口（唇）、鼻、耳五个器官。

1.3.10 五方（宫）　five directions

东、南、中、西、北五个方位。

1.3.11 五行相生　five elements mutually generate

五行之间存在着有序的资生、助长和促进的关系，借以说明事物间有相互协调的一面。其次序是木生火、火生土、土生金、金生水、水生木。任何一行都有"生我"及"我生"两方面的关系，生我者为"母"，我生者为"子"。

1.3.11.1 母行　mother element

五行相生，生我之行。

1.3.11.2 子行　child element

五行相生，我生之行。

1.3.11.3 木生火　wood generating fire

在五行相生中，木资生、助长火的作用，用以说明肝对心的资助关系。

1.3.11.4 火生土　fire generating earth

在五行相生中，火资生、助长土的作用，用以说明心对脾的资助关系。

1.3.11.5 土生金　earth generating metal

在五行相生中，土资生、助长金的作用，用以说明脾对肺的资助关系。

1.3.11.6 金生水 metal generating water
在五行相生中，金资生、助长水的作用，用以说明肺对肾的资助关系。

1.3.11.7 水生木 water generating wood
在五行相生中，水资生、助长木的作用，用以说明肾对肝的资助关系。

1.3.12 五行相克 five elements mutually restrain
五行之间存在着有序的克制和制约关系，借以说明事物间相颉颃的一面。其次序是木克土、土克水、水克火、火克金、金克木。任何一行都有我克（所胜）、克我（所不胜）两方面的关系。

1.3.12.1 所胜 restricted object
五行相克中之我所克制的一行。

1.3.12.2 所不胜 un-restricted object
五行相克中之克制于我的一行。

1.3.12.3 木克土 wood restricting earth
在五行相克中，木制约土的作用，用以说明肝对脾的制约关系。

1.3.12.4 土克水 earth restricting water
在五行相克中，土制约水的作用，用以说明脾对肾的制约关系。

1.3.12.5 水克火 water restricting fire
在五行相克中，水制约火的作用，用以说明肾对心的制约关系。

1.3.12.6 火克金 fire restricting metal
在五行相克中，火制约金的作用，用以说明心对肺的制约关系。

1.3.12.7 金克木 metal restricting wood
在五行相克中，金制约木的作用，用以说明肺对肝的制约关系。
同义词：木逢金而伐

1.3.13 五行相乘 five elements mutually over-restrain one another
五行中某一行对其所胜一行的过度克制，即相克太过，是事物间关系失去相对平衡的另一种表现。其次序同于五行相克的次序。

1.3.13.1 木乘土 wood over-restricting earth
在五行相乘中，木过度克土的现象，用以说明肝对脾的过度制约关系。

1.3.13.2 土乘水 earth over-restricting water
在五行相乘中，土过度克水的现象，用以说明脾对肾的过度制约关系。

1.3.13.3 水乘火 water over-restricting fire
在五行相乘中，水过度克火的现象，用以说明肾对心的过度制约关系。

1.3.13.4 火乘金 fire over-restricting metal
在五行相乘中，火过度克金的现象，用以说明心对肺的过度制约关系。

1.3.13.5 金乘木 metal over-restricting wood
在五行相乘中，金过度克木的现象，用以说明肺对肝的过度制约关系。

1.3.14 五行相侮 five elements mutually counter–restrain one another
五行中某一行对其所不胜一行的反向克制，是事物间关系失去相对平衡的另一种表现。其次序与五行相克的次序相反。

同义词：反克；反侮

1.3.14.1 木侮金 wood counter-restricting metal
在五行相侮中，木反克金的现象，用以说明肝对肺的反克关系。

1.3.14.2 金侮火 metal counter-restricting fire
在五行相侮中，金反克火的现象，用以说明肺对心的反克关系。

1.3.14.3 火侮水 fire counter-restricting water
在五行相侮中，火反克水的现象，用以说明心对肾的反克关系。

1.3.14.4 水侮土 water counter-restricting earth
在五行相侮中，水反克土的现象，用以说明肾对脾的反克关系。

1.3.14.5 土侮木 earth counter-restricting wood
在五行相侮中，土反克木的现象，用以说明脾对肝的反克关系。

1.3.15 五行生克 generation and restriction among five elements
五行相生与五行相克的合称，是自然界（或有机体）的正常（或生理）现象。

1.3.16 五行乘侮 over–restriction and counter–restriction among five elements
五行相乘与五行相侮的合称，是自然界（或有机体）的异常（或病理）现象。

1.3.17 五行制化 restriction and generation among five elements
五行相生与五行相克，相互为用，生中有克，克中有生，平衡协调。

1.3.18 五行胜复 alternate preponderance among five elements
五行相胜相制，克制与反克制。其一般规律是：凡先有胜，后必有复，以报其胜。

1.3.19 亢害承制 harmful hyperactivity checked for harmony
五行之间的盛极必制。一行相胜至极，另一行必从而制之，以维持其平衡。

第二章　脏腑与气血津液精

2.1 脏腑类

2.1.1 藏象 visceral manifestation
动物体内在脏腑生理功能、病理变化表现于外的征象。

2.1.2 脏腑 zang-fu
动物体内脏的总称，包括五脏、六腑、奇恒之腑，是动物体的重要组成部分。

2.1.3 脏腑学说 theory of visceral state
通过观察动物机体外在表现、征象，研究动物体各脏腑的生理功能、病理变化及其相互关系的学说，是中兽医学基本理论的重要组成部分。

2.1.4 五脏 five zang organs
心、肝、脾、肺、肾的合称，是化生和储藏精气的器官，具有藏精气而不泻的特点。

2.1.4.1 五脏所主 those dominated by five zang viscera
心主脉，肺主皮毛，脾主肌肉，肝主筋，肾主骨的总称。

2.1.4.2 五液 five humors
五脏所化生的汗、涕、泪、涎、唾合称。

2.1.4.3 五华 manifestations of five zang viscera
五脏精华之气显露于外的部位。即心华在面，肺华在毛，脾华在唇，肝华在爪，肾华在齿的总称。

2.1.4.4 五体 five body constituents
皮、肉、筋、骨、脉的总称。

2.1.4.5 五窍 five signal apertures on the head
目、耳、鼻、口、舌五个器官。它们分属于五脏，为五脏之外窍，即肝开窍于目、肾开窍于耳、肺开窍于鼻、脾开窍于口和心开窍于舌。

2.1.4.6 五脏所恶 those intolerated by five zang viscera
五脏各随其性能与气化而有所恶，心恶热，肺恶寒，脾恶湿，肝恶风，肾恶燥的总称。

2.1.4.7 藏精气而不泻 house of essence qi without leakage

五脏生理特点。五脏贮藏精气，藏而不泄，满而不实，勿使妄泄。

2.1.4.8 心 heart

位于胸中，有心包护于外，与小肠相表里，主血脉和藏神，开窍于舌，在液为汗，下络于小肠，为动物体生命活动的中心，有统管脏腑功能活动的作用。为五脏之一。

2.1.4.8.1 心气 heart qi

心藏之气，与心血相对，由心精（血）所化，是心脏生理活动的物质基础及其动力来源。

2.1.4.8.2 心血 heart blood

心主之血，与心气相对，起营养和滋润的作用，是维持心脏生理活动的物质基础。

2.1.4.8.3 心阴 heart yin

心之阴精，与心阳相对，是维持心生理功能的四种基本物质之一，其作用是滋养心脏，令心阳潜藏，宁心静神。

2.1.4.8.4 心阳 heart yang

心之阳气，与心阴相对，通过温养保持血液流动状态，合心气推动血液运行，并通过温煦、激发振奋心神，制约心阴。

2.1.4.8.5 心系 heart system

指由心联络的小肠、脉、面、舌等构成的系统。

2.1.4.8.6 心藏神 heart housing mind

心具有主宰动物体五脏六腑、形体官窍的一切生命活动和精神意识活动的功能。

同义词：心主神明；心主神志

2.1.4.8.7 心志喜 joy as heart emotion

心主精神情志之喜。

同义词：心在志为喜

2.1.4.8.8 心主血脉 heart governs the blood and vessels

心有推动和调控血液在脉管内运行，以营养和濡润全身的作用。

2.1.4.8.8.1 心主血 heart dominating blood

心具有总管机体血液运行和生成的作用。

2.1.4.8.8.2 心主脉 heart dominating vessel

心与脉相连，使脉道通利、血行脉中的功能。

同义词：心合脉

2.1.4.8.9 心华在面 heart manifesting in complexion

心的功能状态会在面部色彩、光泽上面表现出来。

2.1.4.8.10 心开窍于舌 heart opens into the tongue

心经的别络上行于舌，心的气血上通于舌，心的生理功能及病理变化最易在舌上反映出来，舌的生理功能直接与心相关。

同义词：心主舌；舌为心之苗

2.1.4.8.11 心在液为汗 fluid of the heart is sweat

汗由津液所化生，津液是血液的重要组成部分，血为心所主，血汗同源。

同义词：心主汗；汗为心之液

2.1.4.8.12 心恶热 heart intolerating heat

心易被火热所伤，过热则病。

2.1.4.8.13 心合小肠 pairing between the heart and small intestine

脏腑相合之一。心与小肠互为表里，有经脉相互络属。心气正常，有利于小肠气血的补充，小肠才能发挥分别清浊的功能；小肠功能的正常，又有助于心气的正常活动。

同义词：心与小肠相表里

2.1.4.8.14 心包络 pericardium

心脏外面的包膜，有保护心脏、代心受邪的作用，与六腑中的三焦互为表里。

2.1.4.8.15 心肾相交 coordination between heart and kidney

心火与肾水之间上下、升降、水火、阴阳之间的协调平衡，又称水火既济。

2.1.4.9 肺 lung

位于胸中，上连气道，下络于大肠，与大肠相表里，主气、司呼吸，主宣发和肃降，通调水道，外合皮毛，朝百脉而助心行血，开窍于鼻，在液为涕。为五脏之一。

2.1.4.9.1 肺气 lung qi

肺藏之气，肺脏生理活动的物质基础及其动力来源。主要由肺津化生，具有推动和调控呼吸、行水等作用。

2.1.4.9.2 肺阴 lung yin

肺之阴精，与肺阳相对。肺功能获得的一种基本物质，具有制约肺阳，滋润肺脏，肃降肺气，调节气机，协同纳气，走表滋汗的作用。

2.1.4.9.3 肺阳 lung yang

肺之阳气，与肺阴相对。具有蒸化肺阴之气，推动呼吸运动；蒸腾津液为雾，调通三焦水道而灌溉全身；内温养脏腑，外温养形体并具抗御外邪的作用。

2.1.4.9.4 肺系 lung system

肺与大肠、皮、毛、鼻等构成的由肺联络的系统。

2.1.4.9.5 肺主气 lung governs qi

肺有主宰呼吸之气和一身之气的生成、出入与代谢的功能。

2.1.4.9.5.1 肺司呼吸 lung governs breathing

肺为体内外气体交换的场所，通过肺的呼吸作用，动物机体吸入自然界的清气，呼出体内的浊气，实现动物机体与外界环境间的气体交换，以维持正常的生命活动。

2.1.4.9.5.2 肺主声 lung dominating voice

肺气鼓动声带而发声的功能。

2.1.4.9.6 肺主宣发 lung governs upward and outward diffusion

肺气宣通与布散的功能,与肺主肃降相对。通过宣发作用将体内代谢后的气体呼出体外;并将脾传输至肺的水谷精微之气布散全身,外达皮毛;宣发卫气,以温分肉、司腠理开合。

2.1.4.9.7 肺主肃降 lung governs descent and purification

肺气清肃与向下、向内的功能,与肺主宣发相对。通过肺的向下、向内作用,吸入自然界清气;将津液和水谷精微布散全身,并将代谢产物和多余水液输布于肾和膀胱,排出体外;保持呼吸道的清洁。

2.1.4.9.8 肺主通调水道 lung regulates waterways

肺的宣发和肃降运动对体内水液的输布、运行和排泄有疏通和调节的作用。

2.1.4.9.8.1 肺为水之上源 lung as upper source of water

肺居上焦而调节动物体的水液代谢。

2.1.4.9.8.2 肺为贮痰之器 lung as container of phlegm

痰饮随气升降,无处不到而主要停聚于肺。

2.1.4.9.9 肺朝百脉 lung presides over the hundred vessels

全身的血液,都要通过经脉而汇聚于肺,经肺的呼吸进行气体交换,而后输布于全身。

2.1.4.9.10 肺主治节 lung dominating management and regulation

肺通过调控气、血、津液而治理调节全身生理活动的功能。

2.1.4.9.11 肺为华盖 lung as canopy

肺在脏腑中位居最高,覆盖诸脏。

2.1.4.9.12 肺为娇脏 lung as delicate zang viscus

肺与外界直接相通,为清虚之体,不耐寒热,易受邪侵。

2.1.4.9.13 肺主皮毛 lung governs skin and body hair

肺与皮毛(一身之表)在生理、病理方面均存在着极为密切的关系。肺之精气具有润泽皮毛、固护肌表、调节汗孔散气和排泄汗液的作用;皮毛有散气作用,参与呼吸调节,而有"宣肺气"的功能。在病理上,肺经病变可反映于皮毛,皮毛受邪也可传之于肺。

同义词:肺外合皮毛

2.1.4.9.14 肺华在毛 lung manifesting in hair

皮毛的焦枯光泽是肺脏的生理功能的反映。

2.1.4.9.15 肺开窍于鼻 lung opens into the nose

鼻为肺窍,有司呼吸和主嗅觉的功能。肺通过呼吸道最上端的鼻与自然界相通以呼吸。

同义词：肺主鼻

2.1.4.9.16 肺在志为悲 grief as lung emotion

肺主精神情志之悲。

同义词：肺志悲

2.1.4.9.17 肺在液为涕 fluid of the lung is nasal discharge

涕由肺津所化，并有赖于肺气的宣发；涕的量和性状可反映肺的受邪和病变情况。

同义词：肺主涕

2.1.4.9.18 肺恶寒 lung intolerating cold

肺本属寒，易被寒邪所伤，过寒则病。

2.1.4.9.19 肺合大肠 pairing between the lung and large intestine

脏腑相合之一，肺与大肠相表里，有经脉互相络属。肺气清肃下行，能促进大肠传导糟粕；大肠传导通畅，肺气才能清肃通利。

同义词：肺与大肠相表里

2.1.4.9.20 肺肾相生 mutual promotion between lung and kidney

肺肾或肺肾阴液之间的相互资生。

同义词：金水相生

2.1.4.10 脾 spleen

位于腹内。经脉络于胃，与胃相表里，主运化，主升清、统血，主肌肉四肢，开窍于口，在液为涎。为五脏之一。

2.1.4.10.1 脾气 spleen qi

脾藏之气，是维持脾脏生理活动的物质基础及其动力来源。

2.1.4.10.2 脾阴 spleen yin

脾之阴精，与脾阳相对。具有宁静、内收、濡养作用。营养于脾且协助脾气、脾阳等运化水谷精微的重要基础物质。

2.1.4.10.3 脾阳 spleen yang

脾之阳气，与脾阴相对。具有振奋、推动、温煦作用，是运化水谷与统摄血液不可缺少的重要因素。

2.1.4.10.4 脾系 spleen system

脾与胃、唇、口等构成的由脾联络的系统。

2.1.4.10.5 脾主运化 spleen governs transportation and transformation

脾有消化、吸收、运输营养物质及水液的功能。动物机体的脏腑经络、四肢百骸、筋肉、皮毛，均有赖于脾的运化以获取营养，为脏中之母。

2.1.4.10.5.1 脾为后天之本 spleen dominating acquirement

脾运化的水谷精微是动物出生后生命物质的主要来源。

2.1.4.10.5.2 脾（胃）为气血生化之源 spleen as source of qi-blood formation

脾（胃）消化吸收的水谷精微是化生气血的主要物质来源。

2.1.4.10.5.3 脾为生痰之源 spleen as source of phlegm

脾运化水液功能失常聚湿成痰饮。

2.1.4.10.5.4 脾藏营 spleen housing nutrient

脾藏纳由水谷精微化生的营血的功能。

2.1.4.10.6 脾主统血 spleen contains blood

脾统摄血液在脉中正常运行，不致溢出脉外。

2.1.4.10.7 脾主升清 spleen ascends the nutrients

脾能将胃肠吸收的水谷精微输布于肺，再通过心、肺的作用化生气血，以营养濡润全身。

2.1.4.10.8 脾旺不受邪 strong spleen being pathogen resistant

脾气健旺，正气充足，充养心肺肝肾，故脾旺四时不受邪。

2.1.4.10.9 脾主四肢 spleen dominating limbs

脾运输水谷精微至四肢以维持其正常的生理活动。

2.1.4.10.10 脾主肌肉 spleen dominating muscle

脾化生的水谷精微充养肌肉维持其正常生理活动。

2.1.4.10.11 脾华在唇 spleen manifesting in lips

脾有经络与唇相通，唇是脾的外应，口唇可以反映出脾运化功能的盛衰。

2.1.4.10.12 脾开窍于口 spleen opens into the mouth

脾主水谷的运化，口是水谷摄入的门户；脾气通于口，与食欲和味觉有直接联系。

同义词：脾主口

2.1.4.10.13 脾在志为思 pensiveness as spleen emotion

脾主精神情志之思虑。

同义词：脾志思

2.1.4.10.14 脾在液为涎 fluid of the spleen is thin saliva

脾气布散脾精上注于口而为涎，以辅助脾胃之消化，但不溢出口外。

同义词：脾主涎

2.1.4.10.15 脾恶湿 spleen intolerating dampness

脾喜燥恶湿，脾为太阴湿土之脏，运化水湿，易受湿邪困扰，水湿过盛则易伤脾。

2.1.4.10.16 脾与胃相表里 pairing between the spleen and stomach

脏腑相合之一。脾与胃同居中焦，互为表里，通过经脉的相互络属，为气血生化之源，同为后天之本。

同义词：脾合胃

2.1.4.11 肝 liver

位于腹腔的前部，胆附于其下（马有胆管、无胆囊），通过脉络于胆，与胆相表里，主藏血，主疏泄，主筋。肝开窍于目，在液为泪。为五脏之一。

2.1.4.11.1 肝气 liver qi

肝气是推动肝进行各种生理活动的物质基础，肝主疏泄，调畅气机，是肝气功能的具体体现；肝统藏血液是调节分配血量的主要职能；主持谋虑，辅佐心神，参与思维活动也是肝气的重要功能之一。

2.1.4.11.2 肝血 liver blood

指肝脏所藏之血，全身血液的组成部分，血藏肝内，一是柔软肝体，制约肝用，防止太过；二是荣养筋、目、爪、冲任二脉等组织器官；三是肝血藏魂守舍，舒畅精神情志。肝脏生理活动的物质基础。

2.1.4.11.3 肝阴 liver yin

肝之阴精，根于肾阴，与肝阳相对。具有滋养、宁静、柔润作用，并能制约过亢的肝阳。

2.1.4.11.4 肝阳 liver yang

肝之阳气，根植于肾阳，与肝阴相对。具有温煦、升发、疏泄作用。

2.1.4.11.5 肝系 liver system

肝与胆、筋、爪、目等构成的由肝联络的系统。

2.1.4.11.6 肝主疏泄 liver governs the free flow of qi

肝具有疏通、畅达、宣泄全身气血、津液的作用。

2.1.4.11.7 肝主谋虑 liver dominating design of strategy

肝参与考虑谋划的思维活动。

2.1.4.11.8 肝在志为怒 anger as liver emotion

肝主精神情志之怒。

同义词：肝志怒

2.1.4.11.9 肝藏血 liver stores blood

肝有贮藏血液及调节血量的功能。

2.1.4.11.10 肝主升发 liver dominating rise and dispersion

肝升腾阳气，条达舒畅，生机不息的特性。

2.1.4.11.11 肝为刚脏 liver being firm-characierized zang viseus

肝喜条达，阳气用事，其气易上亢逆乱的特性。

2.1.4.11.12 肝体阴用阳 liver being yin in substance and yang in function

肝脏本体与功能（特性）的关系。肝主藏血，以血为体，血属阴；肝主疏泄，以气为用，气属阳。肝体阴柔，其用阳刚，阴阳和调，刚柔相济。

2.1.4.11.13 肝主筋 liver dominating tendon

肝有为筋膜提供营养，以维持其正常功能的作用。

2.1.4.11.14 肝华在爪 liver manifesting in nail

爪甲的色泽形态是肝脏功能的反映。

2.1.4.11.15 肝开窍于目 liver opens into the eyes

目主视觉，肝有经脉与之相连，其功能的发挥有赖于五脏六腑之精气，特别是肝血的滋养。

同义词：肝主目

2.1.4.11.16 肝在液为泪 fluid of the liver are tears

肝开窍于目，泪从目出，由肝精、肝血经肝气疏泄于目而化生。

2.1.4.11.17 肝恶风 liver intolerating wind

肝主风，肝风容易化热、化火，风胜则病。

2.1.4.11.18 肝合胆 pairing between the liver and gallbladder

脏腑相合之一。胆附于肝，有经脉相互络属。肝为脏属阴，胆为腑属阳，在五行中均属木，构成了阴阳表里相合关系，在生理上相互联系，在病理上相互影响。

同义词：肝与胆相表里

2.1.4.11.19 肝肾同源 liver and kidney tron same source

肝藏血、肾藏精，精血相互滋生。

2.1.4.12 肾 kidney

位于腰部，左右各一。通过脉络于膀胱，与膀胱相表里。肾藏精，主命门之火，主水，主纳气，主骨、生髓、通于脑，开窍于耳，司二阴，在液为唾。为五脏之一。

2.1.4.12.1 肾精 kidney essence

肾藏之精，包括生殖之精和五脏六腑之精，是肾脏生理活动的物质基础。

2.1.4.12.2 肾气 kidney qi

肾精所化之气。肾脏生理活动的物质基础和维持机体生命活动的基本动力。

2.1.4.12.3 肾阴 kidney yin

肾之阴液，为全身阴液之根本，与肾阳相对。具有宁静、滋润和濡养作用。

2.1.4.12.4 肾阳 kidney yang

肾之阳气，为机体阳气之根本，与肾阴相对。具有温煦、推动、振奋作用。

同义词：肾主命门之火

2.1.4.12.5 肾藏精 kidney stores essence

精的产生、贮藏及转输均由肾所主。

2.1.4.12.5.1 肾主先天之本 kidney dominating innateness

肾精主持胚胎形成和生殖发育的功能。

同义词：肾主先天

2.1.4.12.5.2 肾主生殖 kidney dominating reproduction

肾中精气促进生殖器官成熟、维持繁衍生殖的功能。

2.1.4.12.6 肾主纳气 kidney governs qi reception

肾有摄纳呼吸之气、协助肺司呼吸的功能。

2.1.4.12.7 肾主水 kidney governs water

肾具有主持和调节水液代谢的作用。

2.1.4.12.8 肾主封藏 kidney dominating storage

肾固密、贮藏脏腑之精的特性。

2.1.4.12.9 肾主骨 kidney dominating bone

肾精生髓充养骨骼的功能。

2.1.4.12.10 肾生髓 kidney produces marrow

肾精化生脊髓、骨髓和脑髓的功能。

2.1.4.12.11 肾华在齿 kidney manifesting in teeth

动物牙齿之色泽荣枯是肾脏功能的反映。

2.1.4.12.12 肾开窍于耳 kidney opens into the ears

肾的上窍是耳。

同义词：肾主耳

2.1.4.12.13 肾司二阴 kidney controls two lower orifices

前阴尿生殖道和后阴肛门均为肾之外窍。二阴的功能与肾相关。肾主封藏，固摄下元而主司二阴。

同义词：肾开窍于二阴

2.1.4.12.14 肾在志为恐 fear as kidney emotion

肾主精神情志之恐。过度的恐惧，有时可使肾气不固，气泄于下，导致二便失禁。

同义词：肾志恐

2.1.4.12.15 肾在液为唾 fluid of the kidney is thick saliva

肾主管保护和润泽口腔的唾液。

2.1.4.12.16 肾恶燥 kidney intolerating dryness

肾为水脏，易燥伤阴液为病。

2.1.4.12.17 肾与膀胱相表里 pairing between the kidney and urinary bladder

肾与膀胱相表里，通过脉相互络属。脏腑相合之一。

同义词：肾合膀胱

2.1.4.12.18 命门 life gate

动物体生命的根本，气化的本源，与肾的功能密切相关，对五脏六腑的功能发挥着决定性的作用。其位置说法有二：其一认为命门在两肾之间；其二认为左为肾，右为命门。

2.1.5 六腑 six fu organs

胆、胃、小肠、大肠、膀胱、三焦的统称，是受盛和传化水谷的器官，具有传化浊物、泻而不藏的特点。

2.1.5.1 六腑以通为用 six fu viscera unobistructed in function
六腑气机通畅则传化物而不藏，实而不满的特性。

2.1.5.2 六腑以降为顺 six fu viscera descended in function
六腑气机下降，传导化物，泻而不藏的特性。

2.1.5.3 传化之腑 fuviscera in transportation of transformed products
胃、小肠、大肠、三焦、膀胱五个传导化物之腑的统称。

2.1.5.3.1 传化物而不藏 transportation of transformed products without storing them
六腑的生理特点。受纳和传化水谷，排泄糟粕。

2.1.5.4 胆 gallbladder
附于肝（马有胆管，无胆囊），内藏胆汁。居六腑之首，又为奇恒之腑。主要生理功能是贮藏、排泄胆汁。

2.1.5.4.1 胆者精之腑 gallbladder being essence of fu viscera
胆是六腑之一，贮藏和排出胆汁，而胆汁是促进食物消化的精微物质。

2.1.5.4.2 胆气 gallbladder qi
胆之精气。具有分泌与排泄胆汁及主决断的功能。

2.1.5.4.3 胆汁 bile
由肝之精气化生汇聚而成，贮存于胆囊，排泄进入小肠，参与饮食的消化、吸收。

2.1.5.4.4 胆主决断 gallbladder dominating decision
胆具有助心判断事物、作出决定的功能。

2.1.5.5 胃 stomach
位于膈下，上接食道，下连小肠。有经脉络于脾，与脾相表里，胃主受纳和腐熟水谷。

同义词：太仓

2.1.5.5.1 五脏六腑之海 reservoir of five zang and six fu viscera
胃。受纳腐熟水谷，五脏六腑皆禀气于胃。

2.1.5.5.2 胃脘 stomach cavity
胃腔的统称。其上部为上脘，包括贲门；下部为下脘，包括幽门；上下脘之间为中脘。

2.1.5.5.3 胃气 stomach qi
胃之气。胃的生理功能，以降为顺。胃气是指脾胃消化吸收水谷精微的功能作用。其一是指胃的生理功能和特性；其二是脾胃的消化功能；其三是指胃功能在脉象上的反映；其四是专指胃中之阳气；其五是指机体诸气之后天之本，古有有胃气则生，无胃气则亡之说。

2.1.5.5.4 胃阳 stomach yang

胃之阳气，与胃阴相互为用。共同维持正常的纳食化谷功能。

2.1.5.5.5 胃阴 stomach yin

胃之阴液，与胃阳相互为用。共同维持正常的纳食化谷功能。

2.1.5.5.5.1 胃津 stomach fluid

胃的津液。

2.1.5.6 胃主受纳 stomach governs receiving and holding

胃有接受和容纳饮食的作用。

2.1.5.7 胃主腐熟 stomach governs decomposition

胃的主要功能之一，将饲草料初步消化为食糜的过程。

2.1.5.8 胃主通降 stomach dominating descent

胃以降为顺，胃的气机通畅下降，使初步消化的食糜向下传送至肠道。

2.1.5.8.1 胃主降浊 stomach dominating residue descent

与脾主升清相对。胃气通降使腐熟后的水谷下传至肠道，并将糟粕排出体外，保持胃肠虚实更替状态。

2.1.5.9 胃喜柔润 stomach preferring softness and moisture

胃喜润恶燥，胃为阳明燥土之腑，赖阴液的滋润以维持正常的生理功能。

2.1.5.10 仓廪之本 source of granary supply

脾、胃、大肠、小肠、三焦、膀胱等有出纳、转输、传化水谷的功能。

2.1.6 小肠 small intestine

位于腹部，前连于胃，后接大肠。有经脉络于心，与心相表里。主要生理功能是受盛化物和分别清浊。

2.1.6.1 小肠主受盛 small intestine dominating reception of residue as container

受盛之官，小肠接受和容纳胃腐熟之水谷（食糜）的功能。

2.1.6.2 小肠主化物 small intestine dominating digestion

小肠消化水谷，吸收精微的功能。

2.1.6.3 小肠主液 small intestine governs thick body fluids

小肠在吸收水谷精微的同时，也吸收了大量水液。

2.1.6.4 泌别清浊 separation of the refined from residue

小肠一方面对食糜做进一步消化，并将水谷精华部分吸收以营养全身；另一方面传输食物残渣和水液至大肠和膀胱，变化为粪便和尿液而排出体外。

2.1.7 大肠 large intestine

位于腹部，前连小肠，后通肛门。有经脉络于肺，与肺相表里。主要功能是传送糟粕。

2.1.7.1 大肠主传导 large intestine dominating conveyance

大肠把由小肠传输下来的食物残渣，吸收剩余水分和营养，形成粪便糟粕，排出体外。

2.1.7.2 大肠主津 large intestine governs the thin body fluids

大肠接受食物残渣，吸收水分而参与水液代谢的功能。

2.1.8 膀胱 bladder

位于腹后部，有经脉络于肾，与肾相表里；为津液之腑、州都之官，主藏津液。

2.1.8.1 膀胱主藏津液 bladder dominating fluid storage

膀胱为津液之腑，州都之官。依赖肾的气化而储存尿液的功能。

2.1.8.2 膀胱气化 qi transformation of the urinary bladder

水液经过小肠的吸收后，输布于肾的部分，经肾阳的蒸化成为尿液，渗入膀胱，贮存到一定量后，引起排尿动作，排出体外。

2.1.9 三焦 san jiao

上焦、中焦和下焦的合称。是水液升降出入的道路。属六腑之一。

2.1.9.1 上焦 upper jiao

三焦之一，从咽喉至胸膈部分，包括心、肺等。

2.1.9.1.1 上焦如雾 upper jiao resembles mist

对心肺输布营养至全身的作用形象化的描写与概括，喻指上焦宣发卫气，敷布水谷精微、血和津液的作用，如雾露之灌溉。

2.1.9.1.2 上焦主纳 upper jiao dominates reception

上焦摄纳清气和水谷精微的功能。

2.1.9.2 中焦 middle jiao

三焦之一，位于膈下至脐之间。包括脾、胃、肝、胆等。

2.1.9.2.1 中焦如沤 middle jiao resembles foam

对脾胃、肝胆等脏腑消化饮食作用的形象化描写与概括，喻指中焦消化饮食的作用，如发酵酿造的过程。

2.1.9.2.2 中焦主化 middle jiao resembles transformation

中焦消化水谷，吸收精微，化生营血的功能。

2.1.9.3 下焦 lower jiao

三焦之一，位于脐下。包括大肠、小肠、肾、膀胱等。

2.1.9.3.1 下焦如渎 lower jiao resembles a sluice

对大肠、肾和膀胱排泄糟粕和尿液的作用和形式的描写与概括，喻指肾、膀胱、大肠等脏腑排泄二便的功能，如沟渠之通导。

2.1.9.3.2 下焦主出 lower jiao dominates discharge

下焦主导排泄二便的功能。

2.1.9.4 决渎之官 organ of drainnge

三焦。水液升降出入的道路。

2.1.10 奇恒之腑 extraordinary fu organs

脑、髓、骨、脉、胆、胞宫，因其形态似腑，功能似脏，不同于一般的脏腑。

2.1.10.1 脑 brain

藏于颅腔之中，为脑髓汇聚而成，是神志活动的本原之处。

2.1.10.1.1 脑为髓海 brain as sea of marrow

诸髓皆汇聚于脑。

2.1.10.1.2 元神之府 house of original mentality

脑。神志活动的本原之处。

2.1.10.2 髓 marrow

由肾精所化生，有骨髓和脊髓之分，与脑相通。

2.1.10.3 骨 bone

贮藏骨髓，保护内脏，支撑形体和配合运动。

2.1.10.3.1 百骸 skeleton

全身骨骼的统称。

2.1.10.3.2 齿为骨之余 teeth are the extension of the bone

齿与骨的关系。齿与骨均属肾所主，赖肾精充养。

2.1.10.4 脉 vessel

气血精津液运行的通道。

2.1.10.4.1 脉为血府 vessel as house of blood

血液汇集、循行的居处。

2.1.10.4.2 脉舍神 spirit adhering to vessel

心藏神，主血脉，脉为血府，神寓其中。

2.1.10.4.3 息脉 wheezing and pulse

呼吸喘息和脉搏。

2.1.10.5 胞宫 uterus with its appendages

雌性动物整个内生殖器官，主发情与孕育胎儿。

2.2 气血津液精类

2.2.1 气 qi

构成动物体和维持其生命活动的最基本物质，也泛指脏器组织的机能活动，是物质与功能的统一。

2.2.1.1 气一元论 qi monism

属中国古代哲学范畴，其核心思想是用一元论来认识和阐释物质世界的构成及其运动变化规律。

2.2.1.2 气化 qi transformation

一泛指脏腑器官的生理活动，多用于表示某些器官的特殊功能；二指自然六气的变化。

2.2.1.3 气机 qi movement

气的运动，以升、降、出、入为基本形式。

2.2.1.4 先天之气 innate qi

禀受于父母的先天之精气。

2.2.1.5 后天之气 acquired qi

肺吸入的自然界清气和脾胃所运化的水谷精微之气。

2.2.1.6 元气 genuine qi

根源于肾，包括元阴、元阳（即肾阴、肾阳）之气。它由先天之精所化生，藏之于肾，又赖后天精气的滋养，才能不断地发挥其作用。

同义词：原气；真气

2.2.1.7 宗气 zong-pectoral qi

由脾胃所运化的水谷精微之气和肺所吸入的自然界清气结合而成，积聚于胸中，灌注于心肺，主要功能是出喉咙而司呼吸，灌心脉而行气血。属后天之气。

2.2.1.8 营气 ying-nutrient qi

水谷精微所化生的精气之一，与血并行于脉中，是血液的组成部分，是宗气贯入血脉中的营养之气。具有化生血液、营养周身的功能。

2.2.1.8.1 营出中焦 nutrient qi derived from middle jiao

营气生成于三焦，其运行始于中焦的前肢太阴肺经。

2.2.1.8.2 营行脉中 nutrient qi moving in vessels

营气与卫气相对而言，运行于经脉之中。

2.2.1.9 卫气 wei-defensive qi

源于下焦肾，由水谷所化生，行于脉外，是动物机体阳气的一部分。具有温煦皮肤、腠理、肌肉，司汗孔开阖与护卫肌表、抗御外邪的功能。

2.2.1.9.1 卫出上焦 defense qi coming from upper jiao

卫气由上焦肺气宣发而出，以熏肤充身泽毛。

2.2.1.9.2 卫出中焦 defense qi coming from middle jiao

卫气生成于中焦，为脾胃水谷精微所化。古有脾胃为卫气之本，肺为卫气之主，肾为卫气之根之说。

2.2.1.9.3 卫出下焦 defense qi coming from lower jiao

卫气昼行于阳二十五周，自后肢太阳膀胱经始；夜行于阴二十五周，自后肢少阴肾经始。肾与膀胱皆属下焦。

2.2.1.10 脏腑之气 qi of the zang-fu organs

各脏腑的气。不仅指构成脏腑的最基本物质和各脏各腑的生理功能，而且包括脏腑间功能的协调乃至整体生命功能的表现。

2.2.1.11 中气 qi of the middle jiao

一泛指中焦脾胃之气，即脾胃受纳、运化、升清降浊和统摄血液的生理功能；二指脾气，即脾之升提内脏、防止脱垂的功能。

2.2.1.12 气主煦之 qi governs warming

气具有气化生热、温煦动物机体的作用。

2.2.2 血 blood

行于脉中而循环流注于全身且具有营养和滋润作用的红色液体，主要含有营气和津液。

2.2.2.1 血主濡之 blood governs nourishing and moistening

血具有营养和滋润全身的功能。

2.2.3 津液 body fluids

体内一切正常水液的总称，包括各脏腑组织的内在体液及其分泌物，是构成机体和维持机体生命活动的基本物质之一。

2.2.3.1 津 thin fluids

性质清轻稀薄、流动性较大，分布于体表、皮肤、肌肉和孔窍，并能渗注于血脉，起滋润作用。

2.2.3.2 液 thick fluids

性质重浊、稠厚，流动性较小，灌注于骨节、脏腑、脑髓等组织，起濡养作用。

2.2.4 精 essence

（广义上）气、血、津液、水谷精微等动物机体一切有形的精微物质；（狭义上）生殖之精。

2.2.4.1 先天之精 innate essence

生命的本原物质，由父母生殖之精汇合而成，藏于肾，为构成动物体胚胎和繁衍后代的基本物质。

2.2.4.2 后天之精 acquired essence

由水谷精微升华而来，藏于肾，输布于脏腑，是维持动物机体生命活动的物质基础。

2.2.4.3 精气 essential qi

精与气的总称。气化于精，精生于气，精与气互化互生，共同成为构成动物体和维持生命活动的基本物质及其功能。

2.2.5 气为血帅 qi is the commander of the blood

气对血有推动、统摄和化生等作用，具体表现为气能生血、气能行血、气能摄血。

2.2.5.1 气能生血 qi engenders blood

气参与并促进血液的生成。

2.2.5.2 气能行血 qi circulates blood

气具有推动血液在脉中运行的作用。

2.2.5.3 气能摄血 qi controls blood

气具有统摄血液在脉中正常循行而不逸出脉外的作用，主要体现在脾气统血的生理功能之中。

2.2.6 血为气母 blood is the mother of qi

血为气的物质基础，血能化气，并可作为气运行的载体。

2.2.6.1 血能载气 blood conveying qi

血液具有运载水谷之精气、自然界之清气的功能。

2.2.6.2 血能养气 blood nourishes qi

血液可以充养动物体之气，使气保持旺盛。

2.2.7 气能生津 qi engenders body fluids

气是津液生成的物质基础和动力。

2.2.8 气能行津 qi circulates body fluids

津液的输布和排泄均依赖于气的升降出入和有关脏腑的气化功能。

2.2.9 气能摄津 qi controls body fluids

气有固摄津液以控制其排泄的作用。

2.2.10 津能载气 body fluids carries qi

津液为气的载体之一,气依附于津液而存在,否则就会涣散不定。

2.2.11 津血同源 body fluids and blood share the same source

津液和血都来源于水谷精气,两者有相互滋生、相互转化、同出一源、相互影响的关系。

2.2.12 精血同源 essence and blood share the same source

血由水谷精气化生,精也有赖于水谷精气的培育补充,两者有相互滋生、相互转化、同出一源、相互影响的关系。

2.2.13 血汗同源 blood and sweat share the same source

血为津化,津化为汗,血汗均赖津液所化。

第三章 经 络

3.1 常用经络术语

3.1.1 经络学说 meridian theory
研究动物机体经络系统的组织结构、循行分布、生理功能、病理变化及其与脏腑相互关系的学说，是中兽医学理论体系的重要组成部分。

3.1.1.1 经络 meridians and collaterals
动物体内经脉和络脉的总称，是动物机体联络脏腑、沟通内外和运行气血的通路，是动物体组织结构的重要组成部分。

3.1.1.1.1 经脉 meridians
经络系统中的直行主干，为全身气血运行的主要通道，是十二经脉、奇经八脉以及附属于十二经脉的经别、经筋、皮部的合称。

3.1.1.1.2 络脉 collaterals
从经脉中分出而遍布全身的细小分支，包括十五络脉、浮络和孙络等。

3.1.1.1.3 十四经脉 fourteen meridians
十二正经与任脉、督脉的合称。

3.1.2 经气 qi of meridians
一指经络的功能活动；二指运行于经脉中之气。
同义词：脉气

3.1.3 经隧 meridian passage
潜布于体表之下运行气血的经络通路。

3.1.4 十二经脉 twelve meridians
包括前肢三阳经、前肢三阴经、后肢三阳经、后肢三阴经，为经络系统的主体。
同义词：十二正经

3.1.4.1 前肢三阴经 three-yin meridians of forelimb
十二经脉中循行于前肢内侧的三条阴经，是前肢太阴肺经、前肢厥阴心包经和前肢少阴心经的总称。其循行方向从胸部开始，循行于前肢内侧，止于前肢末端。

3.1.4.1.1 前肢太阴肺经 lung meridian of forelimb-taiyin

十二正经之一。在体内属肺络大肠，并与胃、喉相连；在体表起于胸部外上方，沿前肢上部屈侧前缘向下，止于第一指末端桡侧。

3.1.4.1.2 前肢厥阴心包经 pericardium meridian of forelimb-jueyin

十二正经之一。在体内属心包络三焦，并与横膈膜相连；在体表起于胸，经腋下，沿前肢内侧中线向下，止于第三指末端前端中央。

3.1.4.1.3 前肢少阴心经 heart meridian of forelimb-shaoyin

十二正经之一。在体内属心络小肠，并与咽部及眼相连；在体表起于胸，经腋下，沿前肢内侧后缘向下，止于第五指末端桡侧。

3.1.4.2 前肢三阳经 three-yang meridians of forelimb

十二经脉中循行于前肢外侧的三条经脉，是前肢阳明大肠经、前肢太阳小肠经和前肢少阳三焦经的总称。其循行方向由前肢末端开始，循行于前肢外侧，抵达头部。

3.1.4.2.1 前肢阳明大肠经 large intestine meridian of forelimb-yangming

十二正经之一。在体内属大肠络肺；在体表起于第二指末端桡侧，沿前肢外侧前缘向上，经肩部、颈部、颊部，止于对侧鼻孔旁。

3.1.4.2.2 前肢少阳三焦经 sanjiao meridian of forelimb-shaoyang

十二正经之一。在体内属三焦络心包络，并与耳、眼相连；在体表起于第四指末端尺侧，沿前肢外侧中线向上，经过肩部、侧头部、耳部，止于眼部。

3.1.4.2.3 前肢太阳小肠经 small intestine meridian of forelimb-taiyang

十二正经之一。在体内属小肠络心，并与胃、眼和内耳相连；在体表起于第五指末端尺侧，沿前肢伸侧两骨之间向上，经肘部、肩胛部、颈侧、颜面、眼部，止于耳。

3.1.4.3 后肢三阳经 three-yang meridians of posterior limb

十二经脉中循行于后肢外侧的三条经脉，是后肢阳明胃经、后肢太阳膀胱经和后肢少阳胆经的总称。其循行方向由头部开始，经背腰部，循行于后肢外侧，止于后肢末端。

3.1.4.3.1 后肢阳明胃经 stomach meridian of hindlimbs-yangming

十二正经之一。在体内属胃络脾；在体表起于鼻旁，经侧头部、面部、颈部、胸腹部，沿后肢外侧前缘、胫骨外侧向下，止于第三趾末端外侧。

3.1.4.3.2 后肢少阳胆经 gallbladder meridian of hindlimbs-shaoyang

十二正经之一。在体内属胆络肝；在体表起于外眼角，经头侧部、胸腹部，沿后肢外侧中线向下，止于第四趾末端外侧。

3.1.4.3.3 后肢太阳膀胱经 bladder meridian of hindlimbs-taiyang

十二正经之一。在体内属膀胱络肾，并与脑相连；在体表起于内眼角，向上越过头顶向后，向下经颈部、背部两侧、臀部，沿后肢外侧后缘向下，止于第五趾末端外侧。

3.1.4.4 后肢三阴经 three-yin meridians of posterior limb

十二经脉中循行于后肢内侧的三条经脉，是后肢太阴脾经、后肢厥阴肝经和后肢少阴肾经的总称。其循行方向由后肢末端开始，循行于后肢内侧，经腹达胸。

3.1.4.4.1 后肢太阴脾经 spleen meridian of hindlimbs-taiyin

十二正经之一。在体内属脾络胃，并与心及舌根相连；在体表起于后肢第二趾末端内侧，沿后肢内侧前缘向上，经腹部、胸部，止于侧胸部。

3.1.4.4.2 后肢厥阴肝经 liver meridian of hindlimbs-jueyin

十二正经之一。在体内属肝络胆，并与生殖器、胃、横膈膜、咽喉、眼球相连；在体表起于第二趾末端外侧，沿后肢内侧中线向上，经外阴部、腹部，止于侧胸部。

3.1.4.4.3 后肢少阴肾经 kidney meridian of hindlimbs-shaoyin

十二正经之一。在体内属肾络膀胱，并与脊髓、肝、膈膜、喉部、舌根、肺、心、胸腔等相连；在体表起于足垫前缘正中凹陷处，沿后肢内侧后缘向上，经腹部，止于胸部。

3.1.5 十二经别 twelve divergent meridians

十二正经别行深入体腔的分支，能够到达十二经脉所不到之处，加强脏腑、表里间的联系。

注：在循行过程中六阳经的经别均流回原来的阳经去，六阴经的经别均流入与其相表里的阳经去。

同义词：别行的正经

3.1.6 十二经筋 twelve muscle regions

经脉所连属的筋肉系统，即十二经脉及其络脉中气血所濡养的肌肉、肌腱、筋膜、韧带等。

注：其功能主要是连缀四肢百骸，主司关节运动。

3.1.7 十二皮部 twelve cutaneous regions

十二经脉及其功能活动在皮肤上所分属的部位，也是经络之气在皮肤所散布的部位。

3.1.8 奇经八脉 eight extraordinary meridians

任、督、冲、带、阴维、阳维、阴跷、阳跷八条经脉的总称。

3.1.8.1 督脉 du meridian

行于背正中线，总督一身之阳脉。

同义词：阳脉之海

3.1.8.2 任脉 ren meridian

行于腹正中线，总任一身之阴脉。

同义词：阴脉之海

3.1.8.3 冲脉 chong meridian

行于颈、腹两侧，经后肢内侧达掌或蹄之中心，与后肢少阴经并行，总领一身气血的要冲，能调节十二经气血。

同义词：十二经之海；血海。

3.1.8.4 带脉 dai meridian

起于肋下，横行绕腰部一周，状如束带，约束纵行诸经脉。

3.1.8.5 阴维脉 yinwei meridian

起于后肢下端内侧，沿大腿内侧上行，经腹侧至咽喉与任脉会合，维系三阴经。

3.1.8.6 阳维脉 yangwei meridian

起于后肢下端外侧，沿后肢外侧上行，经背侧前行至颈后与督脉会合，维系三阳经。

3.1.8.7 阴跷脉 yinqiao meridian

起于足跟内侧，随后肢少阴经上行，至内眼角与阳跷脉会合，主一身左右之阴，使行动健捷。

3.1.8.8 阳跷脉 yangqiao meridian

起于足跟外侧，随后肢太阳经上行，至内眼角与阴跷脉会合，主一身左右之阳，使行动健捷。

3.1.9 十五络脉 fifteen collaterals

十二络脉（每一条正经都有一条络脉）加上任脉、督脉的络脉和脾的大络，总共为十五条，它是所有络脉的主体。

3.1.9.1 脾之大络 large splenic collateral channel

名曰大包，从大包穴分出，浅出于渊腋穴下三寸处，散布于胸肋部。

3.1.10 孙络 tertiary collaterals

从络脉中分出的细小分支。

3.1.11 浮络 floating collaterals

浮于体表的络脉。

3.1.12 阳脉 yang meridian

指经脉中的阳经，包括前、后肢三阳经，督脉、阳维脉、阳跷脉等。

3.1.13 阴脉 yin meridian

指经脉中的阴经，包括前、后肢三阴经，任脉、阴维脉、阴跷脉等。

3.1.14 十二经脉流注次序 the order of perfusion in the twelve meridians

十二经脉气血流注从前肢太阴肺经开始，依次流经前肢阳明大肠经，后肢阳明胃经，后肢太阴脾经，前肢少阴心经，前肢太阳小肠经，后肢太阳膀胱经，后肢少阴肾经，前肢厥阴心包经，前肢少阳三焦经，后肢少阳胆经，至后肢厥阴肝经而终，再由后肢厥阴肝经复传于前肢太阴肺经，流注不已，从而构成了周而复始、如环无端的循环传注系统。

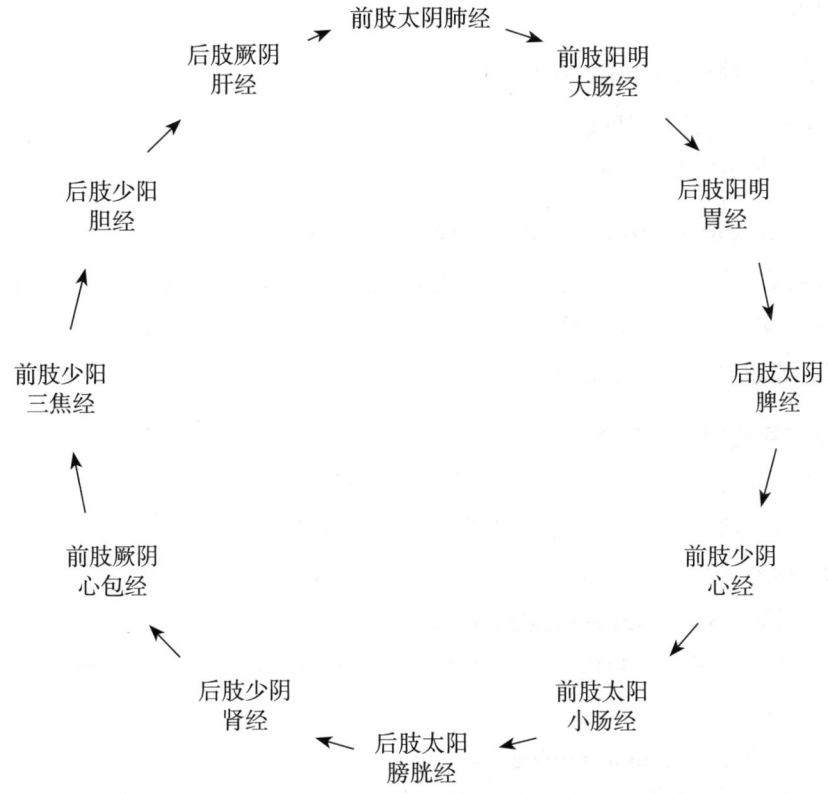

第四章 病因病机

4.1 病因类

4.1.1 病因学说 etiological theory
研究各种病因的概念、形成、性质、致病特点及其所致病证的兽医临床表现理论，是中兽医学理论体系的重要组成部分。

4.1.1.1 辨证求因 differentiate pattern to reveal an etiology
在中兽医理论指导下，以病证的临床表现为依据，通过分析病证的症状体征，推断出发病的病因病机，为治疗提供依据。

同义词：随证求因；审证求因

4.1.1.2 病因 causative factors
引起动物疾病发生的原因。广义上为发病学，研究疾病如何发生；狭义上为致病因素，中兽医称为"邪气"。主要包括外感因素、内伤因素和其他因素。

同义词：病邪；病源

4.1.1.2.1 外感病因 exogenous etiological factors
外感致病因素，来源于自然界，多从皮毛、口鼻侵入动物机体而引发疾病的致病因素，包括六淫和疫病之气。

4.1.1.2.2 内伤病因 endogenous etiological factors
内伤致病因素，主要包括饲喂失宜、劳役之伤、内生五邪以及内伤七情。

4.1.2 六淫 six pathogenic factors
风、寒、暑、湿、燥、火六种外感病邪的统称。

同义词：六邪

4.1.2.1 风邪 pathogenic wind
具有轻扬开泄、善动不居、升发、向上、向外特性的邪气，为六淫之首，百病之长。

4.1.2.1.1 风性开泄 wind attributing to opening and releasing
风邪具有疏通、透泄之性，其致病使皮毛腠理开泄，出现汗出、恶风等症。

4.1.2.1.2 风易伤阳位 wind tending to attack yang portion of body
风邪轻浮，具有向上向外之性，易伤及动物体头部、肌表和阳经。

4.1.2.1.3 风性主动 wind attributing to mobility
风邪具有动摇不定之性，其致病易使肢体震颤摇动。
同义词：风胜则动

4.1.2.1.4 风性善行数变 wind being mobile and changeable
风邪具有行无定处，变幻无常之性，病位游移，发病急，变化多，传变快。

4.1.2.1.5 风为百病之长 wind being leading cause of diseases
风邪为六淫之首，为外感病邪的先导，易与寒、湿、燥、热等邪气相合为病。

4.1.2.2 寒邪 pathogenic cold
具有寒冷、凝滞、收引等特性的邪气。

4.1.2.2.1 寒易伤阳 cold tending to injure yang
寒邪具寒冷之性，其致病最易损伤动物体阳气，使全身或局部出现明显的寒象。

4.1.2.2.2 寒性凝滞 cold attributing to congealing
寒邪具有凝结阻滞之性，其致病易使气血凝滞，经脉不通，不通而痛。

4.1.2.2.3 寒性收引 cold attributing to contraction
寒邪具有收缩牵引之性，其致病易使气机收敛，腠理闭塞，经络、筋脉和肌肉等收缩挛急。

4.1.2.3 暑邪 pathogenic summer heat
夏至以后，立秋之前，具有炎热、升散兼湿等特性的邪气。

4.1.2.3.1 暑性炎热 summer-heat attributing to burning heat
暑为夏季火热之气所化，其致病使动物体表现出一系列阳热症状。

4.1.2.3.2 暑性升散 summer-heat attributing to rise and dispersion
暑邪具有上升发散之性，其致病多直入气分，使腠理开泄而汗多。

4.1.2.3.3 暑易挟湿 summer-heat tending to be mixed with dampness
暑季气候炎热，多雨潮湿，热蒸湿动，暑邪为病，易挟湿邪侵袭动物机体。

4.1.2.3.4 暑易扰心 summer-heat tending to disturb heart
暑邪具有上扰心神之性，其致病易使心神失守或清窍闭塞。

4.1.2.3.5 暑易伤津耗气 summer-heat tending to consume fluid and qi
暑邪具有耗伤动物体津气之性，其致病使汗多伤津，气随津泄，呈现气津（阴）两亏之象。

4.1.2.4 湿邪 pathogenic dampness
在取类比象的思想指导下，具有易阳气机、重浊、黏腻、趋下致病特点的邪气。

4.1.2.4.1 湿阻气机 dampness hampering qi movement
湿邪具有阻滞气机运行，其致病使脏腑气机升降出入失常，经络气血运行不畅。

4.1.2.4.2 湿易伤阳 dampness tending to injure yang

湿邪具有损伤动物体阳气之性，致病以损伤脾阳为甚。

4.1.2.4.3 湿性重浊 dampness attributing to heaviness and turbidity

湿邪具有沉重、秽浊之性，其致病使肢体沉重，分泌物和排泄物污浊。

4.1.2.4.4 湿性黏滞 dampness attributing to viscosity and stagnation

湿邪具有黏腻停滞之性，其致病使分泌物和排泄物黏稠、涩滞，且病程较长，缠绵难愈。

4.1.2.4.5 湿性趋下 dampness characterized by downward going

湿邪具有下趋之性，其致病易侵袭阴位，伤及动物体下部。

4.1.2.5 燥邪 pathogenic dryness

致病具有干燥、收敛、易伤肺等特性的邪气。

4.1.2.5.1 燥性干涩 dryness characterized by aridity and astringency

燥邪具有干燥、枯涩之性，其致病易伤机体津液而致阴津亏损。

4.1.2.5.2 燥易伤肺 dryness tending to injure lung

燥邪具有从口鼻、肌表而入，损伤肺气之性，其致病使肺津受损、宣降失职。

4.1.2.6 火邪 pathogenic fire

具有炎上、易伤津耗气、易生风动血及易扰动心神等特性，导致阳热性病证的邪气。

4.1.2.6.1 热邪 heat pathogen

致病以热性、阳性的实性病理变化为特征的外邪。

4.1.2.6.2 温邪 warm pathogen

春温、风温、暑温、湿温、温燥、冬温、温疫和温毒等各种温热病致病邪气的统称。

4.1.2.6.3 火性炎上 fire characterized by flaming

火邪具有燔灼向上之性，其致病热象显著，多见动物体腰背、头面等部。

4.1.2.6.4 火易生风 fire tending to cause wind

火邪具有扰动肝风之性，其致病为热极生风致肝风内动。

4.1.2.6.5 火易动血 fire tending to cause bleeding

火邪具有灼伤脉络，迫血妄行之性，其致病为出血、发斑。

4.1.2.6.6 火易耗气伤津 fire tending to consume qi and fluid

火邪具有耗伤正气、消烁阴津之性，其致病为气津两亏。

4.1.2.6.7 火易扰心 fire tending to disturb heart

火邪具有扰乱心神之性，其致病为心神失守。

4.1.2.6.8 火邪易致疮痈 fire pathogen tending to cause carbuncle

火热之邪入于血，可聚于局部，腐蚀血肉而发为痈肿疮疡。

4.1.2.7 合邪 combined pathogens

六淫中，两种或两种以上联合侵犯动物体的邪气。如风热、风寒湿等。

4.1.3 疫疠 epidemic pestilence

具有很强传染性的外感致病因素。

同义词：疠气

4.1.4 内生五邪 five endogenous pathogenic factors

由于气血津液、脏腑等生理功能失调从体内产生的类似风、寒、湿、燥、火外邪致病的病因，即"内风""内寒""内湿""内燥"和"内火"。

4.1.4.1 内风 internal wind

脏腑气血失调，体内阳气亢逆的致病因素，主要包括肝阳化风、阴虚生风、血虚生风和热极生风四种。

同义词：肝风内动

4.1.4.1.1 肝阳化风 wind produced by liver yang

由于肝肾阴亏，阴不制阳，导致肝阳亢逆无制，而从体内产生的一种风邪。

4.1.4.1.2 阴虚风动 wind produced by yin deficiency

由于阴液枯竭，筋脉失养，而从体内产生的一种风邪。

同义词：阴虚生风

4.1.4.1.3 血虚生风 wind produced by blood deficiency

由于血液虚少，筋脉失养，而从体内产生的一种风邪。

4.1.4.1.4 血燥生风 wind produced by blood dryness

由于血虚津亏，失润化燥，肌肤失于濡养，而从体内产生的一种风邪。

4.1.4.1.5 热极生风 wind produced by extreme heat

由于邪热炽盛，伤及营血，燔灼肝经，筋脉失养，而从体内产生的一种风邪。

4.1.4.2 内寒 internal cold

由于动物机体机能衰退、阳气不足，而从体内产生的一种寒邪。

4.1.4.3 内湿 endogenous dampness

由于脾肾阳虚，水液代谢失调，而从体内产生的一种湿邪。

4.1.4.4 内燥 endogenous dryness

由于动物机体津液耗伤，而从体内产生的一种燥邪。

4.1.4.5 内火 endogenous fire

由于脏腑阴阳失调，而从体内产生的一种火邪。

4.1.5 内伤七情 seven emotions

喜、怒、忧、思、悲、恐、惊七种主要内伤性致病因素，是动物对客观事物或现象所作出的不同情志反应。

4.1.5.1 大怒伤肝 excessive anger damaging the liver

过度愤怒导致肝血被耗，引起肝气上逆、肝阳上亢或肝火上炎的病证。

4.1.5.2 怒则气上 anger causing the qi to rise

过度愤怒影响肝的疏泄功能，导致肝气上逆，血随气逆，出现目赤舌红、呕血，甚至昏厥、猝倒等症。

4.1.5.3 暴喜伤心 overwhelming joy impairing heart

过度欢喜，会使心气涣散，出现神不守舍的病证。

4.1.5.4 喜则气缓 joy causing the qi to slack

欢喜过度会使心气涣散、神不守舍，出现精神不能集中，甚至失神狂乱的症状。

4.1.5.5 思虑伤脾 pensiveness injuring spleen

思虑过度，损伤脾气，使脾气郁结，运化失职。

4.1.5.6 思则气结 pensiveness leading to qi knotting

思虑过度导致脾气郁结，从而出现食欲减退甚至废绝、肚腹胀满或便溏等症状。

4.1.5.7 悲忧伤肺 melancholy injuring lung

悲忧过度，使肺气积郁，清肃失职，甚则肺气耗伤。

4.1.5.8 悲则气消 sadness consuming the qi

过度悲伤会损伤肺气，出现气短、精神萎靡不振、乏力等症。

4.1.5.9 惊恐伤肾 fear/fright damaging the kidney

惊恐过度会耗伤肾的精气，出现肾虚不固的病证。

4.1.5.10 恐则气下 fear causing the qi to descend

过度恐惧可使肾气不固、气泄于下，出现大小便失禁，甚至昏厥的症状。

4.1.5.11 惊则气乱 fright causing the qi to become chaotic

突然受惊，损伤心气，致使心气紊乱，出现心悸、惊恐不安等症状。

4.1.6 饲喂失宜 improper feeding

饲喂不节、饲喂不洁、饲喂偏嗜等饮食失于正常的致病因素。

4.1.6.1 饲喂不节 irregular eating

饲喂量明显低于或超过动物机体自身的适度采食量和饲喂不规律等。包括过饥、过饱和时饥时饱等。

4.1.6.1.1 过饱　excess feeding
饮喂量太多或动物贪食，超过胃肠受纳及传送的限度，可能损伤胃肠。

4.1.6.1.2 过饥　insufficient feeding
饮喂量不足或食物品质低劣，营养不足，长期可导致气血生化乏源而气血亏虚。

4.1.6.1.3 时饥时饱　irregular excess and insufficient feeding
饮喂不规律，饥饱失时，可影响脾胃运化，导致脾胃气机升降失常和功能紊乱，进而引发疾病。

同义词：饮食无时

4.1.6.2 饲喂不洁　contaminated food
食用不洁净或陈腐变质，甚至有毒的食物而导致疾病发生。

4.1.6.3 饲喂偏嗜　food preference
异常喜好某种食物，或长期偏食某些食物而导致某些疾病的发生。

4.1.7 劳逸失度　work-rest imbalance
过度劳累和过度安逸的总称。

4.1.7.1 过劳　overwork
过度劳伤的致病因素。

4.1.7.1.1 劳力过度　physical exhaustion
动物使役不当或运动不当，超过力所能及的体力活动，能损伤脏腑，使脏气虚少。

4.1.7.1.2 劳神过度　mental exhaustion
断奶、离群等引发过度的精神刺激，能耗伤心脾，甚则殃及诸脏。

4.1.7.1.3 配种过度　sexual exhaustion
过度配种，或过度采精，耗伤肾精。

4.1.7.2 过逸　physical inactivity
过度安逸，可致气血运行不畅、筋骨柔弱、脾胃呆滞，甚则继发他病的致病因素。

4.1.8 可致病的病理产物　pathological product that can cause disease
在初始病因作用下，机体脏腑功能失调所产生并能引起新的病理变化的致病因素。包括痰饮、瘀血、结石等。

4.1.8.1 痰饮　phlegm-fluid retention
痰与饮的合称，因脏腑功能失调，致使体内津液凝聚变化而成的水湿，可阻碍气血运行而成为继发的致病因素。

4.1.8.1.1 痰　phlegm
脏腑病理变化过程中，机体水液代谢障碍形成的质地黏稠的渗出液。属继发性致病因素。

4.1.8.1.1.1 寒痰 cold-phlegm

痰浊与寒邪相合的病邪。

4.1.8.1.1.2 热痰 heat-phlegm

痰浊与热邪相合的病邪。

4.1.8.1.1.3 湿痰 damp-phlegm

脾失健运，聚湿生痰而致痰浊与湿邪相合的病邪。

4.1.8.1.1.4 燥痰 dryness-phlegm

痰浊与燥邪相合的病邪。

4.1.8.1.1.5 风痰 wind-phlegm

痰浊与风邪相合的病邪。

4.1.8.1.2 饮 retained fluid

脏腑病理变化过程中，机体水液代谢障碍形成的质地清稀的渗出液。属继发性致病因素。

4.1.8.2 瘀血 static blood

全身血液运行不畅，或局部血液停滞，或体内存在离经之血。

4.1.8.3 结石 calculus

停滞于脏腑的砂石样物质。属继发性致病因素。

4.1.9 外伤 external injury

常见的外伤性致病因素，包括创伤、挫伤、烫火伤及虫兽伤等。

4.1.10 寄生虫 parasite

能够引起动物发病的内寄生虫和外寄生虫。

4.1.11 胎毒 fetal toxin

在妊娠期间胚胎禀受自母体的热毒，为动物出生后发生疮疹和遗毒等病变的致病因素。

4.1.12 虫兽伤 injury by animal and insect

虫、兽等各类动物咬啮、螫刺所致受伤的因素。

4.1.13 水土不服 non-acclimatization

由于个体迁移，自然环境、气候和生活习惯改变，不能适应其变化而引起发病的因素。

4.1.14 禀赋不足 constitutional insufficiency

先天体质虚弱、气血亏损，为发病的内在因素。

4.2 病机类

4.2.1 病机学说 pathogenesis theory

研究疾病发生、发展、转归规律的理论。

4.2.1.1 病机 pathogenesis

各种病因作用于动物机体，引起疾病发生、发展与转归的机理。论述疾病发生后的演变、传和不传、有无并发症、继发症等。

4.2.1.2 正气 healthy qi

动物体各脏腑组织器官的机能活动，及其对外界环境的适应力，对致病因素的抵抗力以及康复能力。

4.2.1.3 邪气 pathogenic qi

与正气相对。各种致病因素的总称。

4.2.1.4 发病 onset of disease

发病是指疾病的发生过程，是机体处于病邪的损害和正气抗损害之间的矛盾斗争过程。

4.2.1.4.1 新感 new contraction

感邪之后即时而发的发病类型，与伏邪相对。

4.2.1.4.2 伏邪 latent pathogenic qi

感邪之后邪藏体内，逾时而发的发病类型，与新感相对。

4.2.1.4.3 卒发 sudden onset

急暴而病的发病类型。

4.2.1.4.4 徐发 gradual onset

缓慢而病的发病类型。

4.2.1.4.5 继发 occurrence of secondary disease

在原发疾病未愈的基础上，继续发生新的疾病。

4.2.1.4.6 复发 recurrence

疾病已愈，再次发病。

4.2.1.4.6.1 食复 relapse due to improper diet

疾病初愈，因饲喂不当、饮食不洁等因素致疾病复发。

4.2.1.4.6.2 劳复 relapse due to overwork

疾病初愈，因过劳使正气受损，导致疾病复发。

4.2.1.4.6.3 药复 relapse due to drugs
病后用药失当，导致疾病复发。

4.2.1.4.7 两感 damage to paired meridians
表里两经同时受邪而发病。

4.2.1.4.8 直中 direct attack
病邪逾三阳经而径直侵犯三阴经，或外邪直接损伤脏腑而为病。

4.2.1.4.9 合病 disease involving two or more meridians
两经或三经的病证同时出现。

4.2.1.4.10 并病 disease of one meridian involving another meridian
一经病证未罢又出现另一经病证。

4.2.2 八纲病证病机 pathogenesis of eight principal syndromes
从八纲辨证的角度，阐释病证的发病机制。

4.2.2.1 正邪相争 struggle between anti-pathogenic qi and pathogenic factors
致病因素与动物体正气之间的抗争，双方的盛衰消长，决定了病变发展的趋势与速度。

同义词：邪正相搏

4.2.2.2 邪正盛衰 waxing and waning of anti-pathogenic qi and pathogenic factors
在疾病发生、发展过程中，动物机体的正气与致病邪气之间相互斗争所发生的盛衰病机变化，影响着疾病的发展和转归，以及病证的虚实变化。其中，邪气亢盛而正气未衰为实；正气虚衰、抗病力弱为虚。

同义词：邪正消长

4.2.2.2.1 邪气盛则实 exuberance of pathogenic qi leading to excessiveness
在疾病过程中，邪气亢盛，正气未虚，邪正相争导致以邪气盛为主的实性病理变化。

4.2.2.2.1.1 实证病机 pathogenesis of excess syndrome
机体感受外邪，或体内病理性产物蓄积而正气未衰时产生的病理变化。

4.2.2.2.1.2 五实 five zang excess
五脏实热闭阻病变的总称。心受邪而脉洪盛，肺受邪而皮肤灼，脾受邪而腹胀满，肝受邪而闷瞀，肾受邪而二便不通。

4.2.2.2.2 精气夺则虚 loss of essential qi leading to deficiency
在疾病过程中，气过度耗损，邪正相搏，导致以气虚为主的虚性病理变化。

4.2.2.2.2.1 虚证病机 pathogenesis of deficiency syndrome
正气虚衰，抗病力弱，邪正相搏形成各种衰退性的病理变化。

4.2.2.2.2.2 五虚 five zang deficiency
五脏精气虚损病变的总称。心气虚而脉细，肺气虚而皮寒，脾气虚而食欲废绝，肝气虚而气少，肾气虚而二便滑泄。

4.2.2.2.3 虚实错杂 deficiency complicated with excess
虚证和实证同时并存于动物机体的病机。

同义词：虚实夹杂

4.2.2.2.3.1 虚中夹实 deficiency complicated with excess
证候以正虚为主，兼有实邪的病机。

4.2.2.2.3.2 实中夹虚 excess complicated with deficiency
邪实为主，兼有正气虚衰的虚实夹杂的病机。

4.2.2.2.4 虚实真假 true-false of deficiency and excess
动物疾病发展到严重阶段时，虚实外证表现与疾病本质相反的病机。

4.2.2.2.4.1 真虚假实 true deficiency with false excess
本质为虚、表象似实的病机。

同义词：至虚有盛候

4.2.2.2.4.2 真实假虚 true excess with false deficiency
本质为实、表象似虚的病机。

同义词：大实有羸状

4.2.2.2.5 虚实转化 deficiency-excess transformation
实邪久留而损伤正气或正气不足而实邪积聚，导致虚与实之间的相互转换变化，包括由实转虚和因虚致实。

4.2.2.2.5.1 由实转虚 excess transformation into deficiency
以邪气盛为主的实性病变，向以正气虚损为主的虚性病变的转化。

4.2.2.2.5.2 因虚致实 deficiency transforming into excess
疾病本来是以正气虚为矛盾主要方面的虚性病变，因正气虚弱，脏腑功能减退，气、血、水液运行障碍，产生气滞、血瘀、痰饮、水湿等实邪留滞的病机。

4.2.2.2.6 正胜邪退 vital qi increasing with pathogenic qi decreasing
在邪正盛衰的变化过程中，正气旺盛，正气战胜邪气，邪气逐渐消退，疾病趋向好转而痊愈的一种转归。

4.2.2.2.7 邪去正虚 pathogenic qi retreating with deficient vital qi
邪气被祛除，病邪对机体的损害作用已经消失，但正气在疾病的发展变化过程中已被耗伤，而有待恢复的一种转归。

4.2.2.2.8 邪盛正衰 exuberance of pathogenic qi with decrease of vital qi
在邪正盛衰的变化过程中，邪气太盛，正气逐渐衰退，疾病逐渐恶化，甚至死亡的一种转归。

4.2.2.2.9 正虚邪恋 lingering pathogenic qi due to deficient vital qi
在疾病后期，正气已虚，但邪气去而未尽，正气又一时无法祛邪外出，因而病势缠绵，经久而不能痊愈的病理状态。

4.2.2.3 阴阳失调 disharmony between yin and yang

在疾病的发生发展过程中，由于各种致病因素的影响，导致动物机体的阴阳双方失去相对的平衡协调而出现阴阳偏盛、偏衰、互损、格拒、亡失等一系列病机。

4.2.2.3.1 阴阳偏盛 excess yin or yang

阳邪或阴邪侵犯动物体引起阳偏盛或阴有余的病机统称。

4.2.2.3.1.1 阳盛 exuberance of yang

阳热偏盛的病理变化。

4.2.2.3.1.1.1 阳盛则阴病 predominance of yang making yin suffer

阳热偏盛导致各种伤津、伤阴的病理变化。

同义词：阳胜则阴病

4.2.2.3.1.1.2 阳盛则热 yang excess leading to heat

阳邪偏盛导致热且实的病机。

同义词：阳胜则热

4.2.2.3.1.2 阴盛 exuberance of yin

阴邪偏盛的病理变化。

4.2.2.3.1.2.1 阴盛则阳衰 predominance of yin making yang suffer

阴寒偏盛导致阳气衰微的病理变化。

同义词：阴胜阳病

4.2.2.3.1.2.2 阴盛则寒 yin excess leading to cold

阴邪偏盛导致寒且实的病机。

同义词：阴胜则寒

4.2.2.3.2 阴阳偏衰 waning of yin or yang

邪气侵犯动物体引起阴阳之气亏虚的病机统称。

4.2.2.3.2.1 阳衰 decline of yang

阳气偏衰的病理变化。

同义词：阳虚

4.2.2.3.2.2 阳虚则寒 yang deficiency leading to cold

阳气虚弱，温煦功能减退，必然导致寒而虚的病机。

4.2.2.3.2.3 阴衰 decline of yin

机体阴液亏损的病理变化。

同义词：阴虚

4.2.2.3.2.3.1 阴虚则热 yin deficiency leading to heat

动物机体阴液亏损，阴不制阳，阳相对偏亢，必然导致热且虚的病机。

4.2.2.3.2.3.2 阴虚则阳亢 yin deficiency leading to yang hyperactivity

动物机体阴液亏损，阴不敛阳，阳相对亢盛而浮越于上的病机。

4.2.2.3.2.3.3 阴虚火旺 yin deficiency leading to fire hyperactivity
阴液亏损，阴不制阳，阳相对亢盛而致虚火炽盛的病机。
同义词：水不制火

4.2.2.3.2.3.4 虚火上炎 flaming of deficiency-fire
阴液亏虚，阴不制阳，水不制火，致虚火上升的病理变化。

4.2.2.3.3 阴阳互损 mutual impairment between yin and yang
阴或阳任何一方虚损到一定程度，累及另一方使其虚损所导致的阴阳两虚的病机。

4.2.2.3.3.1 阳损及阴 deficiency of yang affecting yin
由于阳虚到一定程度，累及于阴，使阴气生化不足，从而在阳虚的基础上又导致阴虚，形成以阳虚为主的阴阳两虚的病机。

4.2.2.3.3.2 阴损及阳 impairment of yin affecting yang
由于阴虚到一定程度，累及阳气生化不足，或阳气无所依附而耗散，从而在阴虚的基础上又出现阳虚，形成以阴虚为主的阴阳两虚的病机。

4.2.2.3.3.3 阴阳两虚 dual deficiency of yin and yang
阴虚和阳虚并见的病机。

4.2.2.3.4 阴阳胜复 alternative predominance of yin and yang
阴阳对立双方，一方亢盛，导致另一方的报复，出现阴胜阳复或阳胜阴复的病理变化。

4.2.2.3.5 阴阳转化 transformation between yin and yang
在疾病过程中，阴证和阳证在一定的条件下所形成的相互转变的病理变化。

4.2.2.3.5.1 由阳转阴 yang syndrome transforming to yin syndrome
在疾病过程中，一定条件下，阳证转化为阴证的病理变化。

4.2.2.3.5.2 由阴转阳 yin syndrome transforming to yang syndrome
疾病发展过程中，在一定条件下，阴证转化为阳证的病理变化。

4.2.2.3.6 阴阳格拒 yin yang rejection
阴或阳的一方偏盛至极而壅踞于内，将相对的一方阻遏于外，所形成的寒热真假的病机，包括阴盛格阳和阳盛格阴。

4.2.2.3.6.1 阴盛格阳 excessive yin rejecting yang
阳气极虚，导致阴寒之气偏盛，壅闭于里，逼迫阳气浮越于外，而出现内真寒外假热的病机。
同义词：阴极似阳

4.2.2.3.6.2 虚阳上浮 floating upward of deficiency-yang
阴盛格阳所致的阳气浮越于上或阴液亏虚、阴不制阳、阳无所附而浮越于上的病机。

4.2.2.3.6.3 阳盛格阴 excessive yang repelling yin
阳热盛极于内，阳气闭郁、逼阴浮越于外所形成的真热假寒的病机。
同义词：阳极似阴

4.2.2.3.7 阴阳亡失 collapse of yin and yang
机体的阴气或阳气突然大量地亡失，导致生命垂危的病机变化，包括亡阴和亡阳。

4.2.2.3.7.1 亡阴 yin collapse
阴液在短时间内大量亡失，脏腑功能突然严重衰竭，导致生命垂危的病机。
同义词：阴脱

4.2.2.3.7.2 伤阴 injury of yin
阴液耗损的病理变化。

4.2.2.3.7.3 亡阳 yang collapse
阳气在短时间内大量亡失，脏腑功能突然严重衰竭，导致生命垂危的病机。
同义词：阳脱

4.2.2.3.7.4 伤阳 injury of yang
损伤阳气的病理变化。

4.2.2.3.7.5 阴竭阳脱 yin exhaustion and yang collapse
阴液枯竭，阳气衰败，动物体功能衰竭，生命垂危的病机。

4.2.2.3.8 阴阳离决 separation between yin and yang
阴阳之间的关系分离、破裂，导致生命垂危的病理变化。

4.2.2.4 寒热失调 imbalance between cold and heat
机体阴阳失调，寒温失其常度而发生的偏寒、偏热或寒热错杂、真假的病理变化。

4.2.2.4.1 实寒 excess-cold
寒邪偏盛的实而寒的病理变化。

4.2.2.4.2 虚寒 deficiency-cold
阳气偏虚的虚而寒的病理变化。

4.2.2.4.3 实热 excess-heat
阳热亢盛的实而热的病理变化。

4.2.2.4.4 实火 excess-fire
邪火炽盛所致的实性而热极的病理变化。

4.2.2.4.5 虚热 deficiency-heat
阴血不足所致虚而热的病理变化。

4.2.2.4.6 虚火 deficiency-fire
真阴亏损及阴盛格阳所致之热或阴虚阳亢，水不制火，所致的阴虚内热、火热上攻的病理变化。

4.2.2.4.7 寒热错杂 cold and heat in complexity
寒和热交错混杂的病理变化。包括上寒下热、上热下寒、表热里寒、表寒里热等。

4.2.2.4.7.1 上寒下热 upper cold and lower heat
动物机体在同时期之内，上部表现为寒证，下部表现为热证的证候。

4.2.2.4.7.2 上热下寒 upper heat and lower cold
　　动物机体在同时期之内,上部表现为热证,下部表现为寒证的证候。

4.2.2.4.8 寒热真假 true-false of cold and heat
　　由阴阳格拒所致的真寒假热或真热假寒的病理变化。

4.2.2.4.8.1 真热假寒 true heat with false cold
　　内热炽盛而外见假寒之象的病理变化。

4.2.2.4.8.2 真寒假热 true cold with false heat
　　阴寒内盛而外见假热之象的病理变化。

4.2.2.4.9 寒热转化 transformation between cold and heat
　　疾病过程中,病性发生转化,出现由寒化热或由热转寒的病理现象。

4.2.2.4.9.1 由寒化热 cold syndrome transforming to heat syndrome
　　病本为寒证,后出现热证,随热证出现而寒证消失的病理现象。

4.2.2.4.9.2 由热化寒 heat syndrome transforming to cold syndrome
　　病本为热证,后出现寒证,随寒证出现而热证消失的病理现象。

4.2.2.5 表里病机 pathogenesis of exterior and interior
　　病邪侵入动物体而致病位浅深和病势轻重的病理变化。包括表里寒热、表里虚实和表里同病等。

4.2.2.5.1 表里寒热 cold or heat of exterior or interior
　　表里之病寒热的病理变化。

4.2.2.5.1.1 表寒 exterior cold
　　风寒之邪,侵袭肌表,郁遏卫阳,尚未入里的寒性病理变化。

4.2.2.5.1.2 表热 exterior heat
　　风热之邪,侵袭肌表,郁阻卫气,尚未入里的热性病理变化。

4.2.2.5.1.3 里寒 interior cold
　　脏腑阴寒偏胜或阳气衰微的病理变化。

4.2.2.5.1.4 里热 interior heat
　　脏腑阳热亢盛或阴虚内热的病理变化。

4.2.2.5.1.5 郁火 stagnant fire
　　阳气郁滞日久而引起内热的病理变化。

4.2.2.5.2 表里虚实 deficiency or excess of exterior or interior
　　表里之病虚实的病理变化。

4.2.2.5.2.1 表实 exterior excess
　　外邪侵入肌表,邪正相争,腠理闭塞的病理变化。

4.2.2.5.2.2 表虚 exterior deficiency
　　卫外的阳气不足,腠理不固而出现的病理变化。

4.2.2.5.2.3 里实 interior excess
体内寒凝、气滞、血瘀、食滞、虫积、停痰、留饮等，邪盛而正未虚的病理变化，或外邪入里化热，结于胃肠而致阳明腑实的病理变化。

4.2.2.5.2.4 里虚 interior deficiency
脏腑气血阴阳亏虚，机能减退的病理变化。

4.2.2.5.3 表里同病 disease of both exterior and interior
表证和里证同时存在的病理变化。

4.2.2.5.3.1 表寒里热 exterior cold and interior heat
表寒与里热并存的病理变化。

4.2.2.5.3.2 表热里寒 exterior heat and interior cold
表热与里寒并存的病理变化。

4.2.2.5.3.3 表里俱寒 cold in both exterior and interior
表寒与里寒并存的病理变化。

4.2.2.5.3.4 表里俱热 heat in both exterior and interior
表热与里热并存的病理变化。

4.2.2.5.3.5 表实里虚 exterior excess and interior deficiency
表实与里虚并存的病理变化。

4.2.2.5.3.6 表虚里实 exterior deficiency and interior excess
表虚与里实并存的病理变化。

4.2.2.5.3.7 表里俱实 excess in both exterior and interior
表实与里实并存的病理变化。

4.2.2.5.3.8 表里俱虚 deficiency in both exterior and interior
表虚与里虚并存的病理变化。

4.2.2.5.3.9 表里转化 transformation between exterior and interior
表证里证可以互相转化，即表证入里或里证出表。

4.2.2.5.3.10 表证入里 inner invasion of exterior syndrome
表证未解，病势向内发展的病理变化。
同义词：由表及里

4.2.2.5.3.11 里证出表 interior disease involving superficies
病邪本在脏腑之里，由于正气抗邪，病邪由里透达于外的病理变化。
同义词：由里出表

4.2.3 气血津液病证病机 pathogenesis of qi, blood and body fluid syndromes
由外邪或内伤引起的气、血、津液运行失常、输布失度、生成不足、亏损过度的病变机制。

4.2.3.1 气血失调　qi-blood disharmony
气和血的生成、运行和功能异常的病机。

4.2.3.1.1 气失调　disorder of qi
气的生成、运行和生理功能异常的病机。

4.2.3.1.1.1 气虚　qi deficiency
气虚弱而致全身或脏腑功能衰退的病机。

4.2.3.1.1.1.1 气虚中满　flatulence caused by qi deficiency
脾胃气虚，运行无力而致脘腹胀满的病理变化。

4.2.3.1.1.1.2 劳则气耗　overwork leading to qi consumption
劳累过度耗伤真气，而致真气亏虚的病理变化。

4.2.3.1.1.2 气机失调　disorder of movement of qi
气的升降出入失去协调平衡的病理变化。

4.2.3.1.1.2.1 气闭　qi blockage
气闭阻于内，不能外出，以致清窍闭塞、出现昏厥的病机。

4.2.3.1.1.2.2 气逆　qi counterflow
气机升降出入反常，应降不降，气机上逆或横逆的病机。

4.2.3.1.1.3 气陷　qi sinking
气虚升举无力，应升反而下陷的病机。

4.2.3.1.1.4 气脱　qi collapse
气不内守而外逸脱失的病机。

4.2.3.1.1.5 气机不畅　delayed qi movement
气机郁结或阻滞的病理变化。

4.2.3.1.1.6 气郁　qi stagnation
情志不舒、气机郁结所导致的病机。

4.2.3.1.1.7 气郁化火　transformation of qi depression into fire
气郁日久，化生火热而气郁与火热并存的病理变化。

4.2.3.1.1.8 气滞　qi impediment
气于体内运行不畅，产生阻滞的病机。

4.2.3.1.2 血失调　disorder of blood
血液生成、运行以及生理功能异常的病机。

4.2.3.1.2.1 血虚　blood deficiency
血液亏虚，脏腑经络失养所导致的病机。

4.2.3.1.2.2 血瘀　blood stasis
血液运行迟缓、凝聚而停滞的病机。

4.2.3.1.2.3 血寒　blood cold
局部脉络寒凝气滞，血行不畅表现的病机。

4.2.3.1.2.4 血热 blood heat
热邪侵犯血分而引起的病机。

4.2.3.1.2.5 血脱 blood collapse
各种急性大出血而致血脉空虚、血液脱失的危重病机。

4.2.3.1.2.6 血不归经 blood flowing outside channels
血液溢出脉外，不循经脉运行的病理变化。
同义词：血不循经；血溢

4.2.3.1.2.7 血不养筋 failure of blood to nourish tendons
肝血不足，筋失濡养而致筋脉拘急的病理变化。

4.2.3.1.3 气血关系失调 disorder of qi and blood
气为血帅、血为气母的关系异常所导致的病理变化。

4.2.3.1.3.1 气血两虚 dual deficiency of qi and blood
气虚和血虚并存，机体失养、功能减退的病机。

4.2.3.1.3.2 气滞血瘀 blood stasis due to qi stagnation
气机运行阻滞，以致血液运行障碍、气滞与血瘀并存的病机。

4.2.3.1.3.3 气虚血瘀 blood stasis due to qi deficiency
气虚而血行无力，导致血行瘀滞、气虚与血瘀并存的病机。

4.2.3.1.3.4 气不摄血 qi failing to contain blood
气虚固摄无力，统摄血液功能减退，血液逸出脉外的病机。

4.2.3.1.3.5 气随血脱 qi collapse following heavy blood loss
大量出血的同时，气随血液的突然流失而脱散，导致气血并脱的病机。

4.2.3.1.3.6 血随气逆 blood counterflow with qi
因气机上逆使血随之上冲而致呕血或清窍闭塞的病机。

4.2.3.2 津液失常 disturbance of body fluid
津液的生成、输布、排泄所发生的紊乱或障碍现象。

4.2.3.2.1 津液不足 insufficient body fluid
体内津液减少，导致脏腑、皮肤、孔窍缺乏津液，失去滋养濡润，出现以燥化为特征证候的病理变化。

4.2.3.2.2 伤津 consumption of body fluid
津液耗伤所致病理变化的总称。

4.2.3.2.3 津脱 fluid depletion
大汗或呕吐等导致机体津液大量脱失的危重病理变化。

4.2.3.2.4 液脱 liquid depletion
阴液极度亏损而致形体羸瘦，脏腑生理功能衰微，其则生命垂危的病理变化。

4.2.3.2.5 气随液脱 qi prostration following liquid depletion
津液大量丢失，气亦随津液大量外泄，导致严重气虚，全身衰竭的病理变化。

4.2.3.2.6 气化无权 failure in qi transformation
阳气亏虚，气化功能减退而致气、血、津、液等化生及其代谢产物排泄异常或水液代谢障碍的病理变化。

同义词：气化不利

4.2.3.2.7 水不化气 failure of water to transform into qi
气化功能失调而致水液代谢障碍或阴液亏虚，津液枯竭，无以化气导致的病理变化。

4.2.3.2.8 水停气阻 water retention causing qi stagnation
水液停潴体内，阻碍气机运行的病理变化。

4.2.3.2.9 津枯血燥 fluid exhaustion causing blood dryness
津液亏乏，血失充盈濡润而燥热内生，甚则血燥生风的病理变化。

4.2.3.2.10 津亏血瘀 fluid depletion causing blood stasis
津液亏损，血液循环郁滞不畅的病理变化。

4.2.3.2.11 气阴两虚 deficiency of both qi and yin
气虚和阴虚同时并见的病理变化。

4.2.4 脏腑病证病机 pathogenesis of zang-fu viscera syndromes
脏腑病变发生、发展、变化以及相互影响的病理机制。

4.2.4.1 五脏病机 five zang viscera pathogenesis
五脏阴阳气血失调的病变机制。

4.2.4.2 六腑病机 six fu viscera pathogenesis
六腑传化、通降等功能失调的病变机制。

4.2.4.3 心与小肠病证病机 heart and small intestine pathogenesis
心与小肠病变发生、发展、变化以及相互影响的机制。

4.2.4.3.1 心病病机 heart pathogenesis
心阴阳气血失调的病变机制。

4.2.4.3.1.1 心气虚 heart qi deficiency
心气虚损，心脏功能减退，血行无力，心神不安，伴有气虚的病理变化。

同义词：心气不足

4.2.4.3.1.2 心血虚 heart blood deficiency
血液亏虚，血不养心，心神失养，血脉空虚，伴有血虚的病理变化。

同义词：心血不足

4.2.4.3.1.3 心阳虚 heart yang deficiency
阳气不足，虚寒内生，温煦失职，心神失养，血行迟滞的病理变化。

同义词：心阳不振

4.2.4.3.1.4 心阳暴脱 sudden collapse of heart yang
阳气突然衰败而欲脱导致心神失守，血行异常，且伴亡阳的病理变化。

4.2.4.3.1.5 心阴虚 heart yin deficiency
阴液不足，心阴耗伤，心神失养，虚热内扰或阴虚火旺的病理变化。
同义词：心阴不足

4.2.4.3.1.6 心火亢盛 rampancy of heart fire
火热炽盛，内炽于心而扰神、上炎、伤津、动血的病理变化。
同义词：心热内盛

4.2.4.3.1.7 心火上炎 upward flaming of heart fire
心经火旺，循经燔灼，躁扰心神，上炎头面口舌的病理变化。

4.2.4.3.1.8 热扰神明 heat disturbing mind
邪热炽盛，扰于心神致心神失常的病理变化。

4.2.4.3.1.9 心血瘀阻 heart blood stagnation and obstruction
血行迟缓，瘀血阻滞心脉的病理变化。

4.2.4.3.1.10 痰火扰心 phlegm-fire disturbing heart
火热痰浊扰乱心神而致神志不宁的病理变化。

4.2.4.3.1.11 痰蒙心窍 mind confusion by phlegm
痰浊内蕴，蒙蔽心神而致神志昏蒙的病理变化。

4.2.4.3.2 小肠病机 small intestine pathogenesis
小肠受盛化物功能失调、二便异常的病变机制。

4.2.4.3.2.1 小肠虚寒 deficiency-cold in small intestine
阳气亏耗，虚寒内生，小肠受盛化物功能减退而清浊不分的病理变化。

4.2.4.3.2.2 小肠实热 excess-heat in small intestine
心火炽盛，循经下移小肠，小肠泌别清浊功能失调的病理变化。

4.2.4.4 肺与大肠病证病机 lung and large intestine pathogenesis
肺与大肠病变发生、发展、变化以及相互影响的病理机制。

4.2.4.4.1 肺病病机 lung pathogenesis
肺阴阳气血失调的病变机制。

4.2.4.4.1.1 肺气虚 lung qi deficiency
肺气虚弱而致呼吸功能减退，卫外失司的病理变化。
同义词：肺气不足

4.2.4.4.1.2 肺阳虚 lung yang deficiency
阳气亏虚，肺失温煦，虚寒内生而致宣降功能减退的病理变化。
同义词：肺气虚寒

4.2.4.4.1.3 肺阴虚 lung yin deficiency
阴液亏虚，虚热内生而致肺失清润，宣降失职的病理变化。

同义词：肺阴不足

4.2.4.4.1.4 肺气不宣 lung qi failing in dispersion

肺气宣发功能失调的病理变化。

4.2.4.4.1.5 肺失清肃 lung qi failing in purification and descent

肺气清肃下降，功能失调的病理变化。

4.2.4.4.1.6 肺气上逆 counterflow rise of lung qi

肺气清肃失司，气机上逆的病理变化。

4.2.4.4.1.7 风热犯肺 wind-heat invading lung

风热邪气侵袭肺卫，而致肺气宣降功能失常的病理变化。

4.2.4.4.1.8 风寒袭肺 wind-cold attacking lung

风寒邪气侵袭肺卫，而致肺气宣降功能失常的病理变化。

4.2.4.4.1.9 燥热伤肺 dryness-heat injuring lung

燥热邪气袭肺，耗伤阴津而致肺气宣降功能失常的病理变化。

4.2.4.4.1.10 痰浊阻肺 turbid phlegm obstructing lung

痰湿壅盛，内阻于肺，肺失宣降的病理变化。

4.2.4.4.1.11 痰热壅肺 phlegm-heat accumulated in lung

痰热交结，壅积于肺，肺失宣降的病理变化。

4.2.4.4.2 大肠病机 large intestine pathogenesis

大肠传导糟粕和吸收水分功能失常的病变机制。

4.2.4.4.2.1 大肠热结 heat accumulation in large intestine

大肠邪热炽盛，气机阻滞而致燥屎内结或迫津下泄的病理变化。

同义词：大肠燥结

4.2.4.4.2.2 大肠寒结 cold accumulation in large intestine

阴寒凝结，阳气虚衰，大肠传导滞涩无力的病理变化。

4.2.4.4.2.3 大肠湿热 dampness-heat in large intestine

湿热内蕴，损伤肠络，阻滞气机，大肠传导失常的病理变化。

4.2.4.4.2.4 热迫大肠 heat distressing large intestine

热邪下迫，来势急骤，大肠传导失常的病理变化。

4.2.4.4.2.5 大肠虚寒 deficiency-cold in large intestine

阳气虚衰，温煦失职，虚寒内生，大肠传导功能减退的病理变化。

4.2.4.4.2.6 大肠液亏 liquid insufficiency of large intestine

津液不足，大肠失润，气机阻滞，大肠传导失常的病理变化。

4.2.4.5 脾与胃病证病机 spleen and stomach pathogenesis

脾与胃病变发生、发展、变化以及相互影响的病理机制。

4.2.4.5.1 脾病病机 spleen pathogenesis

脾阴阳气血失调的病变机制。

4.2.4.5.1.1 脾气虚 spleen qi deficiency
脾气亏虚，功能减退，运化失职，化源亏乏，形体失养的病理变化。
同义词：脾气不足

4.2.4.5.1.2 脾气不升 spleen qi failing to ascend
脾气虚弱，水谷精微不得上输，心、肺、头目失于滋养的病理变化。
同义词：脾不升清；清阳不升

4.2.4.5.1.3 中气下陷 sinking of spleen qi
脾气虚弱，升举无力，降多升少，气机下陷的病理变化。

4.2.4.5.1.4 脾阴虚 spleen yin deficiency
脾阴亏虚，脾失濡养，运化失常的病理变化。
同义词：脾阴不足

4.2.4.5.1.5 脾阳虚 spleen yang deficiency
脾阳亏虚，虚寒内生，温煦失职，运化失常的病理变化。
同义词：脾阳不振

4.2.4.5.1.6 脾失健运 dysfunction of spleen in transportation
脾运化功能失常的病理变化。

4.2.4.5.1.7 脾不统血 failure of spleen to control blood
脾气统摄血液功能失调而致血溢脉外的病理变化。

4.2.4.5.1.8 脾虚湿困 spleen deficiency with dampness retention
脾气虚衰，运化失司而致湿浊内停的病理变化。

4.2.4.5.1.9 湿热蕴脾 accumulation of dampness-heat in spleen
脾失健运，水湿停滞，湿蕴生热，湿热郁蒸的病理变化。
同义词：湿热困脾

4.2.4.5.1.10 寒湿困脾 retention of cold-dampness in spleen
寒湿内盛，困阻脾阳，运化失职的病理变化。

4.2.4.5.1.11 脾虚生风 production of wind by spleen deficiency
脾气虚弱，化源亏乏，筋脉失养而致虚风内动的病理变化。

4.2.4.5.1.12 脾虚生痰 production of phlegm by spleen deficiency
脾虚失运，水湿停聚痰浊内生的病理变化。

4.2.4.5.2 胃病病机 stomach pathogenesis
胃受纳腐熟水谷及通降失常的病变机制。

4.2.4.5.2.1 胃气虚 stomach qi deficiency
胃气虚弱，胃失濡养，受纳腐熟功能减退，胃失和降的病理变化。
同义词：胃气不足

4.2.4.5.2.2 胃阳虚 stomach yang deficiency
阳虚气弱，虚而有寒，受纳腐熟功能减退，胃失和降的病理变化。

同义词：胃虚寒

4.2.4.5.2.3 胃阴虚 stomach yin deficiency

阴液亏虚，胃失濡润，虚而有热，受纳腐熟功能减退，胃失和降的病理变化。

同义词：胃阴不足

4.2.4.5.2.4 胃寒 stomach cold

脾阳虚衰，过食生冷或寒邪直中所致阴寒凝滞胃腑的病证。

4.2.4.5.2.5 胃热 stomach heat

胃火阳热偏盛的病理变化。

4.2.4.5.2.6 胃火炽盛 blazing of stomach fire

胃中火热炽盛，循经上炎，耗伤津液，胃失和降，功能亢进的病理变化。

4.2.4.5.2.7 胃热消谷 acceleration of digestion by stomach heat

火热炽盛，内炽于胃，腐熟功能亢进的病理变化。

同义词：胃热杀谷

4.2.4.5.2.8 胃气不和 disorder of stomach qi

胃的受纳、腐熟水谷和通降功能失调所致病理变化的统称。

同义词：胃不和

4.2.4.5.2.9 胃气不降 failure of stomach qi to descend

胃的通降功能受阻，腑气不通的病理变化。

同义词：胃失和降

4.2.4.5.2.10 胃气上逆 counterflow rise of stomach qi

胃的气机逆转向上所引发的病证。

4.2.4.5.2.11 胃纳呆滞 poor appetite and digestion

胃的受纳功能停滞，摄食减少的病理变化。

4.2.4.5.2.12 食滞胃脘 food stagnating in the stomach venter

水谷停滞胃脘不化的病理变化。

4.2.4.6 肝与胆病证病机 liver and gallbladder pathogenesis

肝与胆病变发生、发展、变化以及相互影响的病理机制。

4.2.4.6.1 肝病病机 liver pathogenesis

肝阴阳气血失调的病变机制。

4.2.4.6.1.1 肝气虚 liver qi deficiency

肝之阳气不足，功能减退，升发无力，疏泄不及的病理变化。

同义词：肝气不足

4.2.4.6.1.2 肝血虚 liver blood deficiency

肝脏藏血失职，血液亏虚，经脉、头部和眼睛等失于濡养的病理变化。

同义词：肝血不足

4.2.4.6.1.3 肝阴虚 liver yin deficiency

肝阴液亏虚，肝失濡润，虚热内生的病理变化。

同义词：肝阴不足

4.2.4.6.1.4 肝阳虚 liver yang deficiency

肝阳气亏虚，虚寒内生，疏泄和藏血功能低下的病理变化。

4.2.4.6.1.5 肝气郁结 liver qi stagnation

肝失疏泄，气机郁滞，情志抑郁的病理变化。

4.2.4.6.1.6 肝气横逆 sideward counterflow of liver qi

肝失疏泄，气机郁滞，侵犯脾胃的病理变化。

4.2.4.6.1.7 肝火上炎 liver fire flaming

肝火炽盛，气火上冲，气血壅盛脉络的病理变化。

4.2.4.6.1.8 肝经风热 wind-heat in liver meridian

风热邪气，侵袭肝经，上扰头目的病理变化。

4.2.4.6.1.9 肝经郁热 stagnated heat in liver meridian

肝气郁结，郁久化热，气郁与热盛并存的病理变化。

4.2.4.6.1.10 肝经湿热 dampness-heat in liver meridian

湿热邪气，蕴结于肝，肝气郁滞，循经下注的病理变化。

4.2.4.6.1.11 肝阳上亢 hyperactivity of liver yang

肝肾阴虚，阳亢于上，本虚标实，上盛下虚的病理变化。

4.2.4.6.1.12 肝风内动 internal stirring of liver wind

阴阳气血失调，肝阳气亢逆而为风的病理变化。

4.2.4.6.1.13 寒滞肝经 cold accumulated in liver meridian

多因外感寒邪，滞留肝经，致使气血凝滞的病理变化。

4.2.4.6.2 胆病病机 gallbladder pathogenesis

胆功能失调的病变机制。

4.2.4.6.2.1 胆经郁热 stagnation of heat in gallbladder meridian

火热内扰，疏泄失常，胆失宁谧的病理变化。

4.2.4.6.2.2 胆郁痰扰 gallbladder stagnation with phlegm disturbance

痰浊内蕴，胆气郁结，疏泄失常，扰于心神的病理变化。

4.2.4.7 肾与膀胱病证病机 kidney and bladder pathogenesis

肾与膀胱病变发生、发展、变化以及相互影响的病理机制。

4.2.4.7.1 肾病病机 kidney pathogenesis

肾阴阳气血失调的病变机制。

4.2.4.7.1.1 肾气虚 kidney qi deficiency

肾之阳气亏虚，生长生殖功能下降，气化无权，封藏固摄功能失职的病理变化。

同义词：肾气不足

4.2.4.7.1.2 肾气不固 insecurity of kidney qi
肾气亏虚，封藏失职，固摄无权的病理变化。
同义词：下元不固

4.2.4.7.1.3 肾不纳气 failure of kidney to receive qi
肾气虚损，摄纳肺气功能减退，肺气逆上的病理变化。

4.2.4.7.1.4 肾阳虚 kidney yang deficiency
阳气虚衰，温煦失职，虚寒内生，气化无权的病理变化。
同义词：肾阳不足；命门火衰

4.2.4.7.1.5 肾虚水泛 kidney insufficiency with water diffusion
肾阳虚衰，气化无权，水湿泛滥的病理变化。

4.2.4.7.1.6 肾阴虚 kidney yin deficiency
阴液亏损，肾失滋养，虚热内扰，甚则阴虚火旺的病理变化。
同义词：肾阴不足

4.2.4.7.1.7 肾精不足 kidney essence insufficiency
肾失所藏，肾精亏虚，发育迟缓，生殖功能减退的病理变化。

4.2.4.7.1.8 精脱 essence prostration
肾精亏损，甚则脱失，官窍失养的病理变化。

4.2.4.7.2 膀胱病机 bladder pathogenesis
因气化不利，膀胱贮尿、排尿异常的病变机制。

4.2.4.7.2.1 膀胱湿热 dampness–heat in bladder
湿热蕴结，膀胱气化不利的病理变化。

4.2.4.7.2.2 热结膀胱 heat accumulation in bladder
热入膀胱，瘀热相搏，结于少腹，上扰心神的病理变化。

4.2.4.7.2.3 膀胱虚寒 deficiency–cold of bladder
肾阳虚衰，虚寒内生，膀胱气化功能减弱的病理变化。

4.2.4.7.2.4 膀胱气闭 bladder qi blockage
膀胱气化功能障碍，引起小便不畅的病理变化。

4.2.4.8 脏腑兼病病机 combined zang-fu visceral pathogenesis
脏腑关系失调的病变机制。

4.2.4.8.1 心肺气虚 heart–lung qi deficiency
心气与肺气俱虚，血行无力，宣降失常的病理变化。

4.2.4.8.2 心脾两虚 insufficiency of both heart and spleen
心与脾气血阴阳亏损，以血不养心、运化失职为主的病理变化。

4.2.4.8.3 心肝火旺 blazing of heart–liver fire
心肝邪热炽盛，循经上炎，躁扰神明，灼伤脉络的病理变化。

4.2.4.8.4 心肝血虚 heart-liver blood deficiency
心肝血液亏损，头目、筋脉、爪甲、血海失养的病理变化。

4.2.4.8.5 心肾阴虚 heart-kidney yin deficiency
心与肾阴液亏损，虚热内生，阴虚火旺而致心火上炎的病理变化。

4.2.4.8.6 心肾阳虚 heart-kidney yang deficiency
心与肾阳气亏虚，虚寒内生，失于温煦，功能减退，气化失常，血行无力的病理变化。

4.2.4.8.7 心肾不交 non-interaction of heart and kidney
肾阴亏虚，心阳独亢，水火未济的病理变化。
同义词：水火未济

4.2.4.8.8 水气凌心 insult of water to heart
肾阳亏虚，气化无力，水液泛滥，上凌于心，抑遏心阳的病理变化。

4.2.4.8.9 水寒射肺 attack of lung by water-cold
肾阳亏虚，气化无力，水液泛滥，上逆犯肺，肺失宣降的病理变化。

4.2.4.8.10 心胃火燔 outburst of heart and stomach fire
心营热盛，胃火烁津，心神不宁，津伤液耗的病理变化。

4.2.4.8.11 心胆气虚 heart-gallbladder qi deficiency
心气亏虚，胆虚气怯，心神不宁的病理变化。

4.2.4.8.12 肺脾两虚 insufficiency of both lung and spleen
肺气亏虚，脾气亦衰，肺失宣降，脾不健运的病理变化。
同义词：肺脾气虚

4.2.4.8.13 肺肾阴虚 lung-kidney yin deficiency
肺与肾阴液亏虚，虚热内扰，阴虚火旺，肺络灼伤，清肃失司的病理变化。

4.2.4.8.14 肺肾气虚 lung-kidney qi deficiency
肺与肾之气俱虚，肃降与纳气不足的病理变化。

4.2.4.8.15 脾胃虚弱 spleen-stomach deficiency
脾与胃气虚，功能减退，升清降浊失常，化源匮乏的病理变化。

4.2.4.8.16 寒伤脾胃 cold damaging spleen and stomach
寒邪伤及脾胃的病理变化。

4.2.4.8.17 脾胃虚寒 deficiency-cold of spleen and stomach
脾与胃阳气亏虚，阴寒内生，功能减退的病理变化。
同义词：脾胃阳虚

4.2.4.8.18 脾胃湿热 dampness-heat in spleen and stomach
湿热内蕴中焦，纳运升降失常的病理变化。

4.2.4.8.19 脾胃阴虚 spleen-stomach yin deficiency
脾与胃阴液亏损，失于濡润，纳运功能减退，虚热内扰的病理变化。

4.2.4.8.20 脾肾阳虚 spleen-kidney yang deficiency
脾与肾阳气虚衰，虚寒内生，健运失职，气化无权，水谷不化，水液泛滥的病理变化。

4.2.4.8.21 土壅木郁 spleen earth obstructed and hepatic qi stagnated
脾气壅滞而致肝失疏泄的病理变化。

4.2.4.8.22 肝火犯肺 attack of lung by liver fire
肝火炽盛，上逆犯肺，肺失肃降的病理变化。

同义词：木火刑金

4.2.4.8.23 肝气犯脾 attack of spleen by liver qi
肝气郁滞，横逆犯脾，脾失健运的病理变化。

4.2.4.8.24 肝气犯胃 attack of stomach by liver qi
肝气郁滞，横逆犯胃，胃失和降的病理变化。

4.2.4.8.25 肝郁脾虚 liver stagnation with spleen insufficiency
肝失疏泄与脾气虚弱并见，肝气郁结，木不疏土，脾气虚弱，健运失职的病理变化。

4.2.4.8.26 肝脾不和 disharmony between liver and spleen
肝疏泄与脾运化功能不能协调配合的病理变化。多因受惊、失群、离仔使肝气郁结，脾失健运；或因饲养使役不当，损伤脾气，脾不健运影响肝失疏泄所致。

4.2.4.8.27 肝胆湿热 dampness-heat in liver and gallbladder
湿热内蕴，熏蒸肝胆，疏泄失常，胆汁泛滥的病理变化。

4.2.4.8.28 肝胆实热 excessive heat in liver and gallbladder
火热炽盛，蕴结肝胆，疏泄失职，阳热上冲的病理变化。

4.2.4.8.29 肝胆寒湿 cold-dampness in liver and gallbladder
寒湿侵袭肝胆的病理变化。

4.2.4.8.30 水不涵木 failure of water to nourish wood
肾阴亏虚，不能滋养肝木，出现肝阴不足，虚风内动的病理变化。

4.2.5 六经病证病机 pathogenesis of six meridians syndromes
太阳、阳明、少阳、太阴、少阴、厥阴六经病的病变机制。

4.2.5.1 太阳病机 pathogenesis of taiyang meridian
风寒等外邪侵袭太阳，邪正交争于肌表的病变机制。

4.2.5.1.1 营卫不和 disharmony of nutrient-defensive qi
营失内守，卫失外固，营卫失于和调而汗自出的病理变化，包括卫强营弱和卫弱营强。

4.2.5.1.1.1 卫强营弱 strong defensive qi and weak nutrient qi
卫气亢奋，郁于肌表，内迫营阴，营失内守而汗自出的病理变化。

4.2.5.1.1.2 卫弱营强 weak defensive qi and strong nutrient qi
卫气虚弱，肌表不固，营阴失守而汗自出的病理变化。

4.2.5.1.2 卫气不固 insecurity of defensive qi
卫阳虚衰，失于固表，腠理疏松的病理变化。
同义词：表气不固

4.2.5.1.3 营气不从 stagnation of nutrient qi
血脉中营气运行障碍，出现痈肿的病理变化。

4.2.5.2 阳明病机 pathogenesis of yangming meridian
邪入阳明，阳热亢盛，胃肠燥热的病变机制。

4.2.5.2.1 阳明燥热 dryness-heat in yangming meridian
邪热亢盛，充斥阳明，弥漫全身，消烁津液，尚无实滞的病理变化。

4.2.5.2.2 阳明腑实 excessiveness of yangming fu viscera
邪热内盛，与肠中积滞搏结，劫伤津液，燥结成实的病理变化。

4.2.5.2.3 胃家实 stomach-intestinal excessiveness
病邪深入阳明，肠胃燥热亢盛，无形邪热充斥或邪热与肠中积滞搏结的病理变化。

4.2.5.2.4 阳明虚寒 deficiency-cold of yangming meridian
胃阳虚衰，阴寒内盛，纳谷腐熟失司，胃失和降，浊阴上逆的病理变化。

4.2.5.3 少阳病机 pathogenesis of shaoyang meridian
邪犯少阳，枢机不运，经气不利，殃及半表半里的病变机制。

4.2.5.3.1 半表半里病机 pathogenesis of half-exterior and half-interior
病邪已离太阳之表，未入阳明之里，而与正气相持于表里之间的病理变化。

4.2.5.3.2 邪郁少阳 stagnation of pathogen in shaoyang meridian
邪犯少阳，枢机不利，正邪相持于半表半里的病理变化。

4.2.5.4 太阴病机 pathogenesis of taiyin meridian
邪入太阴，脾阳虚衰，寒湿内生的病变机制。

4.2.5.4.1 太阴寒湿 cold-dampness in taiyin meridian
邪在太阴，脾胃阳虚，寒湿中阻的病理变化。

4.2.5.4.2 太阴虚寒 deficiency-cold of taiyin meridian
邪在太阴，脾胃阳虚，虚寒内生的病理变化。

4.2.5.5 少阴病机 pathogenesis of shaoyin meridian
邪犯少阴，病入心肾，心肾阳虚或心肾阴虚的病变机制。

4.2.5.5.1 少阴热化 transformation of pathogen in shaoyin into heat
邪犯少阴，从阳化热，肾水亏于下，心火亢于上，心肾不交的病理变化。

4.2.5.5.2 少阴寒化 transformation of pathogen in shaoyin into cold
邪犯少阴，从阴化寒，心肾阳虚而阴寒内盛的病理变化。

4.2.5.6 厥阴病机 pathogenesis of jueyin meridian
邪入厥阴，阴阳对峙，厥热胜复，寒热夹杂的病变机制。

4.2.5.6.1 厥热胜复 preponderance of cold or heat
邪入厥阴，邪正交争，阴阳消长，寒热交替出现的病理变化。

4.2.5.6.2 热深厥深 the deeper the heat, the severer the limb coldness
邪热深伏，阳气内郁，阴阳之气不相顺接，热郁重而肢厥甚的病理变化。
同义词：厥深热深

4.2.5.6.3 热微厥微 the milder the heat, the milder the limb coldness
邪热深伏，阳气内郁，阴阳之气不相顺接，热郁轻而肢厥微的病理变化。
同义词：厥微热微

4.2.6 卫气营血病证病机 pathogenesis of defense qi nutrient blood syndromes
从卫气营血辨证的角度，阐释病证的发病机制。

4.2.6.1 卫分病机 pathogenesis of defense level
温病初期，温邪侵袭动物体肌表，肺卫功能失调的病变机制。属卫气营血病理演变过程的表浅阶段。

4.2.6.1.1 邪郁肺卫 stagnation of pathogen in lung defense level
温邪袭表，伤及肺卫，卫气郁阻，肺气失宣的病理变化。
同义词：卫气郁阻；邪郁卫表

4.2.6.2 气分病机 pathogenesis of qi level
温邪由卫入里，正邪相争，里热亢盛，耗伤津液的病变机制。

4.2.6.2.1 热盛气分 exuberant heat in qi level
邪入气分，里热蒸腾，伤津耗液的病理变化。
同义词：气分热盛

4.2.6.2.2 热炽阴伤 yin injury by blazing heat
邪入气分，里热壅盛，耗伤阴液而致热结阴亏的病理变化。

4.2.6.3 营分病机 pathogenesis of nutrient level
温热邪气侵入气分与血分之间，营热阴伤，扰于心神，热窜血络的病变机制。

4.2.6.3.1 热陷心包 invasion of pericardium by heat
邪热炽盛，内陷心包，扰乱神明的病理变化。
同义词：热入心包

4.2.6.3.2 逆传心包 abnormal transmission of warm pathogen to pericardium
温邪犯肺之后，不顺传气分，而直入心包。

4.2.6.3.3 心营过耗 over consumption of heart nutrient
温热邪气伤及心营，耗伤阴血，灼伤津液的病理变化。

4.2.6.3.4 营阴损伤 nutrient yin injury
温热邪气侵入营分，耗伤阴津的病理变化。

4.2.6.4 血分病机 pathogenesis of blood level
温热邪气侵入血分，动血耗血，瘀热内阻的病变机制。

4.2.6.4.1 热入血分 entering of heat into blood level
温热邪气侵入血分，迫血妄行，热瘀交结，内扰心神，伤阴动风的病理变化。

4.2.6.4.2 血热妄行 bleeding due to blood heat
热入血分，损伤血络，加速血行而表现出血的病理变化。

4.2.6.4.3 血分瘀热 stagnation of heat in blood level
温热邪气深入营血，瘀热互结的病理变化。

4.2.6.4.4 血分热毒 heat toxin in blood level
热毒深陷血分，耗血伤阴，甚则蒙蔽心神，危及生命的病理变化。

4.2.6.5 卫气同病 disease involving both defense and qi levels
表邪入里化热，气分热势已盛而卫分病变尚未消除的表里俱实的病理变化。

4.2.6.6 气营两燔 blazing of both qi and nutrient levels
气分与营分邪热炽盛的病理变化。

4.2.6.7 气血两燔 blazing of both qi and blood levels
气分与血分邪热炽盛，扰及心神，灼伤血络，迫血妄行的病理变化。

4.2.6.8 卫营同病 disease involving both defense and nutrient levels
卫分病变和营分病变并存的病理变化。

4.2.7 三焦病证病机 pathogenesis of triple energizer syndromes
温病上、中、下三焦病变的机制。

4.2.7.1 上焦湿热 dampness-heat in upper energizer
湿热侵袭上焦，困遏卫阳，肺失宣降的病理变化。

4.2.7.2 中焦湿热 dampness-heat in middle energizer
湿热病邪传入中焦，困阻脾胃，脾失健运的病理变化。

4.2.7.3 下焦湿热 dampness-heat in lower energizer
湿热流注下焦、大肠、膀胱，气机阻滞，气化失司的病理变化。

4.2.7.4 三焦湿热 dampness-heat in triple energizer
湿热弥漫，累及上、中、下三焦的病理变化。

第五章 治 则

5.1 治未病类

5.1.1 治未病 preventive treatment

预防法则。采取一定的措施防止疾病发生和发展的治疗原则,包括未病先防、既病防变两方面。

5.1.1.1 未病先防 preventing a disease before it arises

预防法则。在动物未发病之前,采取各种有效措施,预防疾病的发生。

5.1.1.2 既病防变 prevention of progress of disease

预防法则。在疾病发生以后,应早期诊断、早期治疗,以防止疾病的发展与传变。

5.1.1.3 愈后防复 post-recovery prevention

预防法则。在疾病治疗完成后,采取一系列的预防措施,以防止疾病复发。

5.1.2 四季药 seasonal formula

按不同季节灌服,用以调理动物机体,预防疾病的药方。

5.1.3 针刺六脉血 bleeding at the six points

对膘肥体壮的马等动物,应用血针刺胸堂、眼脉、带脉、肾堂、鹘脉、尾本六个穴位放血,能调理气血阴阳,疏通经络,并能泻热和增强对一些热性病的抵抗力,从而可以预防疾病。

5.2 治则类

5.2.1 治则 treatment principles

在治疗动物疾病时需遵循的基本原则,对兽医临床的具体理法、处方、用药等具有普遍的指导意义。

5.2.1.1 治病求本 treating disease under the root

治疗疾病时必须通过辨证探究其本质,根据本质进行治疗。

5.2.2 正治 routine treatment

针对疾病的本质,从正面进行治疗,即逆着疾病症象而治的治疗原则。

同义词：逆治

5.2.2.1 寒者热之 treat cold with heat

针对寒性的病证需使用温热方药或具有温热功效的措施进行治疗的原则。治疗寒证宜采用温法，根据病情，或辛温解表，或温中散寒，或温肾壮阳。

5.2.2.2 热者寒之 treat heat with cold

针对热性的病证需使用寒凉方药或具有寒凉功效的措施进行治疗的原则。治疗热证宜用清法，根据病情或辛凉解表，或清热泻火，或壮水滋阴。

5.2.2.3 虚则补之 treat deficiency with reinforcement

针对虚弱性的病证采用补益方药或具有补益功效的措施进行治疗的原则。

同义词：虚者补之

5.2.2.4 实则泻之 treat excess with purgation/reduction

针对性质属实的病证采用攻泻方药或具有攻伐功效的措施进行治疗的原则。

同义词：实者泻之

5.2.3 反治 paradoxical treatment

针对疾病出现假象，或大寒证、大热证用正治法发生格拒的情况，采用与表面症状性质相同的药物进行治疗，亦即顺从疾病假象性质的治疗原则。

同义词：从治

5.2.3.1 寒因寒用 treat false cold with cold

针对真热假寒证用寒凉方药或具有寒凉功效的措施治疗的原则。

5.2.3.2 热因热用 treat false heat with heat

针对真寒假热证用温热方药或具有温热功效的措施治疗的原则。

5.2.3.3 塞因塞用 treating obstructive syndrome with tonifying methods

针对正气虚损、本虚标实所致闭塞不通的病证用补益、固涩方药进行治疗的原则。

5.2.3.4 通因通用 treat uncontrolled discharge by unblocking

针对因邪实所致内有积滞而外见泄泻、崩漏等病证，采用通利的方药或具有通利功效的措施进行治疗的原则。

5.2.3.5 调理阴阳 regulating yin and yang

利用药物、食物、针灸等纠正阴阳的偏胜偏衰，使之恢复相对平衡协调的治疗原则。

5.2.4 标本缓急 symptom, root-cause, non-urgency and urgency

应用标本关系分析病证的主次先后、轻重缓急而确定治疗步骤的原则。

5.2.4.1 急则治标　treat the tip first in acute conditions

与缓则治本相对而言，在大出血、暴泻、剧痛等标症紧急的情况下，及时采取止血、止泻、止痛等救治标病，然后治其本病的治疗原则。

5.2.4.2 缓则治本　treat the root in remissive stages

与急则治标相对而言，在病势缓和、病情缓慢的情况下，针对本病的病机治疗或采取调理、补益为主的治疗原则。

5.2.4.3 标本同治　treat both the tip and root

标病本病同时俱急，在时间与条件上皆不宜单治标或单治本时，采取同治的方法。

5.2.5 扶正祛邪　reinforce healthy qi to eliminate pathogenic factors

对于正虚为主、因虚致实的病证，采取扶助正气为主，使正气加强，增强机体的抗病能力而达到祛除病邪目的的治疗原则。

5.2.5.1 扶正固本　reinforce healthy qi to strengthen the body

使用补益正气的方药及加强患病动物护养等方法，以扶助动物机体正气，提高动物机体抵抗力，达到祛除邪气、战胜疾病、恢复健康目的的治疗原则。

同义词：扶正

5.2.5.2 祛邪扶正　eliminating pathogen to support vital qi

祛逐病邪并扶助机体正气的治疗原则。对于邪气实而正气偏虚的病证，采取以消除病邪为主，扶助正气为辅，从而达到邪祛正安的目的。

同义词：祛邪安正

5.2.5.3 先补后攻　elimination after reinforcement

正虚而邪实，正气虚衰难以耐受攻伐，先扶正后祛邪。

同义词：先扶正后祛邪

5.2.5.4 先攻后补　reinforcement after elimination

邪盛而正虚，正气尚耐攻伐，祛邪后扶正。

同义词：先祛邪后扶正

5.2.5.5 攻补兼施　reinforce healthy qi and remove pathogenic factors simultaneously

对于虚实夹杂，或虚实病情相当的病证，扶正与祛邪同时使用的治疗原则，即攻邪与扶正并重的治疗原则。

5.2.5.6 寓攻于补　reinforcement containing elimination

治疗虚实夹杂病证的一种治疗原则。指将攻邪目的寓寄于补正之中，通过补正为主，待正气来复而能驱邪以愈病。或指在补益方药中，适当加入攻邪之品，使扶正而不留邪。

同义词：补中寓攻

5.2.5.7 寓补于攻 elimination containing reinforcement

治疗虚实夹杂病证的一种治疗原则。一指通过攻邪，使邪去而正安，最后达到补益的目的。二指将少量补益之品夹于大队攻伐药之内，使攻邪而不伤正。

同义词：攻中寓补

5.2.6 同病异治 different treatments for the same disease

相同的疾病，可因动物、因时、因地的不同，或因病情的发展、病机的变化、正邪的消长等差异，采取不同治法的治疗原则。

5.2.7 异病同治 same treatment for different diseases

不同的疾病，由于病机相同或处于同一性质的病变阶段（证候相同），采取相同治法的治疗原则。

5.2.8 三因制宜 treatment in accordance with three factors

根据不同气候、地理环境和动物自身特点考虑用药的治疗原则，包括因时制宜、因地制宜和因畜（禽）制宜。

5.2.8.1 因时制宜 treat according to time

根据不同季节气候的寒热燥湿等特点选择适宜的治疗方法和用药的治疗原则。

5.2.8.2 因地制宜 treat according to place

根据不同地区的地理环境特点选择适宜的治疗方法和用药的治疗原则。

5.2.8.3 因畜（禽）制宜 treatment in accordance with animals individuality

根据不同动物的种类、体重、年龄、强弱、公母及病情的轻重、缓急和用药方式的不同特点选择适宜的治疗方法和用药的治疗原则。

5.2.9 调整阴阳 regulate and balance yin and yang

根据动物机体阴阳盛衰的变化而损其有余或补其不足，使之重归于和谐平衡的治疗原则。

5.2.9.1 阳病治阴 treatment of yin for yang disease

阴虚而致虚热之证，治宜滋阴以抑阳，即阳病而从阴论治。

5.2.9.2 阴病治阳 treatment of yang for yin disease

阳虚而致虚寒之证，治宜扶阳以制阴，即阴病而从阳论治。

5.2.9.3 阳中求阴 seeking yin from yang

根据阴阳互根理论，治疗阴虚之证宜滋阴之中佐以扶阳。

5.2.9.4 阴中求阳 seeking yang from yin

根据阴阳互根理论，治疗阳虚之证宜扶阳之中佐以滋阴。

5.2.10 调和气血 regulate and harmonize qi and blood

用具有理气和血作用的方药治疗气血不和病证的治疗原则。

5.2.10.1 气病治血 treating blood for qi disorder

气血互相依存，气病血亦病，治气必兼理血。

5.2.10.2 血病治气 treating qi for blood disorder

气血互相依存，血病气亦病，治血必兼理气。

5.2.10.3 虚者补其母 reinforcing mother-element in deficiency condition

根据五行相生和五脏母子关系理论，五脏虚证可补其属母（生我）之脏以治。

5.2.10.4 实者泻其子 reducing child-element in excess condition

根据五行相生和五脏母子关系理论，五脏实证可泻其属子（我生）之脏以治。

5.2.10.5 脏病治腑 treating fu for zang viscus disease

脏腑相合，脏病可治其相合之腑。

5.2.10.6 腑病治脏 treating zang for fu viscus disease

脏腑相合，腑病可治其相合之脏。

第六章 疾 病

6.1 心系病

6.1.1 口舌生疮 mouth and tongue ulcers

因心火炽盛或阴虚火旺，上攻于口所致。以舌、唇、颊、齿龈红肿、水疱、溃烂及流涎为主要表现的疾病。

同义词：口疮；舌疮；牙疮

6.1.2 血证 bleeding or hemorrhagic syndrome

因血不循经脉运行，溢于脉管外所致。以鼻衄、咯血、吐血、尿血、便血和皮下出血等为主要表现的疾病。

6.1.3 汗证 sweating syndrome

因表卫亏虚、腠理开阖失司，津不内守所致。以皮毛被汗液湿润或大汗淋漓不止为主要表现的疾病。

6.1.4 木舌 swollen and rigid tongue

因心经积热或外伤，气血停聚于舌所致。以舌体肿胀发硬、形如木杆为主要表现的疾病。

6.1.5 心虚 heart deficiency

因心经气血不足所致。以神疲、自汗、易惊、运动能力减退为主要表现的疾病。

6.1.5.1 心气虚 qi deficiency of heart

因运动或使役过度，损伤心气，致心气不足而无力鼓动血脉正常运行，使心失所养所致。以心悸动、气短、倦怠无力等为主要表现的疾病。

6.1.5.2 心血虚 blood deficiency of heart

因心脾虚弱，血液生化之源不足或因失血较多所致。以心悸动、躁动、易惊、口色淡白或苍白等为主要表现的疾病。

6.1.6 心冷吐水 hydroptysis spleen and heart deficiency

因心脾两虚而遇寒湿之邪，使胃受冷，令水湿上泛所致。以口吐清水为主要表现的疾病。

6.1.7 心热风邪 apoplexy

因热邪内聚，化火生痰，干扰心神所致。以发病急速、猝然倒地、汗出肉颤、两目上翻为主要表现的疾病。

6.1.8 中暑 heat stroke

因暑热或暑湿内郁，侵扰心神所致。以突然发病、高热、心肺机能急剧衰竭为主要表现的疾病。

6.1.8.1 热痛 mild heat stroke

因阳气太盛，心、肺、胃三经热极相攻所致。以精神恍惚、头低眼闭、站立如痴、口红身热为主要表现的疾病。

6.1.8.2 黑汗风 severe heat stroke

因暑热而致的瘀血不通、血脉壅滞、积于心胸所致。以突然神昏头低、眼急呆痴、行如酒醉、浑身肉颤、汗出如浆、气促喘粗为主要表现的疾病。

6.1.9 痫病 epilepsy

因痰火积聚，迷乱心神，或外感暑热，热积三焦，使肝、脾、肾、心功能失调所致。以阵发性、神志丧失、四肢抽搐和流涎为主要表现的疾病。

6.1.10 心黄 xin huang

因料毒、疫气布满胃肠，痰血郁结，迷乱其心所致。以烦躁不安、眼急惊狂、自啃其身、咬物伤人为主要表现的疾病。

同义词：心风黄

6.1.11 心痛 cardialgia

因痞气冲塞心胸，瘀痰凝于罗膈，气血凝滞，血流不畅所致。以两目直视、眼闭头低、四肢难移、鼻乍喘粗、时起时卧、前肢跪地、前胸出汗为主要表现的疾病。

6.1.12 脑黄 meningo encephalitis

因暑月炎天，运动或劳役过度，或受烈日暴晒，厩舍闷热，热积于心肺而上注于脑所致。以惊狂或痴呆为主要表现的疾病。

6.1.13 中风 stroke

因气血逆乱，脑脉痹阻或血溢于脑所致。以共济失调、肢体麻木等为主要表现的疾病。

6.2 肺系病

6.2.1 感冒 cold

因风寒或风热之邪侵袭所致。以恶寒、发热、鼻塞、流涕或咳嗽为主要表现的疾病。

6.2.1.1 风寒感冒 common cold of wind cold

因运动或劳役汗出，气候突变，受到风寒雨淋所致。以恶寒、发热、无汗，被毛逆立，外流清涕为主要表现的疾病。

6.2.1.2 风热感冒 common cold of wind heat

因风热之邪侵袭肌表，肺气失和所致。以肌表发热，微恶风寒，鼻流黏涕，口渴欲饮为主要表现的疾病。

6.2.1.3 体虚感冒 body deficiency cold

因机体素虚，腠理不固或过劳出汗，又感受外邪，使卫气受损，营液外泄，营卫不和所致。以恶寒发热，热势不盛，形寒恶风，缠绵难愈为主要表现的疾病。

6.2.2 咳嗽 cough

因肺系外感内伤、宣降失常所致。以肺气上逆作声、鼻流涕、咳痰为主要表现的疾病。

同义词：咔嗽；哐嗽

6.2.2.1 外感咳嗽 exogenous cough

因风、寒、湿等邪气入侵，引发肺气宣降失常，导致肺气上逆而引起的咳嗽，包括风寒咳嗽、风热咳嗽和肺热咳嗽。

6.2.2.2 风寒咳嗽 common cold of wind cold

因外感风寒，肺气受遏，肺气失宣所致。以恶寒，耳鼻俱凉，鼻流清涕，喷嚏，遇寒咳重为主要表现的疾病。

6.2.2.3 风热咳嗽 common cold of wind heat

因外感风热，肺失清肃所致。以体表热，干咳声大，鼻液黏稠，呼出气热，或有微喘主要表现的疾病。

6.2.2.4 肺热咳嗽 body deficiency cold

因肺内壅热，肺津受灼，肺失宣降所致。以鼻乍气粗，咳嗽连声，日重夜轻，咳声宏亮，或有黏性鼻液，粪干尿黄为主要表现的疾病。

6.2.2.5 内伤咳嗽 cough with internal injury

因脏腑失调引起的咳嗽，包括肺虚咳嗽、痰湿咳嗽、肾虚咳嗽。

6.2.2.6 肺气虚咳嗽 lung qi deficiency cough

因运动或劳役过度，或运动过量，饥饱不匀，损伤肺气所致。以久咳不已，咳声低弱，鼻流黏涕或脓涕为主要表现的疾病。

6.2.2.7 肺阴虚咳嗽 lung yin deficiency cough

因肺阴亏损，失于清润，气逆于上，或肺气不足，清肃无权，主气无能所致。以频频干咳，痰少津干为主要表现的疾病。

6.2.2.8 劳伤咳嗽 over exertion cough

因长期劳役或运动过度伤肺所致。以鼻流白色脓涕，咳声低弱，日轻夜重，久咳不愈为主要表现的疾病。

6.2.3 喘证 asthma

因肺气上逆所致。以呼吸喘促、鼻翼和肋扇动，全身晃动为主要表现的疾病。

6.2.3.1 实喘 sthenic dyspnea

因热邪侵肺，痰饮壅阻，肺气失宣所致。以发病急骤，病程较短，呼吸喘粗为主要表现的疾病。

6.2.3.2 虚喘 asthenic dyspnea

因长期劳役或运动过度，肺气受损，肾不纳气所致。以发病较缓，病程较长，呼吸气短难续，声音低微，动则喘甚为主要表现的疾病。

6.2.4 流鼻 nasal discharge

因肺经疾病所致。以鼻孔流出大量鼻液为主要表现的疾病。

同义词：吊鼻

6.2.4.1 肺火流鼻 nasal discharge due to lung fire

因火热之邪壅阻或邪毒侵袭于肺所致。以鼻流黄涕，呼吸有鼻塞音，喘粗为主要表现的疾病。

6.2.4.2 劳伤流鼻 nasal discharge due to over exertion

因长期劳役或运动过度，营养不良所致。以鼻流灰白色鼻涕，咳声低弱，日轻夜重为主要表现的疾病。

6.2.4.3 脑颡流鼻 empyema of frontal sinus

因风热邪毒侵袭，或湿热内盛，上移于脑颡，使气血郁滞所致。以鼻孔流出恶臭难闻，呈豆腐脑样鼻涕为主要表现的疾病。

6.2.5 肺痈 pulmonary abscess

因风热外侵，痰热内结，热毒瘀血壅结于肺所致。以鼻流腥臭而稠或带脓血的浊涕为主要表现的疾病。

同义词：肺壅

6.2.6 胸痛 chest pain

因外感风邪热毒，凝注于肺窍，致肺气不宣，气血郁滞所致。以胸部疼痛为主要表现的疾病。

同义词：肺家痛

6.2.7 肺风毛燥 pruritus and alopecia

因皮毛孔窍受邪雍闭所致。以浑身瘙痒、毛燥脱落为主要表现的疾病。

同义词：肺经疮；肺风燥热

6.2.8 肺败 deterioration of the lung

属肺劳之严重者。因劳伤过度，气血壅结于肺所致。以鼻流脓涕，颌下淋巴结化脓，气喘为主要表现的疾病。

6.2.9 鼻血 epistaxis

因火热内盛或阴虚火旺或外伤鼻络所致。以血从鼻孔流出为主要表现的疾病。

同义词：鼻衄；鼻出血

6.2.9.1 肺热鼻血 lung heat epistaxis

因外感风热，或奔走太急，热邪侵肺所致。以鼻孔流血，血色鲜红，不含气泡，或伴有咳嗽，发热为主要表现的疾病。

6.2.9.2 肝热鼻血 liver heat epistaxis

因运动或劳役过度伤肝或使役中粗暴鞭打，肝火旺盛循经上行于鼻络所致。以鼻流血液，躁动易惊，目赤多眵为主要表现的疾病。

6.2.9.3 胃热鼻血 stomach heat epistaxis

因运动或役后体热未散，又多喂热草热料，或热天运动、劳役过甚，胃热熏蒸所致。以鼻孔流血，口渴喜饮，大便干燥为主要表现的疾病。

6.2.9.4 阴虚鼻血 yin deficiency epistaxis

因肾阴亏虚，虚火上炎，循经上逆于鼻窍所致。以鼻腔流血常兼齿衄，时出时止，久而不愈为主要表现的疾病。

6.2.9.5 外伤鼻血 traumatic epistaxis

因异物击伤鼻部，或投胃管时失误所致。以鼻腔流血为主要表现的疾病。

6.2.10 肺寒吐沫 salivation due to pulmonary cold evil

因外感风寒或内伤阴冷，肺失宣降，津液输布障碍所致。以磨牙锉齿，口吐白沫为主要表现的疾病。

6.2.11 肺把胸膊痛 pain on chest and foreleg

因过劳伤肺，气血凝滞于胸膈所致。以胸膊疼痛为主要表现的疾病。

6.2.12 肺黄 fei huang

因风热之邪犯肺，肺失宣降所致。以发热、咳嗽、气喘、流鼻为主要表现的疾病。

6.2.13 胸水 hydrothorax

因体内水液运化失常而积聚于胸腔所致。以咳逆喘息、前肢开张站立为主要表现的疾病。

同义词：前槽停水

6.2.14 异物呛肺 aspiration or deglutition pneumonia

因饮水、饲料、药物等异物误入于肺所致。以咳嗽不安，出汗发抖，呼吸困难为主要表现的疾病。

6.2.15 锁喉风 locked throat

因气候炎热，热积于心肺，热毒上冲咽喉所致。以咽喉肿胀、呼吸困难犹如锁喉为主要表现的疾病。

同义词：锁喉黄

6.3 脾系病

6.3.1 呕吐 vomiting

因脾胃不和，胃气上逆所致。以胃内容物从口鼻流出为主要表现的疾病。

6.3.1.1 伤食呕吐 vomiting due to improper diet

因过食伤胃所致。以肚腹胀满，嗳气腐臭，呕物酸腐为主要表现的疾病。

6.3.1.2 湿热呕吐 hygropyretic vomiting

因暑湿热燥内侵胃肠所致。以体热，口渴欲饮，食后即吐，呕物清稀、色黄为主要表现的疾病。

6.3.1.3 虚寒呕吐 vomiting due to cold of insufficiency type

因脾阳不足而阴寒内盛所致。以食后呕吐，呕物气味不明显，吐后口内多涎为主要表现的疾病。

6.3.2 草噎 oesophagus obstruction

因咽或食道被食物团块或异物急性阻塞所致。以突然发生咽下障碍为主要表现的疾病。

6.3.3 料伤 indigestion due to improper feeding

因过食精料，使役或运动过少，谷气凝于脾胃，料毒内聚入血所致。以口气酸臭，恶料，拘行束步，四肢如攒为主要表现的疾病。

同义词：伤料；伤食

6.3.4 水伤 indigestion due to improper watering

因乘骑失于牵遛，立即饮水过多，或久渴不饮，饮冷水太过，停滞不散等，水结胃肠所致。以肚腹疼痛，肠鸣泄泻为主要表现的疾病。

6.3.5 脾虚慢草 inappetence due to spleen deficiency

因脾胃亏虚，脾阳不振，脾运失健所致。以饮食减退、日益消瘦为主要表现的疾病。

6.3.6 百叶干 impaction of omasum

因长期饲喂粗劣饲料，使役或运动过度，缺少饮水，胃失濡润所致。以瓣胃干燥，料草不能正常消化并向后运转为主要表现的疾病。

6.3.7 异物伤胃 traumatic reticuloperitonitis

因采食时误将金属等异物吞入胃内所致。以肚腹疼痛、脾胃功能紊乱为主要表现的疾病。

6.3.8 脾虚不磨 indigestion due to spleen deficiency

因脾虚胃实，运化无力所致。以草料停滞胃腑、饮食欲减退为主要表现的疾病。

6.3.9 脾虚浮肿 edema due to spleen deficiency

因劳役或运动过度，寒湿内侵，营养失调或久病而使脾阳不振，脾失健运所致水湿内停。以四肢、腹下及眼睑浮肿，粪渣粗大或有泄泻为主要表现的疾病。

6.3.10 起卧症 equine colic

因饲养管理、使役或运动不当、气候突变而使马属动物胃肠道生理功能障碍和肠变位所致。以腹痛起卧为主要表现的疾病。

6.3.11 结症 impaction of intestines

因肠蠕动和分泌功能障碍，粪便积滞于肠管不能移动所致。以阻塞不通和腹痛起卧为主要表现的疾病。

同义词：肠阻塞

6.3.12 肠积沙　intestinal sabulous

因马骡异嗜或误食大量泥沙,沉积于大肠形成沙包所致。以腹痛为主要表现的疾病。
同义词:肠结沙;沙石结

6.3.13 痢疾　dysentery

因外感时邪疫毒,暑湿热毒侵于肠胃或寒湿内郁大肠或内伤饮食不洁所致。以排粪次数增多而量少、腹痛、里急后重、赤白脓血为主要表现的疾病。

6.3.13.1 湿热痢　damp heat dysentery

因采食发霉饲料或过食精料损伤脾胃,湿热内生所致。以下痢稀糊,赤白相杂,或呈白色胶冻状为主要表现的疾病。

6.3.13.2 疫毒痢　epidemic toxic dysentery

因疫毒、湿热侵及肠道,使气血凝滞,传导失职,使湿热下注,热毒弥漫所致。以泻粪黏腻、夹杂脓血为主要表现的疾病。

6.3.13.3 虚寒痢　cold damp dysentery

因体虚久泻,正气不足,寒湿内郁大肠,虚寒之邪内生,使中焦阳气不足,脾肾阳虚,胃肠气机衰弱所致。以泻粪呈灰色,或泡沫状为主要表现的疾病。

6.3.14 肠黄　enteritis

因动物外感湿热或食生冷不洁草料或饮水,热毒蕴结于肠,肠腑受损而生黄肿所致。以发热、腹痛、暴泻、粪腥臭为主要表现的疾病。

6.3.14.1 急肠黄　acute enteritis

因湿热或热毒壅结肠中,热毒内陷所致。以急性腹泻、腹痛不安、粪便恶臭为主要表现的疾病。

6.3.14.2 慢肠黄　chronic enteritis

因急肠黄未愈,脾胃功能失调所致。以毛焦欣吊、泻粪如浆或夹有黏液或未消化食物、粪水齐下为主要表现的疾病。

6.3.15 肠痈　appendicitis

因湿热邪毒内聚,瘀结肠中所致。以发热、少腹疼痛拘急、触及包块为主要表现的疾病。

6.3.16 肠风下血　intestinal wind bleeding

因风邪侵袭肠道所致。以大便带血、血色鲜红为主要表现的疾病。

6.3.17 马肠臌气 equine intestinal tympany

因马属动物中焦气机不畅，浊气不能外排所致。以肠内积有大量气体，腹胀如鼓，发生腹痛为主要表现的疾病。

6.3.18 牛羊气胀 cattle and sheep bloating

因牛、羊采食青绿多汁的牧草或易于发酵的草料，引起前胃积聚大量气体所致。以腹部急性胀大为主要表现的疾病。

6.3.19 泄泻 diarrhea

因感受外邪，或被饲料所伤，或脾胃虚弱，清浊不分，运化失常所致。以排便次数增多，粪便稀软，或粪稀如水为主要表现的疾病。

6.3.19.1 伤食泄泻 diarrhea due to improper feeding

因食积胃肠，运化失职所致。以肚胀、泄泻、完谷不化、粪便酸臭为主要表现的疾病。

同义词：伤食泻

6.3.19.2 脾虚泄泻 diarrhea due to spleen asthenia

因脾气亏虚、运化失常、湿气注入肠道所致。以腹泻、夹有未消化的饲料、粪渣粗大为主要表现的疾病。

同义词：脾虚泻；脾泄

6.3.19.3 湿热泄泻 damp heat diarrhea

因肠道受湿热之邪侵袭所致。以泄泻为主要表现的疾病。

6.3.19.4 寒湿泄泻 cold damp diarrhea

因寒湿内盛，脾失健运所致。以泻粪如水为主要表现的疾病。

同义词：冷肠泄泻；寒泻；冷肠泻

6.3.19.5 肾虚泄泻 diarrhea due to deficiency of kidney

因脾肾阳气亏虚，运化失常所致。以凌晨腹泻为主要表现的疾病。

同义词：肾虚泻；五更泻

6.3.20 便秘 constipation

因长期饲喂难以消化的饲料或饮水不足、气候炎热损伤津液等所致。以粪便干而坚硬，排粪困难或伴有腹痛为主要表现的疾病。

6.3.20.1 实秘 sthenia constipation

因脾胃不能正常运化传递，糟粕滞于肠内所致。以粪干难下，时有腹痛，肠音弱为主要表现的疾病。

6.3.20.2 虚秘 deficient constipation

因气血亏虚，津液不足所致。以体表寒热不均，频作排粪姿势，而不见粪便排出为主要表现的疾病。

6.3.20.3 热秘 heat constipation

因热邪蕴结肠道，津液被灼，糟粕干结难下所致。以体温升高，拱腰努责，粪球干小，口干喜饮为主要表现的疾病。

6.3.20.4 寒秘 cold constipation

因寒邪入肠胃，阳气不运，津液不行，肠道难以传送所致。以腹痛，倦卧，排便艰涩，恶寒肢冷，耳鼻不温为主要表现的疾病。

6.3.21 便血 bloody stool

因湿热下注，或脾虚气弱不能统摄血液所致。以迫血下利，血随便出为主要表现的疾病。

同义词：粪血；泻血；大便拉血

6.3.21.1 湿热便血 hygropyretic bloody stool

因风热或湿热蕴结，脉络受损所致。以大便黏稠不爽，或伴有黏液脓血，气味腥臭，里急后重为主要表现的疾病。

6.3.21.2 气虚便血 qi deficiency bloody stool

因脾胃虚寒，中气不足，血失统摄，滥于肠内所致。以身瘦毛焦、便溏、粪中带血为主要表现的疾病。

6.3.22 宿水停脐 ascites

因脾肾阳气亏虚，水湿运化失常，肠中水液渗出肠外，潴留腹腔之内，积于脐下所致。以腹下大，呈对称性，触诊有波动感或拍水音为主要表现的疾病。

同义词：腹水病；腹水胀；水胀

6.3.23 宿草不转 impaction of rumen

因脾胃食滞，草料不能正常消化与转输运转所致。以腹部胀痛，触如面团样为主要表现的疾病。

同义词：宿草不消

6.3.24 胃寒 stomach cold

因寒伤胃腑，胃的受纳腐熟及降浊功能减退所致。以腹痛、食欲减退或口流清涎为主要表现的疾病。

6.3.25 胃热 stomach heat

因邪热积于胃腑，耗伤胃津，胃的受纳腐熟功能减退所致。以慢草不食、上腭肿胀、口臭喜饮为主要表现的疾病。

同义词：胃火

6.3.26 大肚结 gastric dilatation

因脾胃功能失调或肠管阻塞不通使食物停滞于胃所致。以腹痛起卧为主要表现的疾病。

同义词：胃结

6.3.27 冷痛 spasmodic colic

因内伤阴冷或外寒直中肠胃，气血不畅所致。以突然发生急性、阵发性腹痛、肠鸣音增强为主要表现的疾病。

同义词：脾气痛

6.3.28 肠入阴 scrotal hernia

因小肠或小结肠通过腹股沟环窜入腹股沟管及阴囊内，引起肠管运转机能障碍所致。以急性腹痛为主要表现的疾病。

6.3.29 肠绞痛 torsion incarceration and intussusception

因肠管蠕动失常或被挤压引起肠管自然位置改变，肠腔闭塞不通，肠管血行瘀滞所致。以剧烈腹痛为主要表现的疾病。

同义词：肠变位

6.3.30 盘肠结 obstruction of colon

因粪便或纤维团等阻塞盘肠（结肠），不能后送所致。以急性腹痛为主要表现的疾病。

6.3.31 翻胃吐草 osteomalacia

因牛马脾虚胃弱或脾肾阳虚所致。以日渐消瘦、咀嚼困难、随吃随吐、面骨肿胀、四肢骨节肿大、卧地难起为主要表现的疾病。

6.3.32 肠嵌闭 incarceration or strangulation

因马的一节肠道，通过某一天然孔隙或意外造成的组织破损口，进入另一组织或器官而引起该肠段的绞窄和急性阻塞，使气血供应受阻所致。以急性、连续性腹痛为主要表现的疾病。

6.3.33 肚胀 abdominal fullness

因大量气体蓄积肠中，肠管气滞血瘀所致，以肚腹胀满、剧烈腹痛为主要表现的疾病。

同义词：肠膨胀

6.3.34 脾虚带下 over discharge in vagina due to spleen defcience

因脾失健运，湿聚下注，伤及冲任所致。以母畜阴户流出白色黏稠浊液，形如带状为主要表现的疾病。

6.3.35 困水膈痰 phlegm retention of diaphragm

因湿寒阻阳，脾运失健，痰饮阻滞于中膈而使皱胃食积，饮水停留于瘤胃或肠中所致。以喜卧懒动，口吐胃水，瘤胃及瓣胃蠕动音微弱为主要表现的疾病。

6.4 肝系病

6.4.1 黄疸 jaundice

因湿热或寒湿内阻中焦，脾失健运，肝失疏泄，胆液外溢于表所致。以口、结膜、鼻黏膜颜色变黄为主要表现的疾病。

6.4.1.1 阳黄 jaundice due to heat damp

因湿热蕴蒸肝胆所致。以发热，口、眼、鼻黏膜鲜黄为主要表现的疾病。

6.4.1.2 阴黄 jaundice due to cold damp

因寒湿侵袭，或阳黄迁延日久所致。以口、眼、鼻黏膜淡黄，晦暗无光，恶寒喜温为主要表现的疾病。

6.4.2 胆胀 gallbladder enlargement

因热毒或湿热侵袭肝胆，气机不利，疏泄失常，胆汁排泄不畅，郁滞胆内所致。以胆囊胀大，兼两眼流泪，颈部抽搐，甚或躁动颠走为主要表现的疾病。

6.4.3 肝热传眼 acute conjunctivitis due to liver heat

因肝经受热邪侵袭传之于眼所致。以结膜红肿、疼痛，眵盛难睁，羞明流泪，视物不见为主要表现的疾病。

6.4.4 云翳遮睛 nebula over the eyes

因肝热冲目或外伤所致。以睛生翳膜、视力减退或失明为主要表现的疾病。

同义词：火蒙眼

6.4.5 肝经风热 wind heat accumulated in the liver channel

因肝经受风热侵袭、上传于眼所致。以眼睑肿胀、瘀血，羞明流泪为主要表现的疾病。

6.4.6 肝胀 liver enlargement

因肝血内瘀，疏泄失常所致。以肝脏肿大、疼痛、四肢僵硬、食少为主要表现的疾病。

6.4.7 肝胆风 hepatobiliary wind

因气血瘀滞于肝胆，生风所致。以狂乱、肌肤颤抖、闻声惊惶为主要表现的疾病。

6.4.8 肝黄 liver huang

因热毒或湿热壅积肝经所致。以黄疸为主要表现的疾病。

6.4.9 月盲 moon blindness

因肝经热毒上攻于目所致。以周期性目赤肿痛、羞明流泪、翳膜遮睛为主要表现的疾病。

同义词：月发眼；月盲眼

6.4.10 夜盲 nyctalopia

因阴血不足，肝肾亏虚所致。以白昼视物如常，在暗处或夜间视物不见为主要表现的疾病。

同义词：雀目；雀蒙眼

6.5 肾系病

6.5.1 肾痛 nephralgia

因寒湿侵袭，奔走太急，运动或使役过度，跳跃篱笆或沟渠，突然起立等因素，引起腰肾部位气血瘀滞所致。以精神倦怠，食欲减少，起卧困难，腰脊僵硬，后肢难移，捏压背脊疼痛为主要表现的疾病。

6.5.2 肾寒 kidney cold

多发于寒冷季节、地区，老弱马、牛受寒邪侵袭引起的肾阳衰弱。以腰肢寒冷，大便泄泻，小便短少为主要表现的疾病。

6.5.3 肾虚腿肿 edema of posterior limbs due to kidney asthenia

因肾阳虚火衰，肾气不能蒸化，水道不通，溢出所致。以腿部肿胀，按之凹陷为主要表现的疾病。

6.5.4 肾虚带下 leucorrhea due to kidney asthenia

因肾虚，任脉不固，带脉失约，引起母畜阴道流出异物所致。以阴道流出白色或红色、形如带状的浊液为主要表现的肾疾病。

6.5.5 肾虚骨痿 bone impotence due to kidney asthenia

因肾虚精亏而骨质脆软，筋软无力所致。以行走困难为主要表现的疾病。

同义词：肾亏骨痿；骨痿

6.5.6 肾厥 renal syncope

因肾脏的严重病变，致肾气衰竭，气化失司，湿浊尿毒内蕴，上泛而蒙闭心神所致。以肾病症状和意识障碍为主要表现的疾病。

6.5.7 淋证 stranguria syndrome

因湿热之邪，蕴结下焦，肾与膀胱经气化不利，致使排尿困难所致。以排尿频急涩痛，淋漓不尽为主要表现的疾病。

6.5.7.1 热淋 stranguria due to heat

因染湿热之邪，使膀胱气化不利所致。以新起尿频、尿痛、尿急、尿血为主要表现的疾病。

6.5.7.2 劳淋 stranguria due to overstrain

因劳倦、外感，热淋等迁延日久或反复发作，邪毒蕴结，气阴亏损所致。以经常腰痛，小便频急、淋漓隐痛等为主要表现的疾病。

6.5.7.3 石淋 stony stranguria

因湿热之邪蕴结下焦，煎熬尿浊杂质，结为砂石，停阻于肾系所致。以腰痛、尿血，或尿出砂石，或经检查发现结石为主要表现的疾病。

6.5.7.4 血淋 bloody stranguria

因湿热蕴结，热重于湿，伤及脉络，使血液外溢所致。以尿血，排尿频急涩痛，淋漓不尽为主要表现的疾病。

6.5.7.5 膏淋 stranguria due to chyluria

因湿热蕴久，阻滞经脉，致使脂液不循常道所致。以尿液浑浊，状如米泔，脂腻似膏，排尿淋漓痛苦为主要表现的疾病。

6.5.8 尿浊 turbid urine

因湿热下注，蕴结于膀胱，或脾肾亏虚，固摄失常，不能蒸化和制约脂液所致。以尿液浑浊，淋漓不断，排尿无痛为主要表现的疾病。

6.5.9 胞转 torsion of the urinary bladder

因排尿受惊，乘热骤饮冷水或打滚翻转等不当运动所致的膀胱流转。以病畜踏地蹲腰，欲卧不卧，时作排尿状，尿淋或无尿，腹胀不安为主要表现的疾病。

6.5.10 癃闭 obstructive dysuria

因败精阻塞、阴部手术等，使膀胱气化失司，水道不利所致。以尿量少、点滴而出，甚至闭塞不通为主要表现的疾病。

6.5.11 尿不禁 urinary incontinence

因肾气亏虚，精元不固，或因尿路损伤所致。以清醒状态下小便不能控制而自行流出为主要表现的疾病。

6.5.12 尿崩 diabetes insipidus

因肾虚下元不固，或脑神等病变及肾，肾之气化失司，水津直趋膀胱而下泄所致。以尿多如崩、尿清如水、烦渴多饮为主要表现的疾病。

6.5.13 肾火症 yellow urine due to renal meridian fever

因肾经积热，虚火上炎，气化失常所致。以尿液短黄、听力减退或烦躁不安为主要表现的疾病。

6.5.14 风水 wind edema

因外界风邪刺激，脉络挛急，血瘀水停所致。以突发局限性水肿为主要表现的疾病。

6.5.15 皮水 puffiness by wind cold wet and hot

因风寒湿热毒邪侵袭，使肺失宣降，水道不利，水液潴留所致。以新起浮肿，尿少，蛋白尿为主要表现的疾病。

6.5.16 石水 proteinuric edema

因皮水等迁延日久，正气渐虚所致。以反复发作且腹部胀痛而硬的浮肿，蛋白尿为主要表现的疾病。

6.5.17 肾水 chronic edema

由多种原因损及肾脏，肾阳虚不能化气行水所致。以长期浮肿，形寒肢冷，腰胯无力为主要表现的疾病。

6.5.18 正水 progressive edema

因风热湿毒之邪伤肾所致。以浮肿，尿少等呈进行性发展为主要表现的疾病。

6.5.19 外肾黄 swelling of the scrotum

因寒湿邪气侵入肾经，流注外肾（睾丸），或肾经受热，湿热下注，凝于外肾所致。以阴囊肿胀为主要表现的疾病。

同义词：木肾黄

6.5.19.1 阴肾黄 testitis due to yin evil

因寒湿所致，以阴囊水肿、增大，但触诊局部无热，运动后可减轻，久则可见水肿扩展至腹下一带为主要表现的疾病。

6.5.19.2 阳肾黄 testitis due to yang evil

多因热邪所致，以睾丸和附睾肿胀、疼痛拒按，患部发红、发热，行走时后肢开张或牵行不动，重者腰背拱起，后腿难移为主要表现的疾病。

6.5.20 内肾黄 lumbar swelling

因喂养太盛，谷气壅阻中焦，热毒积于腰间，肾脏血郁所致。以耳耷头低，水草大减，行立无神，腰旁或腰胯部肿胀为主要表现的疾病。

6.5.21 阳痿 impotence

因公畜器质性损伤或神志因素，阴精耗损过度，命门火衰引起性功能障碍所致。以交配时阴茎不举，或举而不坚为主要表现的疾病。

6.5.22 垂缕不收 penial prolapse

因肾水亏损，肾阳衰亏，或外受损伤致下元不固所致。以阴茎不收，弛垂于外为主要表现的疾病。

6.5.22.1 滑精 premature ejaculation or spermatorrhea

多因配种过多，营养不良，运动或劳役过度，致使肾阳亏虚，精关不固；或因肾阴不足，相火偏盛而早泄。以配种时未交配或刚开始交配而泄精为主要表现的疾病。

同义词：滑泄

6.5.22.2 腰腿风 lumbago and paralysis of the posterior limbs

常因多喂少役、瘀痰料毒内聚生风所致。以后肢无力，摇摆不稳，腰腿瘫痪，四肢筋挛，卧地不起为主要表现的疾病。

6.5.23 胞黄 dysuria due to hot and wet

因火热湿毒壅结膀胱而生黄肿所致。以小便淋漓，尿色黄赤或尿血，排尿痛苦为主要表现的疾病。

6.5.24 胞虚 stranguria due to deficiency of qi and blood

因瘦弱的老马、牛及产后母猪等气血亏虚，肾气不固，膀胱气化失常，不能贮纳尿液所致。以小便淋漓，排尿困难为主要表现的疾病。

6.5.25 尿闭 anuria

因湿热内蕴膀胱，或腰胯挫伤等引起膀胱失职，尿路不通所致。以相当时间内不排尿为主要表现的疾病。

6.5.25.1 损伤尿闭 anuria due to damage

多因损伤或挫伤腰胯所致。以站立困难，腰胯及后躯知觉部分消失或完全消失，肚腹胀满，肛门松弛，尿液难出为主要表现的疾病。

6.5.25.2 实热尿闭 anuria due to heat

多因暑月炎天，运动或使役过急，暑湿热毒内侵，致三焦积热，下焦被热气闭塞，膀胱气化受阻，使水道不得通利所致。以排尿困难、疼痛为主要表现的疾病。

6.5.25.3 瘀血败精尿闭 anuria due to stasis or abortive obstruction

因跌打损伤，经络瘀阻，或败精阻塞尿道所致。以排尿困难或相当长时间内不排尿为主要表现的疾病。

6.5.26 尿血 hematuria

因膀胱积热，损伤脉络，或脾肾亏虚，统摄失司所致。以血随尿出为主要表现的疾病。

6.5.27 肾虚腿肿 swollen of leg due to deficiency of kidney

因肾阳虚衰，水湿停滞，引起腿部浮肿所致。以后腿浮肿为主要表现的疾病。

6.6 外伤及疮疡病

6.6.1 创伤 wounds

因尖锐物体刺伤，刀斧砍伤，弹片击伤，畜角顶伤，虫兽咬伤，踢伤，打伤，挫伤，压伤等筋脉断裂。以皮肉破裂、出血、疼痛、肿胀，重则筋断骨折为主要表现的疾病。

同义词：金疮

6.6.2 豁鼻 laceration of the muzzle

因穿鼻时位置选择不当，拴鼻工具不良，直接用铁丝或粗绳拴牛鼻，在管理和使役中强行牵拉拴牛鼻绳，硬性撕裂牛鼻中隔等所致。以牛上部鼻端和下部鼻镜分离为主要表现的疾病。

6.6.3 水火烫伤 scalds and burns

因燃烧物及灼热的液体、固体、气体以及电流等直接作用于动物，引起肌肤烫伤或烧伤，甚至火毒内攻脏腑所致。以伤处红肿灼痛、起泡、结焦痂，伴发热烦躁，口干尿黄，甚至神昏等为主要表现的疾病。

6.6.4 闪伤 sudden sprain

因急剧运动，腰部筋肌受到突然牵拉引起损伤所致。以腰部不能自由活动，动则痛剧为主要表现的疾病。

6.6.5 闪挫 sprain and contusion

因跌、闪、挫伤而引起腰部疼痛所致。以行动困难，转侧难移，疼痛显著为主要表现的疾病。

6.6.6 瘘管 fistula of inner and outer orifice

因胚胎发育异常，感染，手术等引起溃疡形成管道所致。以疮孔处流脓经久淋漓不断，体表与脏腔之间相通，局部红肿、疼痛，伴有分泌物流出，具有内口和外口为主要表现的疾病。

6.6.7 窦道 fistula of outer orifice

因机体组织感染、坏死，溃疡形成管道所致，以疮孔处流脓经久淋漓不断，体表与深部组织相通，只有外口为主要表现的疾病。

6.6.8 骨折 fracture

因跌扑猛闪、跳涧、滑倒、蹴踢等原因，使动物骨骼受到过强外力作用引起损伤所致。以骨骼部分或全部破裂、折断或粉碎为主要表现的疾病。

6.6.9 角折 horn fracture

因外伤引起有角类动物的犄角损伤所致。以角的一部分断折或全部从基部断折为主要表现的疾病。

6.6.10 疮黄疔毒 furuncle or pyogenic infection on body surface

因外伤或疮疡形成局部化脓性感染所致。以皮肤和肌肉组织发生肿胀或化脓性感染为主要表现的疾病。

6.6.11 黄 stasis and swelling

多因运动或劳役过度、饮喂失时、气候炎热、奔走太急、外感风邪、内伤饮食，致使热邪积于脏腑，循经外传，郁于体表肌腠而成黄肿；或因跌扑挫伤，外物所伤，使气血运行不畅，瘀血凝聚于肌腠所致。以皮肤完整性未被破坏的软组织肿胀为主要表现的疾病。

6.6.12 锁口黄 stomatitis due to hot and poison

因热毒郁结，上冲于口所致。以口角发生肿胀，口难张开为主要表现的疾病。

同义词：箍嘴黄；束口黄

6.6.13 鼻黄 nasal swelling due to hot evil

因热邪积于肺经，上攻于鼻所致。以单侧或双侧鼻部肿胀，软而不痛，久之破流黄水，鼻孔内亦微有肿胀为主要表现的疾病。

6.6.14 颊黄 cheek swelling due to hot evil

因心肺积热，上攻于颊畔所致。以颊部一侧或双侧发生软肿，压之不痛，初期肿胀较小，后逐渐扩大，甚至牵延到食槽，口流涎水，咀嚼困难为主要表现的疾病。

6.6.15 耳黄 ear swelling due to hot and poison

因热毒积于肾经外传于耳所致。以单耳或双耳的耳根肿胀、下垂为主要表现的疾病。

6.6.16 腮黄 gill swelling due to hot evil

因热邪积于脾肺，上冲腮颊所致。以腮部一侧或双侧发生肿胀为主要表现的疾病。

6.6.17 背黄 back inflammation due to hot evil

因热毒破血积聚于背部所致。以背部患部热痛肿硬，日久软化，内有黄水为主要表现的疾病。

6.6.18 胸黄 chest swelling due to hot evil

因心肺壅极，热毒蕴胸所致。以胸前发生黄肿为主要表现的疾病。

6.6.19 肚底黄 abdominal swelling

因湿热、损伤和脾虚等所致。以腹下发生肿胀为主要表现的疾病。

6.6.20 疖 furuncle

因肌肤浅表部位感受火热毒邪所致。以局部红肿热痛，根浅、脓出即愈为主要表现的疾病。

6.6.21 疮 sores

因外伤引起皮肤浅表肿胀、疼痛、溃烂，甚至化脓。以局部发红、肿胀、溃烂为主要表现的疾病。

6.6.22 褥疮 bedsore

因动物久病卧地，气血运行失畅，肌肤失养，长期摩擦，皮肤破损所致。以局限性浅表皮肤破损，疮口经久不愈为主要表现的疾病。

同义词：席疮；压疮；压力性溃疡

6.6.23 疔疮 nail like boil

因竹木刺伤，或感受疫毒、疠毒、火毒等邪所致。以病灶坚硬根深，形如钉状，肿痛灼热，反应剧烈，易于走黄、损筋伤骨等为主要表现的疾病。

同义词：疔

6.6.24 黑疔 black nail like boil

因皮肤浅层组织受伤所致。以疮面覆盖有血样分泌物，后则变干，形成黑色痂皮，形似钉盖，坚硬色黑，不红不肿，无血无脓为主要表现的疾病。

6.6.25 筋疔 nail like boil of exposed membrane

因脊间皮肤组织破溃所致。以疮面溃烂无痂，显露出灰白色而略带黄色的肌膜，流出淡黄色水液为主要表现的疾病。

6.6.26 气疔 nail like boil of foamy pus

因疮面溃烂，局部色白，或因坏死组织分解所致。以疮面发白，或带有泡沫状的脓汁，或流出黄白色的渗出物为主要表现的疾病。

6.6.27 水疔 nail like boil of exudate

因皮肤热毒蕴结所致。以患部红肿疼痛，光亮多水，严重者伴有全身症状为主要表现的疾病。

6.6.28 血疔 nail like boil of pus and blood

因三焦及大肠火毒所致。以皮肤组织破溃，久不结痂，色赤，常流脓血为主要表现的疾病。

6.6.29 丹毒 erysipelas

因皮肤、黏膜破损，外受火毒与血热搏结，蕴阻肌肤，不得外泄所致。以患部突然皮肤红成片、色如涂丹，灼热肿胀，迅速蔓延为主要表现的疾病。

6.6.30 痈 acute suppurative disease

因热毒蕴蒸，气血壅滞所致。以肌肤患病部位红肿热痛，光软无头，伴寒热口渴，易肿、易脓、易溃、易敛为主要表现的疾病。

6.6.31 疽 cellulitis or phlegmon

因疮深而恶，毒邪阻滞气血，发于肌肉筋骨间所致的疮肿。以未成脓者难消，已成脓者难溃，破后状如蜂窝，难于收敛封口为主要表现的疾病。

6.6.32 有头疽 cellulitis in skin and muscle

因外感风热、湿热、火毒之邪，气血瘀滞，结聚于肌肤间所致。以局部红肿热痛，易向深部及周围扩散，有多个脓栓堆积，溃后形如蜂窝，病损面积较大，易致疽毒内陷为主要表现的疾病。

6.6.33 无头疽 cellulitis in joints and bones

因毒邪深陷，寒凝气滞于骨与关节所致。以患部漫肿、皮色不变、疼痛彻骨、难消、难溃、难敛，溃后多损伤筋骨为主要表现的疾病。

6.6.34 项痈 carbuncle of neck

因外感六淫或内伤劳役，致使营卫不和，热毒郁滞项间；或使役用具不当，磨损项部，气血凝聚所致。以项部肿胀，热感，硬而多痛，患畜伸头直项，不敢回顾，日久肿胀溃破，流出邪热壅聚项间为主要表现的疾病。

6.6.35 肩痈 carbuncle of shoulder

因踢打跌扑、挽具磨伤肩部；或因运动、劳役过度，营卫不和，感受风邪，传于肌腠所致。以肩胛部肿块、红肿热痛、疼痛彻骨为主要表现的疾病。

6.6.36 脑颡黄 nasal discharge of pus and blood

因外感风热或风寒，伏郁化热，邪热熏蒸，热毒攻于鼻窦，进而鼻窦痈肿蓄脓；或面额受外邪暴力损伤，额骨或上颌骨骨折、牛角折等造成瘀血停滞鼻窦，久腐化脓所致。以鼻流脓血为主要表现的疾病。

同义词：脑漏；气毒

6.6.37 无名肿毒 inflammation due to wind cold or hot evil

因外感风邪，或寒热客于经络所致。以局部骤发肿痛为主要表现的疾病。

6.6.38 流注 carbuncle of pus

因感染邪毒，流窜血络，阻于肌肉深部所致。以一处或数处漫肿，微热疼痛，皮色如常，内有脓液为主要表现的疾病。

6.6.39 流痰 phthisis of bones and joints

因先天不足，肾亏骨弱，复感痨虫，痰浊凝聚，蚀伤关节所致。以发生于骨与关节，起病缓，化脓迟，溃后流脓清稀或夹败絮样物，不易愈合，多损伤筋骨，形成脓肿或窦道等为主要表现的疾病。

同义词：骨痨

6.6.40 瘰疬 scrofula

因肝郁气滞，痰湿凝聚；或阴虚火旺，感染痨虫，痰火凝结所致。以颈部缓慢出现大小不等的圆滑肿块，累累如串珠，不红不痛，溃后脓水清稀，夹有败絮状物，易成瘘管为主要表现的疾病。

6.6.41 漏瘘 fistula

因疮疡经久不愈，并继续向深处腐烂化脓所致，形成一个反常的、内部向外表开口的管道。以流脓、疼痛、瘙痒为主要表现的疾病。

6.7 肢蹄病

6.7.1 肩膊痛 pain in shoulder and up arm

因寒伤及闪伤等所致。以肩臂部筋肉、关节疼痛为主要表现的疾病。

6.7.1.1 寒伤肩膊痛 shoulder and up arm pain due to cold evil

因寒风外袭，或汗后当风，或夜露风霜，阴雨苦淋、久卧湿地等风寒湿邪侵袭肌表，传于经络，致使经络不通，气血凝滞，引起的肩膊痛。

6.7.1.2 闪伤肩膊痛 shoulder and up arm pain due to sudden sprain

因肩臂部挫伤或剧伸，致使气血瘀滞，引起的肩膊痛。

6.7.2 膊尖痛 pains of the shoulder joint

因膊尖部的直接损伤，或来自肢体下部的冲力所致。以肩关节及其周围组织疼痛为主要表现的疾病。

6.7.3 传经痛 gout

因外感风寒湿邪凝于经络，引起气血瘀滞所致。以四肢轮流疼痛为主要表现的疾病。

同义词：痛风

6.7.4 脱臼 dislocation

因道路不平，泥泞暗坑，踩踏不稳；或跳越沟渠，猛驰失足等所致。以关节头脱出臼窝不能自行复位为主要表现的疾病。

6.7.5 脱膊 shoulder dislocation

因动物在滑跌、急转弯、扭闪时发生肩胛软组织损伤，进而引起跛行所致。以病肢肩胛骨紧贴于躯干，肩胛软骨与健侧相比明显向后下方移位为主要表现的疾病。

6.7.6 抢风痛 paralysis of the radial nerve

因抢风穴周围组织受机械性压迫，致使局部气血不通，经络受阻所致。以前肢麻痹、运步缓慢、难移前肢为主要表现的疾病。

6.7.7 四肢神经麻痹症 quadriplegia

因四肢神经感传机能减弱或丧失所致。以四肢感觉和运动机能障碍为主要表现的疾病。

6.7.8 夹气痛 sprain of triceps

因动物前肢闪伤使肩胛与躯干之间气血瘀滞所致。以驻立时患侧肩胛部低于健侧，患肢外踏或前伸，减负体重，着地时有痛感，运步时抬举困难，向外划弧，随运动疼痛加剧，日久肩胛部外侧肌肉萎缩为主要表现的疾病。

6.7.9 乘重痛 contusion of the elbow joint

因马、骡等闪伤或挫伤所致。以肘关节部位肿痛为主要表现的疾病。

6.7.10 攒筋痛 flexor tendinitis

因疾跑、闪跌、拉挽过力、跳跃障碍等所致。以屈腱的急性或慢性肿痛为主要表现的疾病。

6.7.11 掌骨痛 metacarpal pains

因长期奔走过急，在不平道路上运动或劳役过度，使掌腕骨承受过重的机械压迫和对其骨间韧带的过度牵引，日久引起骨质增生所致；或外力损伤后逐渐形成。以掌骨肿胀疼痛为主要表现的疾病。

同义词：掌骨瘤

6.7.12 缠腕痛 contusion of the fetlock joint

因马、骡等球节（关节）受到过度屈伸、挫伤、扭转，引起扭挫伤所致。以站立时球节稍屈曲，蹄尖着地，运动时轻度支跛，速步运动时跛行明显为主要表现的疾病。

6.7.13 肾冷拖腰 lumbago pains due to wind cold and wet evil

因外感风寒湿邪，如夜露寒霜，久卧湿地，带汗卸鞍，寒湿之邪侵入肾经所致。以精神不振，耳耷头低，毛焦欣吊，腰背拱起，腰拖胯簸，后肢难移，牵行不动，起立困难，重则卧地难起，脉象沉涩，口色青暗，按压腰部时，皮紧腰硬为主要表现的疾病。

6.7.14 腰胯痛 pains of the lumbus and hip

因风寒湿邪侵袭肌表，肾经受寒，传于腰胯；或跌扑闪伤等所致。以站立时表现腰背板硬，触压反应迟钝，牵行不动，后肢难移，起卧困难，久则肌肉萎缩为主要表现的疾病。

6.7.14.1 寒伤腰胯 pains of the lumbus and hip due to cold evil

由于阴雨苦淋，夜露风霜，久卧湿地，久拴阴冷之处，或乘热过河，带汗卸鞍，汗后当风等，风寒湿邪侵袭肌表，肾经受寒，传于腰胯，引起的腰胯疾病。

6.7.14.2 闪伤腰胯 pains of the lumbus and hip due to sudden sprain

因在不平道路上运动或使役，奔跑、滑走、跌倒、翻车、蹴踢、急剧转弯，蹬空失足，嵌挟拔腿，跳越障碍，保定不当等，闪扭腰胯，致使滞气凝于胯内，瘀血注积腰间，引起的腰胯疾病。

6.7.15 胯瓦痛 sprain of the hip joint

因在不平道路或泥泞冰滑路上使役或奔跑、跌倒、踢蹴、打扑等，闪挫胯关节所致。以站立时蹄尖着地，患肢膝关节及跗关节屈曲为主要表现的疾病。

6.7.16 掠草痛 pain of the stifle joint

因在不平道路或泥泞冰滑路上使役或奔驰而闪伤；或跌倒、踢蹴、打扑、冲撞等暴力使掠草穴部受到损伤；或感受风寒湿邪侵袭所致。以膝关节疼痛为主要表现的疾病。

同义词：迎风痛

6.7.17 冷拖竿 string halt

因外感寒湿侵入肾经，传至后肢筋骨所致。以后肢痉挛，腿直如竿，不能屈曲，行步困难为主要表现的疾病。

6.7.18 合子骨肿痛 bone spavin

因在不平道路上过度劳役，奔走过急，急剧转弯，跳越障碍，以及跗关节部挫伤所致。以跗关节肿胀为主要表现的疾病。

6.7.19 蹄伤 injuries to the hoof

因外力作用（如坚硬锐利物体扎刺等）损伤蹄部筋肉所致。以跛行为主要表现的疾病。

6.7.19.1 踏伤 hoof injure due to stampede

因蹄底及蹄叉被尖锐物体扎伤，造成蹄底或蹄叉真皮，甚至蹄骨、屈腱、籽骨及其黏液囊损伤，以跛行和蹄掌肿痛为主要表现的疾病。

6.7.19.2 钉伤 hoof injure due to trim foot

因装蹄时蹄钉误入真皮部，导致蹄底、蹄壁真皮损伤，以跛行，患肢不敢着地为主要表现的疾病。

6.7.20 筋断 tendon rupture

因屈腱过度牵引，如急跑、跌倒、滑跌，或受刀斧等锐物切割伤，或骨软症、屈腱的腱鞘化脓性坏死疾病，以及营养不良等所致。以屈腱断裂，站立时蹄尖翘起，球节下沉，蹄踵着地，运动时出现支跛等为主要表现的疾病。

6.7.21 风蹄 sandcrack

多因家畜肝血不足，肾水不滋所致。以蹄壁角质裂开为主要表现的疾病。

同义词：裂蹄

6.7.22 滚蹄 contraction of flexor tendons

因多种蹄病治疗不及时或运动、劳役过度，久役伤筋，致使板筋肿痛，日久短缩所致。以寸腕部板筋缩短，屈腱挛缩，蹄向后翻，蹄底向上，行走时蹄向前滚动为主要表现的疾病。

6.7.23 毛边漏 quittor

因蹄冠部外伤，厩舍不洁，蹄冠部长时间被粪尿浸渍，或久立湿地，日久湿毒侵入等所致。以蹄甲上部有毛与无毛交界处、蹄冠缘肿胀破溃、流出脓液为主要表现的疾病。

同义词：蹄冠炎

6.7.24 蹄头痛 pain in the head of the hoof

因膘肥肉重，多立少骑，久拴久系，致使血凝蹄头，蹄甲渐长，失于修削，日久蹄胎骨硬所致。以血凝蹄头引起蹄头部疼痛为主要表现的疾病。

6.7.25 五攒痛 laminitis

因饲喂精料过多或运动、劳役之后立即拴系，气血凝滞，致使蹄部疼痛，指动脉亢进，束步难行所致。以站立时腰曲头低，四肢攒于腹下为主要表现的疾病。

同义词：蹄叶炎

6.7.26 败血凝蹄 chronic laminitis due to stasis

因运动不足，久立久拴；或蹄甲久不修削，障碍蹄机；或长途使役，奔走太急，急停立拴，失于牵散，以致血瘀下注，凝集于蹄踵所致。以蹄甲焦枯，并有裂纹，把前把后，腰曲头低，卧多立少，行走如攒为主要表现的蹄病。

6.7.27 腐蹄病 foot rot

因患畜立于湿地，湿毒入侵，或修蹄失宜，刺伤蹄底；或过削蹄底，日久腐烂成漏；或长久舍饲，久不使役，磨损蹄底；或筋失所养，角质萎弱所致。饲料中缺少矿物质、维生素等所致。以牛蹄角质腐败分解、蹄底腐烂为主要表现的疾病。

同义词：漏蹄

6.8 皮肤病

6.8.1 遍身黄 urticaria

多因运动或劳役过度，身体不洁，汗出当风，腠理开泄，外邪贼风乘虚而入，正邪相搏，卫气被郁，营卫不和所致。以遍体瘙痒，皮肤出现大小不一的疙瘩为主要表现的疾病。

同义词：荨麻疹；肺风黄；风疹块

6.8.2 肺风毛燥 pruritus and alopecia

因运动或劳役出汗，汗沉于毛窍，垢尘迷塞肌肤；或营养太盛，多喂少骑，长期失于洗刷所致。以畜体浑身瘙痒、被毛脱落为主要表现的疾病。

同义词：肺经疮；肺风燥热

6.8.3 湿毒 eczema due to wet poison

因湿毒侵袭，或化学药品、昆虫叮咬等刺激皮肤，或饲养管理不良等所致。以皮肤瘙痒呈弥漫性潮红、丘疹、水泡、糜烂、渗液、结痂、脱屑等多种损害为主要表现的疾病。

6.8.4 热气疮 herpetic dermatosis due to hot evil

因外感风热，或肺胃内热，或热病伴发所致。以皮肤黏膜交界处发生成簇水泡、糜烂、破溃、结痂，痒痛相兼为主要表现的疾病。

同义词：火燎疮

6.8.5 皮肤瘙痒症 pruritus

因外受风邪侵袭，血分有热，外受风邪侵袭，郁于肌肤不得外泄所致。以皮肤瘙痒为主要表现的疾病。

同义词：痒风

6.8.6 脱毛症 alopecia

因肺气久虚不能温煦皮毛，或饮喂失调，引起心肺衰弱、气血亏损，致使被毛焦枯脱落所致。以脱毛为主要表现的疾病。

6.8.7 疣 warts
因风湿热邪蕴结，兼感邪毒所致。以皮肤浅表处发生良性赘生物为主要表现的疾病。

6.8.8 臊疣 warts of perineum
因内蕴湿热，外感邪毒所致。以发生于会阴等处皮肤黏膜交界处的疣状突起，呈菜花状，表面湿润，易出血为主要表现的疾病。

6.8.9 黄水疮 sores of pus
因脾肺湿热与外邪相挟所致。以患部皮肤出现脓疱、结痂、流黄水，浸淫成片，瘙痒为主要表现的疾病。

同义词：滴脓疮；脓窝疮

6.8.10 圆癣 ringworm
因湿热内蕴，外感邪毒所致。以皮肤平滑处起红疹、水疱、结痂、脱屑，呈环状有框廓，瘙痒为主要表现的疾病。

同义词：钱癣；金钱癣

6.8.11 阴癣 tinea of pudendum
因阴部湿热，染受邪毒所致，好发于沟股尾端及臀部、大腿内侧。以皮肤丘疹、水疱、结痂、瘙痒为主要表现的疾病。

6.8.12 疥疮 scabies
因疥虫侵袭皮肤所致。以趾缝、腕、肘窝、脐周、阴股部等处皮肤发生疱疹，脱毛，可找到疥虫为主要表现的疾病。

6.8.13 湿疮 sores of wet
因外感风湿热邪，或脾失健运，湿热内生，内外合邪，浸淫肌肤所致。以皮肤呈多型性皮疹，渗液，结痂，瘙痒为主要表现的疾病。

同义词：湿疡

6.8.14 风土疮 sores due to unacclimatization
因水土不服，染受湿热虫邪所致。以皮肤现丘疹、风团、水疱，乍发乍瘥为主要表现的疾病。

同义词：土风疮

6.8.15 顽湿结聚 sores due to wet evil
因风湿聚于肤表所致。以四肢散发豆粒大灰褐色坚实结节，瘙痒，破后有血痂为主要表现的疾病。

6.8.16 面游风 facial wandering wind

因脾肺湿热，感受风邪所致。以面部红斑、脱屑、瘙痒，甚至肿胀、糜烂为主要表现的疾病。

同义词：白屑风；纽扣风

6.8.17 药毒 drug allergy

因内用或外敷药物所致。以皮肤斑疹、水疱、瘙痒等为主要表现的疾病。

同义词：药疹；中药毒

6.8.18 晒疮 sunburn

因日光暴晒，暑热邪毒内侵，疏泄不畅，郁于肌肤所致。以出现红斑、疹、水疱、脱皮，灼痛瘙痒为主要表现的疾病。

6.8.19 恶虫叮咬伤 bite wound of insect

因蚊子、臭虫、跳蚤等叮咬，虫毒侵袭肌肤所致。以皮肤见红色疹点、瘙痒等为主要表现的疾病。

6.8.20 茧唇 callus like disease of the lips

因脾胃湿热结聚所致。以唇部赘生豆粒样物，渗液，结痂增厚，渐至下唇外翻似茧为主要表现的疾病。

6.8.21 紫癜风 purpura

因阴虚内热，或湿热凝滞，复感风邪所致。以皮肤出现紫红色扁平皮疹，瘙痒为主要表现的疾病，可发生于全身各处，常累及口腔。

6.8.22 火赤疮 vesicular disease due to hot evil

因心火炽盛，或脾虚湿盛所致。以皮肤红斑、大疱壁薄，或水疱成群、呈环状排列，瘙痒难忍为主要表现的疾病。

6.8.23 松皮癣 psoriasis

因湿热郁肤，留滞不去所致。以四肢皮肤结节、肥厚，触之硬固，皮色暗褐如松皮，瘙痒为主要表现的疾病。

6.8.24 流皮漏 spreading skin ulcer

因湿热瘀阻，气血亏虚所致。以头面及身体其他部位出现褐红色小结节、融合成片、结痂出脓、坏死溃疡，愈后形成萎缩性瘢痕为主要表现的疾病。

6.8.25 石疽 hard skin nodule

因寒气客于经络，痰凝湿热蕴结，气血瘀滞，日久坚积不散所致。以肌肤结块坚硬不消，隐痛或不痛为主要表现的疾病。

6.9 胎产病

6.9.1 胎动不安 excessive fetal moovement

因肾虚、气血虚弱，或血热、血瘀等引起冲任损伤，胎元不固所致。以母畜妊娠期未满，阴道持续流出血水、浊液或兼有努责、腹痛等胎儿欲坠为主要表现的疾病。

同义词：胎动

6.9.1.1 体虚胎动 fetal restlessness due to deficency

因妊娠期间运动或使役过度，喂养失调，营养不足，致使气血虚弱，冲任不固，胎失所养，导致的胎儿欲坠为主要表现的疾病。

6.9.1.2 血热胎动 fetal restlessness due to blood heat

因损伤或误投伤胎药物，或外感热邪，致阳盛血热，热扰冲任，损伤胎气，导致的胎儿欲坠为主要表现的疾病。

6.9.1.3 外伤胎动 fetal restlessness due to trauma

因跌扑闪挫，或运动、劳役过度，损伤冲任，气血扰乱，不足以养胎载胎，导致的胎儿欲坠为主要表现的疾病。

6.9.2 流产 abortion

因在怀孕期间母畜肾虚、血热、气血虚，发生胎病；或因意外损伤等所致。以阵痛起卧、外阴微红、肿胀、阴门流出羊水，分娩出未足月胎儿为主要表现的疾病。

同义词：小产

6.9.2.1 带下 over discharge in vagina

因体质虚寒，或感受湿热之邪所致。以母畜阴道分泌物过多，并从阴门流出白色、黄色或赤白相杂、带有异味的分泌物，形如带状为主要表现的疾病。

6.9.3 产后恶露不尽 prolonged lochia

因脏腑气血虚弱或感受外邪使恶露排出时间延长或性质改变所致。以产后由子宫内持续流出污浊液体为主要表现的疾病。

6.9.4 胎衣不下 retention of placenta

因畜体羸弱，气血虚弱，或运行不畅所致。以母畜产后，胎衣在正常时间内不能自行排出为主要表现的疾病。

同义词：胎盘滞留

6.9.5 产后腹痛 postpartum abdominal pain

因产前运动或劳役过度，营养不良，畜体虚弱，气血运行不畅；或产后失于调护，风寒乘虚侵入所致。以母畜产后腹中疼痛为主要表现的疾病。

6.9.6 难产 dystocia

因母畜妊娠期间饮喂失调，体质虚弱，气血亏损，致使产力不足；或因产道异常、胎儿过大等所致。以母畜妊娠期满，胎儿不能顺利产出为主要表现的疾病。

6.9.7 阴道脱出 prolapse of vagina

因母畜在妊娠期间饲养失调，营养不良，运动或劳役过度，以致气血亏损所致。以母畜阴道部分或全部外翻脱出于阴门之外为主要表现的疾病。

6.9.8 乳痈 mastitis

因环境卫生不良，外感湿热等邪，或内伤饥饱劳役，或情志所伤，或外力损伤所致。以乳腺气血郁滞，乳络受阻，泌乳减少或停滞，进而乳腺发生红、肿、热、痛以致化脓为主要表现的疾病。

同义词：奶肿；奶黄；乳腺炎

6.9.9 产后发热 postpartum fever

因母畜分娩产道受损，邪毒内侵所致。以母畜分娩后全身持续发热，恶露增加，甚至出现神昏、出血为主要表现的疾病。

同义词：产褥热

6.9.10 缺乳 hypogalactia

因产前运动或劳役过度，营养不良，体质瘦弱；或气血壅滞，经络不畅；或分娩时间过长，气血亏耗；或因早产，冲任空虚所致。以母畜产后乳汁缺少或全无为主要表现的疾病。

同义词：少乳；乳汁不行

6.9.11 胎风 postpartum paralysis

因母畜产后气血亏损，肌肉筋骨失于濡养，引起后肢不能站立所致。雌性动物产后发生以步行拘紧、腰腿疼痛，甚至四肢瘫痪或痉挛、卧地不起为主要表现的疾病。

同义词：产后风；产后瘫痪；趴窝病

6.9.12 胎气 pregnant edema

因妊娠后期，胎儿过度发育，或胎儿过大，母畜营养不足，运动或使役过度等所致。以妊娠母畜四肢、腹下、乳腺及会阴等处出现浮肿甚至腰瘫腿痪为主要表现的疾病。

6.9.13 流产与死胎 early death of the embryo and abortion

因饲养管理不良，运动或劳役过度，过饮冷水，气血亏虚影响胎儿的正常生长；亦可因意外事故伤及胎儿，如跌、挫、跳跃等所致。以妊娠母畜尚未到预产期娩出胎儿或死胎为主要表现的疾病。

6.10 瘟病

6.10.1 温病 warm disease

感受温邪所引起的外感急性热病的总称。
同义词：温热病

6.10.2 时疫 seasonal pestilence

因感受季节性疫疠之气所致。病气常从口鼻入侵，具有强烈传染性的疾病。
同义词：时行疫病

6.10.3 时行感冒 influenza

因气候骤变，冷热失常，且饲养管理不当，时邪疫毒乘虚侵袭肺卫所致。以起病急、传染性强、精神沉郁，食欲减退或废绝，发热、恶寒等为主要表现的疾病。

6.10.4 破伤风 tetanus

因肌肤损破，染受风毒而发。以全身肌肉强直性、阵发性抽搐，牙关紧闭，角弓反张为主要表现的疾病。

6.10.5 三喉症 three kinds of laryngo-pharyngeal diseases

因病邪侵袭咽喉部所致。以咽喉部炎性肿胀或具传染性为主要表现的三种疾病，包括喉骨胀、槽结、颡黄。

6.10.5.1 喉骨胀 laryngeal swelling

因热毒积聚心肺，郁结于咽喉发生肿胀的疾病，多见于幼驹。

6.10.5.2 槽结 strangles

因病气经口鼻入侵，以幼驹高烧、口色鲜红、食槽内结有结节硬肿和鼻流脓涕为主要表现的疾病。
同义词：颌下腺肿

6.10.5.3 颡黄 acute pharyngolaryngitis

因热毒结于咽喉而致肿痛，水草难咽，呼吸受阻的急性疾病。

6.10.6 牛红眼病 bovine acute conjunctivitis

因高温潮湿，空气污浊，病气上攻于目所致。以牛羞明流泪、结膜红肿、云翳遮睛为主要表现的疾病。

6.10.7 兔流涎病 infectious vesicular stomatitis of rabbits

因疠气经口传入所致。以致幼兔发生口舌生疮、大量流涎为主要表现的疾病。

6.11 其他病症

6.11.1 寒结 cold constipation

因寒邪凝滞胃肠所致。以大便秘结、腹痛起卧为主要表现的疾病。

6.11.2 水肿 edema

因脾失健运，肺失宣发，肾气化失常导致体内水湿停留所致。以颌下、胸前、腹下、阴囊、会阴部、四肢或全身水肿为主要表现的疾病。

6.11.2.1 阴水 yin edema

因脾阳不振，肾阳虚衰，不能运化水湿所致的疾病。

6.11.2.2 阳水 yang edema

因感受风邪，湿热等而引起肺气失宣，不能通调水道下输膀胱所致的疾病。

6.11.3 虚劳病 consumptive diseases

因脏腑亏损、气血阴阳不足所致。以病程较长，缠绵难愈为主要表现的疾病。

6.11.4 项脊恅 neck rheumatism

因风寒湿邪侵入项脊所致。以项颈强直、低头困难为主要表现的疾病。

同义词：低头难

6.11.5 疝 hernia

因运动或劳役过度，上坡爬岭，跨越障碍，腹痛打滚，气虚下陷，外伤或先天性缺陷所致。以腹腔内容物通过体壁的天然孔或异常孔突入皮下为主要表现的疾病。

6.11.5.1 阴囊疝 scrotal hernia

小肠或小结肠通过腹股沟窜入阴囊所致的疾病。

6.11.5.2 脐疝 umbilical hernia

腹腔内容物通过脐部突出的腹外疝，在脐部形成明显的局限性半圆形柔软肿胀。

6.11.5.3 鞘管疝 sheath tube hernia

腹腔脏器落入鞘管内所致的疾病。

6.11.6 直肠脱出 retum prolapse

因中气不足，固摄无力，肛门松弛所致。以直肠脱出肛门外为主要表现的疾病。

同义词：脱肛

6.11.7 幼畜惊风 eclampsia of young stock

因外感风寒、内伤乳食、邪热疫毒、阴血耗损或肝阴亏损导致体内生风所致，以幼畜颈项强直、四肢抽搐、角弓反张、眼球震颤为主要表现的疾病。

6.11.7.1 急惊风 acute eclampsia

幼畜为稚阳之体，气血未全，经脉未充，肌腠不密，形体未坚，感受风寒邪热疫毒，由表入里，阴血耗损，火热上炎，化火动风所致。以抽搐或昏迷为主要表现的疾病。

6.11.7.2 慢惊风 chronic eclampsia

多因外感风寒，内伤乳食，时发泄泻，脾胃阳虚气衰，失于温养，或因幼畜先天不足，肾气未充，水不涵木，肝阴亏损，肝阳上亢，虚风内动所致的疾病。以精神沉郁，神志不清，卧地不起，肢体颤抖，耳鼻四肢发凉，频频抽搐而乏力，呼吸微弱，泻粪清稀为主要表现的疾病。

6.11.8 新驹奶泻 diarrhoea in foal

未满月幼驹以消化紊乱、大便作泻为主要表现的疾病。

6.11.8.1 伤乳泻 diarrhoea due to milk

多因喂养过盛，乳汁过浓；或母马运动、劳役归来，喘息未定，幼驹乘饥暴吮热乳，损伤幼驹脾胃致泻。以泻下乳白或灰白、夹有乳块的黏滞粪便，其味酸臭，肠鸣腹胀，吮乳减少，精神稍差为主要表现的疾病。

6.11.8.2 湿热泻 diarrhoea due to dampness and heat evil

多因母马产后瘀血未尽，久而化热，热毒下注于乳汁；或因幼驹吮入患乳痈的病乳，损伤脾胃而引发的疾病。以泻下暴注，日泻频繁，粪稀如水，色黄腥秽，有时带血，肠鸣腹痛，发热口渴，精神淡漠，甚则高热，呼吸加快，日渐衰弱为主要表现的疾病。

6.11.8.3 脾虚泻 diarrhoea due to spleen deficiency

多因幼驹体弱，受风过食，脾虚不能运化乳食谷，内积胃肠，清浊不分而泻。以腹泻日久不愈，粪便稀薄，混有气泡、乳块或饲料块，毛焦欣吊，卧地不起为主要表现的疾病。

6.11.9 幼畜胎粪不下 constipation of young stock

因妊娠期间胎儿热毒壅结，气滞不畅，津液不足；或因母畜体弱分娩后奶少，幼畜吮吸初乳太迟；或因初产母畜泌乳量少，新生幼畜得不到足够的初乳以促进胃肠的蠕动所致。以幼畜出生一天后胎粪排出很少或未排出为主要表现的疾病。

同义词：胎粪难下

6.11.10 幼驹尿血 hematuria of foal

因公马（驴）和母马的血型不合或酷热暑天，幼驹感受热毒所致。以初生幼驹出现红黄色或粉红色尿液为主要表现的疾病。

6.11.10.1 血滚毒症 hematuria due to blood type incompatibility

因公马或驴和母马的血型不合，在母马妊娠期间，血中疫毒传于胎儿所致。以初生幼驹小便短涩，排尿痛苦，卧地不起，呼吸喘促，心脏衰竭为主要表现的疾病。

6.11.10.2 幼驹血尿 hematuria of foal

多因酷热暑天，幼驹感受热毒，心经积热，流注于小肠、膀胱，损伤脉络，迫血妄行，血随尿出所致。以精神倦怠，头低耳耷，食欲减少，体温升高，尿呈红色为主要表现的疾病。

6.11.11 跳肷 spasm of diaphragm

因外感内伤阴冷或内热亢盛，气逆不舒所致。以两肷部有节律跳动或震颤为主要表现的疾病。

同义词：罗膈损；撞膈症

6.11.11.1 虚寒跳肷 spasm of diaphragm due to cold and deficiency

多因风寒侵袭，阴雨浇淋或运动、劳役过度，趁热过饮冷水，冷热相击，逆气积胸，致胸膈发生痉挛。

6.11.11.2 实热跳肷 spasm of diaphragm due to heat evil

多因风热侵袭，暑气熏蒸，内热亢盛、逆气积胸，或罗膈受损，致胸膈发生痉挛。

6.11.12 风瘫 paralysis due to wind evil

因风寒湿邪壅滞经脉，或肝肾亏虚所致。以腰肢疼痛、卧地难起或不起为主要表现的疾病。

6.11.13 痹证 arthralgia syndrome

因风寒湿邪侵袭肢体，传于经络所致。以肌肉关节疼痛、麻木、屈伸不利，甚至关节肿大灼热为主要表现的疾病。

6.11.13.1 行痹 migratory arthralgia
风邪偏盛，肢体酸痛而游走无定位的痹症。

6.11.13.2 痛痹 arthritis
寒邪偏重的痹症。

6.11.13.3 着痹 arthralgia
湿邪偏盛的痹症。

6.11.13.4 热痹 bi syndrome due to heat pathogen heat arthralgia
热邪偏盛的痹症。

6.11.14 痿证 flaccidity syndrome
因肺热伤津，肝肾亏虚或慢性中毒所致。以肢体筋脉迟缓、四肢痿软无力，肌肉萎缩，终至卧地不能起立为主要表现的疾病。

6.11.15 骨眼 protrusion of the swollen third eyelid
因运动或劳役过度，内外受热，突受风寒侵袭，冷热冲击，肝受之而传注于眼所致。以闪骨，第三眼睑（瞬膜）瘀肿为主要表现的疾病。

6.11.16 内障眼 glaucoma
因暑月炎天，运动或劳役过重，热毒冲于心、肝经，或因肝肾阴虚，肝阳偏亢，虚火上炎或先天因素所致。以翳障凝于瞳内不能视物为主要表现的疾病。
同义词：青光眼

6.11.17 外障眼 external oculopathy
因外感六淫，或内有郁热、积滞等引起肝经积热，外传于眼，或肝肾阴虚、虚火上炎以及眼外伤所致。以见眼泡肿胀，结膜潮红，羞明流泪，眵多难睁，有时角膜混浊，或生白色或蓝色云翳，导致视力减退，甚至失明为主要表现的疾病。

6.11.18 口僻 facial paralysis
多由风邪入中面部，痰浊阻滞经络所致。以突发面部麻木，口眼歪斜为主要表现的疾病。
同义词：面瘫

第七章 证 候

7.1 证

7.1.1 证

证是对疾病在某一阶段的病因、病位、病性和邪正盛衰情况的概括。

7.2 基本虚证类

7.2.1 气虚证 qi deficiency pattern

主要指肺脾气虚，表现为动则气喘、咳嗽声低、劳役即汗、大便清稀、完谷不化或水粪齐下、口舌淡白、舌软无力。

7.2.2 气陷证 qi sinking pattern

气虚无力升举，清阳之气应升反下陷，以极度瘦弱，少气倦怠，肛门松弛、久泻脱肛、直肠脱出，内脏、子宫下垂，舌淡苔白，脉弱等为常见症的证候。

7.2.3 气脱证 qi desertion pattern

元气因某种原因而急骤外泄，元气亏虚至极，以突然可视黏膜苍白，口唇青紫，汗出肢冷，呼吸微弱，舌淡脉细数为常见症的危重证候。

同义词：元气虚脱证；元气衰败证

7.2.4 血虚证 blood deficiency pattern

主要指心肝血虚。特点为口色、结膜淡白无华，脉细弱，双目无光。

7.2.4.1 血虚动风证 blood deficiency stirring wind pattern

血液亏虚，形体失养，虚风内动，以站立不稳，时欲倒地，蹄壳干枯皲裂，口色淡白，脉细弱，肢体反应迟钝、震颤，四肢拘挛抽搐等为常见症的证候。

同义词：血虚生风证

7.2.4.2 血虚风燥证 blood deficiency and wind dryness pattern

血虚风胜化燥，皮毛、筋脉失养，以皮肤粗糙、干燥脱屑、瘙痒，或枯皱皲裂，被毛失荣脱落，肌肤反应迟钝，四肢痉挛拘急，站立不稳，舌淡脉细等为常见症的证候。

同义词：血虚肤燥生风证；血虚风盛证

7.2.4.3 血虚津亏证 anemia of blood and body fluids pattern

津血亏虚，形体、脏腑失于营养、滋润，以皮毛枯槁，黏膜淡白，鼻燥，眼干少泪，小便短少，大便干结，舌红少津，脉细而涩等为常见症的证候。

7.2.5 阴虚证 yin deficiency pattern

主要指肺肾阴虚。表现为虚热不退、午后热盛、不劳而汗、口色红、少苔、脉细数无力。干咳无痰，咳声低微或有气喘，或腰拖胯趿，公畜举阳滑精，母畜不孕。

同义词：阴液亏虚证

7.2.5.1 阴虚阳亢证 yin deficiency and yang floating pattern

阴液亏虚，阴不制阳，阳气亢盛，以低热，形瘦，盗汗，滑精，性欲亢进，舌红少苔，脉细数等为常见症的证候。

7.2.5.2 阴虚血燥证 yin deficiency and blood dryness pattern

阴液亏虚，津血被耗，以站立不稳，口干咽燥，皮枯毛燥、瘙痒，午后潮热，盗汗，舌红少津，脉细数等为常见症的证候。

7.2.5.3 阴虚动血证 yin deficiency moving blood pattern

阴液亏虚，虚热迫血妄行，以咳血、吐血、衄血、尿血、便血，午后潮热，盗汗，舌红少苔，脉细数等为常见症的证候。

7.2.5.4 阴虚动风证 yin deficiency moving wind pattern

阴液亏虚，经脉失养，虚风内动，以肢体震颤，拘挛抽搐，形体消瘦，午后潮热，口燥，小便短黄，大便干结，舌红少苔，脉细数等为常见症的证候。

7.2.5.5 阴虚津亏证 deficiency of yin and body fluids pattern

阴津亏耗，形体失养，以口渴喜饮，皮肤干涩，眼眶凹陷，小便短黄，大便干结，午后潮热，形瘦盗汗，舌红苔少而干，脉细数等为常见症的证候。

同义词：阴津亏虚证

7.2.6 亡阴证 yin fluids exhausting pattern

体液大量耗损，阴精欲竭，以身热汗出如油，口渴喜饮，精神兴奋，躁动不安，气促喘粗，口干舌红，脉数无力或脉大而虚等为常见症的危重证候。

同义词：阴脱证

7.2.7 阳虚证 yang deficiency pattern

主要指脾肾阳虚。证见畏寒怕冷，耳鼻四肢发凉，腰膝萎软，阳痿滑精，慢草或不食，瘦弱无力，久泄不止，四肢浮肿。口色淡白，脉象细弱。

同义词：阳虚内寒证；阳气亏虚证

7.2.8 亡阳证 yang qi exhausting pattern

阳气衰竭而欲脱，以冷汗淋漓，四肢厥冷，精神不振、喘息低微，脉微欲绝，舌淡苔润等为常见症的危重证候。

同义词：阳脱证

7.2.9 虚阳浮越证 deficiency yang floating upward pattern

阳虚阴盛，格阳于外，以口燥，皮肤发热，后肢厥冷，尿清长，大便溏，脉浮大无力等为常见症的证候。

7.2.10 气血两虚证 deficiency of both qi and blood pattern

气血亏虚，形体失养，以体瘦毛焦，精神倦怠、四肢无力，站立不稳，心区跳动明显，搏动加快，易惊，舌淡白，脉细弱等为常见症的证候。

7.2.10.1 气随血脱证 qi loss due to blood depletion pattern

因大量出血，气无所附而随之暴脱，以黏膜苍白，四肢厥冷，大汗淋漓，气息微弱，甚至昏厥，脉微欲绝，或虚大无力等为常见症的危重证候。

7.2.10.2 气不摄血证 qi failing control blood pattern

气虚不能统摄血液，以便血、鼻衄、齿衄，精神不振、四肢无力，气短少鸣，鼻黏膜色淡，舌淡脉弱等为常见症的证候。

同义词：气不统血证

7.2.10.3 气血两虚动风证 both qi and blood deficiency moving wind pattern

气血亏虚，形体失养，虚风内动，以精神不振、四肢无力，气短少鸣，唇甲色淡，站立不稳，肢体反应迟钝，四肢挛急，舌淡白，脉弱等为常见症的证候。

7.2.11 气阴两虚证 both qi and yin fluids deficiency pattern

元气不足，阴津亏损，气虚与津液亏虚并见，以精神不振、四肢无力，气短少鸣，口燥，口渴喜饮，午后潮热，小便短少，大便干结，舌体瘦薄红绛，舌苔少而干，脉细数无力等为常见症的证候。

同义词：气阴两亏证

7.2.12 阴血亏虚证 deficiency of yin and blood pattern

阴液精血亏虚，形体失养，以形体消瘦，低烧不退，肢体反应迟钝，站立不稳，心区跳动明显，搏动加快，易惊，舌红苔少，脉细数等为常见症的证候。

7.2.13 阴阳两虚证 deficiency of yin and yang pattern

脏腑阴液阳气俱虚，以摇晃欲倒，精神不振，畏寒肢凉，低烧不退，躁动不安，舌淡少津，脉弱而数等为常见证的症候。

同义词：阴阳两亏证

7.2.13.1 阴损及阳证 yin fluids deficiency leading to yang deficiency pattern

阴精或阴气亏损，日久累及阳气生化不足或无所依附而耗散，从而在阴虚的基础上导致阳虚，终致阴阳俱虚的证候。

7.2.13.2 阳损及阴证 yang deficiency leading to yin fluids deficiency pattern

阳气虚损，日久及阴，终致阴阳俱虚的证候。

7.2.13.3 阴竭阳脱证 both exhausting of yin and yang pattern

阴精亏损，阳无所附，随之而脱所表现的危重证候。

7.2.14 津液亏虚证 body fluids deficiency pattern

津液亏少，脏腑组织失去濡养，以口鼻干燥，皮毛干枯，口干欲饮，粪便干结，尿短少，舌红少津，干咳无痰，脉细数无力等为常见症的证候。

7.2.14.1 津亏证 body fluids deficiency pattern

津液亏虚之轻者，以口鼻、唇舌、皮肤干燥，大便干结，脉细数等为常见症的证候。

同义词：津伤证

7.2.14.2 液亏证 body fluids exhausting pattern

津液亏虚之甚者，以形体消瘦，口唇焦裂，皮毛干枯，眼窝凹陷，关节不利，小便短少，大便干结等为常见症的证候。

同义词：液脱证

7.2.15 津气亏虚证 both body fluids and qi deficiency pattern

津液不足，正气亏虚，以精神不振、呼吸短促，口渴欲饮，皮肤干燥，眼窝凹陷，或汗出量多，舌红苔干，脉细无力等为常见症的证候。

7.2.16 精气亏虚证 deficiency of vital essence pattern

精气亏少，以形体消瘦，站立不稳，体型矮小，动作迟钝，或精少精稀、阳痿早泄等为常见症的证候。

同义词：精亏证；精气不足证

7.2.17 精血亏虚证 both vital essence and blood deficiency pattern

病久体弱，或生化不足，精亏血少，以心区跳动明显，搏动加快，黏膜苍白，损伤久不能复等为常见症的证候。

7.2.18 卫虚证 defensive qi deficiency pattern

卫气亏虚，卫外不固，以恶风汗出，容易感冒，脉浮无力等为常见症的证候。

同义词：卫表不固证；卫气亏虚证

7.2.19 营虚证 deficiencies of ying and qi pattern

营血亏虚，机体失养，以四肢无力，消瘦，自汗，脉弱等为常见症的证候。

同义词：营气亏虚证

7.3 基本实证类

7.3.1 外风证 external wind evil pattern

风邪或夹湿热疫毒等侵袭肌表，卫外机能失常所致的证候。

7.3.1.1 风邪犯表证 wind evil attacking exterior pattern

风邪侵袭肌表，卫外机能失常，以恶风、发热、汗出、脉浮，或皮肤瘙痒、水肿，或咳嗽、咽喉肿痛等为常见症的证候。

同义词：风邪外袭证

7.3.1.2 风湿犯表证 wind dampness attacking exterior pattern

风湿侵袭肌表，卫外机能失常，以恶寒发热，多卧懒动，头低耳耷，舌苔白腻等为常见症的证候。

同义词：风湿证；风湿外袭证

7.3.1.3 风热外袭证 wind heat assailing the outer body pattern

风热侵袭肌表，卫外机能失常，以发热，微恶寒，汗出，舌尖红，苔薄黄，脉浮数，或皮肤瘙痒等为常见症的证候。

同义词：风热证；风热外侵证

7.3.1.4 风热痰毒证 wind heat and sputum toxin pattern

风热痰毒壅滞，气血不畅，以发热，可视黏膜发红，咳嗽，痰多而黄稠，气喘，咽喉肿痛，或疮疡肿痛质硬、难溃难消，舌红苔黄腻，脉滑数等为常见症的证候。

同义词：风热挟痰证；风热痰凝证

7.3.1.5 风湿挟毒证 wind dampness carrying toxin pattern

风湿毒邪侵溃肌肤，以后肢浮肿、溃疡，阴部湿疹、瘙痒、流黄水，或足趾间奇痒等为常见症的证候。

同义词：风湿毒聚证

7.3.1.6 风热挟湿证 wind heat carrying dampness pattern

风热湿邪侵袭肌肤，以发热，渴不多饮，肢体困重，或目赤肿痛、睑缘湿烂，或皮肤湿疹、水疱、瘙痒、流水，舌红苔黄腻等为常见症的证候。

7.3.1.7 风湿化热证 wind dampness transform to fire pattern

风湿之邪郁久而化热，以肢体酸胀困重，关节肿痛、活动不利，或皮肤瘙痒、渗液，发热口渴，舌红苔黄白而干等为常见症的证候。

同义词：风湿郁热证；风湿化火证

7.3.1.8 风寒化热证 wind cold transforming heat pattern

风寒之邪郁久化热，以恶寒发热，咳嗽痰稠，舌尖红，苔黄白而干，脉数等为常见症的证候。

同义词：风寒郁热证

7.3.2 寒凝证 cold coagulation pattern

寒邪侵袭机体，阳气被遏，以恶寒甚，无汗，头身或胸腹疼痛，苔白，脉弦紧等为常见症的证候。

同义词：外寒证

7.3.2.1 寒湿阻滞证 cold dampness causing stagnation pattern

寒湿之邪侵袭，阻滞气机，以头低耳耷，多卧懒动，屈伸不利，无汗，或面浮肢肿，大便稀溏，小便不利，舌苔白润，脉濡或滑等为常见症的证候。

同义词：寒湿证；寒湿凝滞证

7.3.2.2 寒凝气滞证 cold coagulation and qi stagnation pattern

寒邪凝滞气机，多卧懒动，屈伸不利，四肢拘急，或脘腹胀满冷痛，泛吐清水，肠鸣腹泻，苔白润，脉弦紧等为常见症的证候。

7.3.2.3 寒凝血瘀证 cold stagnation and blood stasis pattern

寒邪凝滞气机，血行瘀阻，以畏寒冷痛，得温痛减，肢冷色青，舌紫暗，苔白，脉沉迟而涩等为常见症的证候。

7.3.2.4 寒湿化热证 cold dampness transforming heat pattern

寒湿之邪郁久化热，以发热，肢体沉重，关节红肿痛，口渴不多饮，欲呕，腹痛腹泻，舌红苔黄白，脉滑数等为常见症的证候。

7.3.2.5 寒湿瘀滞证 cold dampness causing stasis and stagnation pattern

寒湿内蕴，血行瘀滞，以形寒肢冷、肢体沉重、疼痛，得温痛减，唇舌紫暗，舌有斑点，舌苔白滑，脉沉细涩等为常见症的证候。

7.3.2.6 真寒假热证 cold pattern with pseudo heat pattern

阴寒内盛，格阳于外，真寒证反见热象的一种证型，以既有四肢厥冷，小便清长，下利清谷，大便清冷，舌淡苔白等真寒症状，又有热不烫手，口渴欲热饮等假热表现的证候。

7.3.2.7 寒热错杂证 cold heat mixed pattern

同一病畜、同一时期，既有寒证又有热证的表现，寒热交错并见的证候。

同义词：寒热挟杂证

7.3.2.8 血寒证 blood cold stasis pattern

寒邪客于血脉，凝滞气机，血行不畅，以形寒肢冷，喜暖恶寒，四肢疼痛发凉，得温痛减，可视黏膜紫暗，舌淡暗，苔白，脉沉迟等为常见症的证候。

同义词：血寒凝滞证

7.3.3 暑热证 summer heat pattern

暑热侵袭，耗气伤津，以发热口渴，精神不振，呼吸短促，躁动不安，站立不稳，汗出，小便短黄，舌红苔黄干等为常见症的证候。

同义词：暑热内郁证

7.3.3.1 伤暑证 summer heat hurt pattern

暑热或暑湿内郁、侵扰心神，以精神倦怠、食欲减退、四肢无力、呆立不动、身热有汗、呼吸气粗、口色发红、口干、频饮冷水、脉洪数等为常见症的证候。

7.3.3.2 中暑证 heatstroke pattern

暑热或暑湿内郁、侵扰心神，以发病急骤，病程短，高热神昏，行走如醉，精神极度衰沉，汗出如浆，气促喘粗，甚则卧地不起，肢体抽搐，口色赤红或赤紫，脉洪数或细数等为常见症的急性证候。

7.3.3.3 暑湿证 summer dampness pattern

暑湿之邪交阻内蕴，以口渴，精神不振，四肢无力，肢体困重，关节痛，躁动不安，汗出量小，舌红苔黄腻，脉滑数等为常见症的证候。

同义词：暑湿热郁证；暑湿内蕴证

7.3.3.4 暑热动风证 summer heat moving wind pattern

暑热炽盛，引动肝风，以高热，四肢抽搐，甚至角弓反张、牙关紧闭等为常见症的证候。

7.3.3.5 暑闭气机证 summer heat blocking qi pattern

暑热卒中，闭阻气机，以突然昏倒，身热汗少，四肢厥冷，气喘不鸣，牙关紧闭等为常见症的证候。

7.3.4 湿阻证 dampness blocking qi pattern

湿浊邪气阻滞气机，以身体困重，屈伸不利，腹胀腹泻，食欲不振，苔滑脉濡等为常见症的证候。

同义同：湿邪阻滞证；湿浊困阻证

7.3.4.1 湿阻气滞证 dampness blocking causing qi stagnation pattern

湿邪阻困，气机郁滞，以身体困重，脘腹胸胁等处疼痛，苔滑或腻，脉弦等为常见症的证候。

7.3.4.2 湿热证 dampness heat tangling pattern

湿热互结，热不得越，湿不得泄，以身热不扬，口渴不欲多饮，大便泄泻，小便短黄，舌红苔黄腻，脉滑数等为常见症的证候。

同义词：湿热内蕴证；湿热壅盛证

7.3.4.3 气分湿热证 qi fen dampness heat pattern

湿热侵迫气分，以身热不扬，腹胀，皮肤和可视黏膜发黄，肢体困倦，呕吐，尿黄，舌红苔黄腻，脉濡数或滑数等为常见症的证候。

7.3.4.4 湿热壅滞证 dampness heat blocking qi pattern

湿热邪气壅滞气机，以身热口渴，胸腹胀满疼痛，不时作呕，便溏糊肛，舌红苔黄腻，脉濡数或滑数等为常见症的证候。

7.3.4.5 湿热侵淫证 dampness heat attacking body pattern

湿热邪气侵浸，以睑缘、耳、鼻、口角、足趾等处红肿湿烂、瘙痒，溃破流水，舌红苔黄腻，脉滑数等为常见症的证候。

7.3.4.6 湿热毒蕴证 dampness heat and toxin accumulation pattern

湿热毒邪蕴结，以四肢、耳、鼻、头面、阴部等处红肿溃烂、瘙痒流水，或发热身黄，甚至昏迷，斑疹，小便闭涩，舌红苔黄腻，脉濡数等为常见症的证候。

同义词：湿热疫毒证

7.3.4.7 湿热瘀阻证 dampness heat causing congestion pattern

湿热蕴结，血行瘀滞，以身热口渴，肢体疼痛，胁下痞块，小便不利，便溏，舌质紫红，苔黄而腻，脉滑数或涩等为常见症的证候。

7.3.5 外燥证 external dryness pattern

秋季感受燥邪，耗伤津液，以皮肤干燥，口鼻干燥等为常见症的证候。

7.3.5.1 温燥证 warm dryness pattern

温燥之邪侵袭，耗伤阴津，以发热，微恶风寒，干咳痰少，躁动不安，口渴，皮肤及鼻干燥，小便短黄，舌苔薄黄，脉浮数等为常见症的证候。

同义词：燥热证

7.3.5.2 凉燥证 cold dryness pattern

凉燥之邪犯表伤肺，以恶寒重，发热轻，无汗，口干、鼻燥，咳嗽痰少，舌苔薄白而干，脉浮紧等为常见症的证候。

同义词：寒燥证

7.3.6 火热炽盛证 severe heat pattern

火热内盛，以发热，口渴饮冷，胸腹热，目赤，大便秘结，小便短黄，舌红苔黄而干，脉数或洪等为常见症的证候。

同义词：实火证；实热证

7.3.6.1 气分证 qi phase pattern

温热病邪由卫入里，邪热亢盛，正邪交争剧烈的证候。

同义词：热炽气分证；热盛证

7.3.6.2 营分证 ying phase pattern

温热邪气由气分深入，侵入气分与血分之间，热陷心包，心神被扰，热窜血络，伤津动血的证候。

同义词：热炽营分证；营热炽盛证

7.3.6.3 血分证 blood phase pattern

温热病最深重的病理阶段。邪热壅盛，迫血妄行，扰乱心神，动血耗血，伤阴动风，瘀热内阻的证候。

同义词：血热证；热炽血分证

7.3.6.4 热入营血证 heat attacking ying blood pattern

温热病邪深入营血分，伤耗阴血，扰乱心神，以身热夜甚，躁动不安，渴不多饮，斑疹隐隐，或出血，便结尿黄，舌绛，脉细数等为常见症的证候。

7.3.6.5 卫气同病证 involving both wei-defence and qi phases pattern

表邪入里化热，气分热势已盛而卫分证候未解，卫分、气分病变同在的证候。

7.3.6.6 气营两燔证 heat blazing in both qi and ying phases pattern

气分与营分邪热炽盛的证候。

7.3.6.7 气血两燔证 heat blazing in both qi and blood phases pattern

气分与血分邪热炽盛，扰及心神，灼伤血络，迫血妄行的证候。

7.3.6.8 热盛动风证 severe heat moving wind pattern

邪热炽盛，引动肝风，以壮热口渴，神志昏迷，四肢抽搐，颈项强直，角弓反张，牙关紧闭，舌红绛，苔黄，脉弦数等为常见症的证候。

7.3.6.9 热盛动血证 severe heat moving blood pattern

邪热炽盛，迫血妄行，以壮热口渴，目赤，便血、尿血、衄血，或斑疹显露，舌红绛，苔黄，脉洪数等为常见症的证候。

7.3.6.10 热盛气滞证 severe heat causing qi stagnation pattern

邪热炽盛，气机郁滞，以发热口渴，胸腹等处疼痛、便秘尿黄，脉弦数等为常见症的证候。

同义词：热郁气滞证

7.3.6.11 热厥证 heat syncope pattern

热蕴于内，阻阴于外的真热假寒证。证见四肢厥冷、口色红、恶热、口腔干燥、尿短赤。

同义词：真热假寒证；热极肢厥证

7.3.6.12 热盛酿脓证 severe heat causing suppuration pattern

邪热壅积，血肉腐败，酿成痈脓，以发热口渴，局部红肿灼痛、拒按，溃破流脓，舌红苔黄腻或黄腐，脉滑数等为常见症的证候。

同义词：热盛肉腐证

7.3.7 痰证 sputum pattern

痰浊内阻，以咳嗽气喘，咯痰量多，呕恶，或局部有圆滑肿块，苔腻脉弦滑等为常见症的证候。

同义词：痰浊阻滞证；痰浊凝聚证

7.3.7.1 风痰证 wind sputum pattern

外风挟痰浊为患，或肝风痰浊内扰，以咯吐泡沫痰涎，或喉中痰鸣，口眼歪斜，苔白腻，脉弦滑等为常见症的证候。

7.3.7.2 寒痰证 cold sputum pattern

寒邪与痰浊凝滞，以咯吐白痰，脘痞，气喘，肢冷，苔白腻，脉弦滑或弦紧等为常见症的证候。

7.3.7.3 湿痰证 dampness sputum pattern

痰湿内阻，以咯吐多量黏稠痰，痰滑易咯，肢体困重，食欲减退，苔白腻，脉濡缓或滑等为常见症的证候。

同义词：痰湿证；痰湿内阻证

7.3.7.4 热痰证 heat sputum pattern

痰浊与邪热互结，以咯吐黄痰，发热口渴，舌红苔黄腻，脉滑数等为常见症的证候。

7.3.7.5 燥痰证 dryness sputum stagnation pattern

燥热痰浊内蕴，以咳嗽，咯浓痰，或痰黏成块，或痰中带血，口鼻干燥，舌干少津，苔腻，脉涩等为常见症的证候。

同义词：燥痰蕴结证

7.3.7.6 痰气互结证 sputum qi tangling pattern

痰气相互阻结，以头低耳耷、眼闭呆立、状如睡眠，有时头顶墙壁，呆立不动，或嘴含饲草而不知咀嚼，痰多，苔白腻，脉滑数等为常见症的证候。

同义词：痰气互郁证

7.3.7.7 痰瘀互结证 sputum stasis tangling pattern

痰浊瘀血相互搏结，以局部肿块疼痛，或肢体反应迟钝、痿废，痰多，或痰中带紫暗血块，舌紫暗或有斑点，苔腻，脉弦涩等为常见症的证候。

同义词：瘀痰内阻证；血瘀痰凝证；痰瘀互搏证

7.3.7.8 痰瘀化热证 sputum stasis transforming heat pattern

痰瘀互结，日久化热，以患处肿硬、热、痛，咯痰色黄或夹血块，舌暗红或有斑点，苔黄腻，脉弦涩等为常见症的证候。

7.3.7.9 痰热气滞证 sputum heat causing qi stagnation pattern

痰热内蕴，阻滞气机，以躁动不安，胸胁疼痛，咳嗽气喘，咯痰黄稠，发热口渴，舌红苔黄腻，脉弦数等为常见症的证候。

7.3.7.10 痰热内扰证 sputum heat interference pattern

痰热内盛，扰乱心神、气机，以咳嗽气喘，咯痰黄稠，发热口渴，躁动不宁，舌红苔黄腻，脉滑数等为常见症的证候。

同义词：痰热搏结证；痰火郁结证

7.3.7.11 痰热内闭证 sputum heat blocking mind pattern

痰热内蕴，阻闭心神，以胸胁疼痛，咳嗽气喘，咯痰黄稠，或有哮鸣，发热口渴，或神志昏迷，狂乱，或喉中痰鸣，舌红苔黄腻，脉滑数等为常见症的证候。

7.3.7.12 痰热动风证 sputum heat moving wind pattern

痰热内盛，引动肝风，以胸胁胀满，咳嗽气喘，发热口渴，咯痰黄稠，或喉中痰鸣，四肢抽搐，或呕吐，舌红苔黄腻，脉滑数等为常见症的证候。

7.3.7.13 痰结毒滞证 sputum and toxin stagnation pattern

痰浊与邪毒蕴结，以咳嗽痰多，局部包块、触之无感，硬痛不移，或溃后流脓水腥腐秽臭，苔垢腻等为常见症的证候。

7.3.7.14 痰食互结证 sputum food tangling pattern

痰浊与宿食互结，阻滞气机，以胸腹胀满疼痛，咳嗽吐痰，食欲减退，腹胀，呕吐痰涎宿食，苔腐腻，脉弦滑等为常见症的证候。

7.3.7.15 痰虫互结证 sputum parasite tangling pattern

痰浊与虫体搏结，蕴聚体内，以咯痰，或突然昏仆、口吐痰涎，或肢体触及包块，苔腻脉滑等为常见症的证候。

同义词：痰虫互搏证

7.3.7.16 痰湿瘀滞证 sputum dampness stagnation pattern

痰湿内阻，气血瘀滞，以胸腹胀满疼痛，精神沉郁、嗜睡，或肌肤肿硬、触之无感，舌淡紫或有斑点，苔滑腻，脉弦涩等为常见症的证候。

7.3.7.17 痰核留结证 sputum forming nuclear lump pattern

痰浊结块，留滞不消，以颈项等处皮下生核，甚或成串，圆滑质硬、推之可移、不红不热不痛，苔腻，脉弦滑等为常见症的证候。

7.3.8 水饮内停证 excessive fluid collecting internally pattern

水饮停聚体腔，以站立不稳，呕吐清水、涎液，苔滑，脉弦等为常见症的证候。

同义词：饮证

7.3.9 水停证 water dampness stagnation pattern

水湿停聚体内，以肢体浮肿，小便不利，或腹大痞胀，舌淡胖，苔白滑，脉濡缓等为常见症的证候。

同义词：水湿内停证；水湿停聚证

7.3.10 气滞证 qi stagnation pattern

某些脏腑或局部气机阻滞，以胸腹胀满疼痛，时轻时重，常随嗳气、肠鸣、矢气而减，脉弦等为常见症的证候。

同义词：气机阻滞证；气机郁结证

7.3.10.1 气滞血瘀证 qi stagnation and blood stasis pattern

气机阻滞，血行不畅，以胸腹胀满疼痛，或有痞块、时散时聚，舌紫或有斑点，脉弦涩等为常见症的证候。

同义词：气血瘀滞证

7.3.10.2 气滞湿阻证 qi stagnation causing dampness stagnation pattern

气机郁滞，湿浊内阻，以胸腹胀满疼痛，欲吐，肢体困重，嗜睡，或有浮肿，苔白腻，脉弦滑或濡缓等为常见症的证候。

同义词：气滞湿困证

7.3.10.3 气滞热壅证 qi stagnation causing severe heat pattern

气机郁滞，邪热壅盛，以胸腹胀满疼痛，发热口渴，舌红苔黄，脉弦数等为常见症的证候。

7.3.10.4 气郁化火证 qi stagnation transforming fire pattern

气机阻滞，日久化火，以精神沉郁，躁动不安，胸腹胀满疼痛，口干，舌红苔黄，脉弦数等为常见症的证候。

同义词：气滞化火证；气滞化热证

7.3.10.5 气滞水停证 qi stagnation blocking body fluid pattern

气机阻滞，水液内停，以肢体浮肿，小便不利，头身困重，胸腹胀满疼痛，舌淡苔白滑，脉弦缓等为常见症的证候。

7.3.11 气逆证 qi upwards reverse pattern

气机逆乱向上，以咳嗽气喘，或呕吐、呃逆、嗳气为常见症的证候。

同义词：气机上逆证

7.3.12 气闭证 qi blocked pattern

气机闭塞不通，以腹痛，无肠鸣矢气，二便不通，或突然昏厥，牙关紧闭，肢体强直等为常见症的证候。

同义词：气机壅闭证；气机闭塞证

7.3.13 血瘀证 blood stasis pattern

瘀血内阻，血行不畅，以局部出现肿块、疼痛拒按，或腹内硬块、疼痛不移、拒按，或出血紫暗成块，舌紫或有斑点，脉弦涩等为常见症的证候。

同义词：瘀血内阻证；瘀血阻滞证

7.3.13.1 血瘀气滞证 blood stasis and qi stagnation pattern

瘀血内阻，气机郁滞，以腹内硬块、疼痛、拒按，或局部青紫肿胀，舌紫或有斑点，脉弦涩等为常见症的证候。

同义间：瘀滞证

7.3.13.2 血瘀动血证 blood stasis moving blood pattern

瘀血阻塞，血溢脉外，以出血色紫暗、夹块，或局部疼痛、固定，或见青紫肿块，舌紫或有斑点，脉涩等为常见症的证候。

7.3.13.3 血瘀化热证 blood stasis transforming heat pattern

血瘀日久化热，以局部疼痛、发热，或有青紫肿块，午后或夜间发热，口干，舌暗红或有斑点，脉涩而数等为常见症的证候。

同义词：瘀热内郁证；瘀滞化热证

7.3.13.4 血瘀水停证 blood stasis blocking body fluids pattern

瘀血内阻，水液停聚，以腹内有癥块，腹痛，腹大而胀，小便不利，或局部青紫、漫肿疼痛，舌淡紫或有斑点，脉涩等为常见症的证候。

7.3.14 邪毒炽盛证 severe evil toxin pattern

各种邪毒壅盛所致证候，其症因毒邪的不同而各具特征。

7.3.14.1 毒邪流窜证 toxin moving randomly pattern

火热毒、风毒、湿毒等流窜为患，以局部肿胀、疼痛拒按、化脓、溃烂等为常见症的证候。

7.3.14.2 风毒证 severe wind toxin pattern

风毒侵袭，以突然肌肤水肿，或起风团，或肢体抽搐，牙关紧闭，或头面、口鼻、两眼赤红肿痛，脉浮数等为常见症的证候。

同义词：风毒炽盛证

7.3.14.3 火毒证 severe heat toxin pattern

火热壅盛成毒，以肌肤等处生疮疖疔痈，红肿，化脓溃烂，发热口渴，舌红苔黄，脉数等为常见症的证候。

同义词：热毒蕴结证；火毒炽盛证

7.3.14.4 火毒流窜证 fire toxin moving randomly pattern

火热毒邪走散流窜，以多处生疮疖疔痈，红肿，化脓溃烂，发热口渴，便秘尿黄，舌红苔黄，脉数等为常见症的证候。

7.3.14.5 热毒入营证 fire toxin attacking ying pattern

火热毒邪侵入营血，以身热夜甚，斑疹隐隐，神昏，渴不多饮，便结尿黄，舌绛，脉细数等为常见症的证候。

同义词：热毒陷营证；火毒陷营证

7.3.14.6 热毒内陷证 heat toxin inward penetration pattern

火热毒邪炽盛，内陷脏腑，以壮热口渴，神昏，可视黏膜暗红，便秘尿黄，舌红绛，苔黄，脉沉数等为常见症的证候。

同义词：火毒内陷证；热毒内陷证

7.3.14.7 湿毒证 dampness toxin stagnation pattern

湿邪蕴结成毒，以肌肤、阴股、后肢、趾间等处生疮疡、湿疹，糜烂流水，痒麻疼痛，苔腻，脉濡缓等为常见症的证候。

同义词：湿毒蕴结证

7.3.14.8 风火热毒证 wind heat toxin pattern

风火热毒壅滞肌肤，以肌肤生疮疖疔痈，红肿、痒麻，化脓溃烂，或头面、口鼻、两目鲜红肿痛，发热口渴，神昏，便秘尿黄，舌红绛，苔黄焦，脉洪数等为常见症的证候。

7.3.14.9 毒入营血证 toxin attacking ying blood pattern

火热等邪毒侵入营血，以壮热口渴，神昏，斑疹紫暗，或出血色暗红，舌绛脉数等为常见症的证候。

7.3.14.10 虫毒证 parasite toxin accumulating pattern

虫毒侵袭，结聚肌肤，以皮肤痒麻、疼痛，水肿、溃烂，或局部存暗红出血点等为常见症的证候。

同义词：虫毒结聚证

7.3.14.11 疫毒证 epidemic toxin attacking pattern

天行疫毒侵袭所致的证候，因疫毒性质不同，而其症各异。

同义词：疫毒侵袭证

7.3.14.12 疫毒内闭证 epidemic toxin blocking pattern

疫毒侵袭，内闭心神，以神昏，喉间痰鸣，呼吸不利，大便秘结等为常见症的证候。

7.3.14.13 脓毒证 sepsis toxin accumulating pattern

脓毒蓄积于内，以痈疡流脓，气味腥臭，发热口渴，苔腐腻，脉滑数等为常见症的证候。

同义词：脓毒蕴结证

7.3.14.14 食毒证 food toxin pattern

食入有毒物所致，以脘腹剧痛，呕吐腹泻，甚或出现厥脱等为常见症的证候。

7.3.14.15 燥毒证 dryness toxin pattern

燥热毒邪侵袭，以咽喉红肿、溃烂，或目赤溃烂，便秘尿黄，口渴欲饮，鼻干肤燥，舌干少津，脉细涩等为常见症的证候。

7.3.14.16 胎毒证 fetal poisoning pattern

幼畜因母体遗毒所致，以咬牙、吮乳困难，颈肌痉挛，头颈弯向一侧或反张或向下弯曲，眼球颤动，呼吸急促，四肢痉挛抽搐，数十分钟后，恢复正常，口舌干燥，舌质红绛，脉数有力等为常见症的证候。

同义词：胎毒蕴结证；胎毒内蕴证

7.3.14.17 阴毒证 yin toxin pattern

阴寒邪毒结聚，以局部漫肿、喜温、不红、难溃、不易成脓，或脓稀气腥，肢冷，苔白，脉沉迟等为常见症的证候。

同义词：寒毒证

7.3.14.18 蛇毒内攻证 snake venom attacking pattern

毒蛇咬伤，蛇毒内攻脏腑，以呼吸困难，或见斑疹、衄血，或冷汗肢厥，或神识不清等为常见症的证候。

7.3.15 食积证 dyspepsia pattern

食物停积胃肠，以肚腹胀满，呕吐酸臭，废食，大便黏滞，臭如败卵，苔腐腻，脉弦滑等为常见症的证候。

7.3.16 虫积证 parasite clumping pattern

寄生虫积聚体内，以腹胀腹痛，贪食易饥，体瘦，四肢无力，大便稀溏等为常见症的证候。

7.3.17 石阻证 lithiasis blocking pattern

体内有结石阻滞，以肚腹胀满，或腰腹压痛等为常见症状，因结石停滞部位不同而其症各具特征。

7.3.17.1 石阻气机证 lithiasis blocking qiji pattern

结石阻滞气机，以肚腹胀满疼痛，或身目发黄，或小便不畅、涩痛，或尿血等为常见症的证候。

7.3.17.2 石阻气闭证 lithiasis blocking qi stagnation pattern

结石阻闭气机，以突发右胁下、脘腹部压痛，欲呕，或腰腹绞痛，血尿等为常见症的证候。

7.3.18 真实假虚证 excess pattern with pseudo deficiency pattern

疾病本质为实证，而表现出某些类似虚弱症状的证候。

7.4 虚实夹杂证类

7.4.1 气虚挟实证 qi deficiency mixed excess pattern

气虚兼有痰湿、水饮、瘀血等邪的证候，其症除有气虚表现外，又因实邪不同而各具特征。

7.4.1.1 气虚发热证 qi deficiency mixed heat pattern

正气亏虚，阳气浮动，以低热不退，劳累更显，食欲减退，四肢无力，气短不鸣，舌淡脉虚为常见症的证候。

7.4.1.2 气虚痰结证 qi deficiency mixed sputum stagnation pattern

气虚而痰浊留结脏腑形体，以气短乏力，咳喘咯痰，瘿瘤瘰疬，苔腻脉滑等为常见症的证候。

同义词：气虚痰阻证

7.4.1.3 气虚湿困证 qi deficiency mixed dampness pattern

正气亏虚，湿邪困阻，以食欲减退、气短，精神不振，四肢无力，或腹胀腹泻等为常见症的证候。

同义词：气虚湿阻证

7.4.1.4 气虚水停证 qi deficiency mixed body fluids stagnation pattern

正气亏虚，水液停聚，以气短，精神不振、四肢无力，肢体浮肿，小便短少等为常见症的证候。

7.4.1.5 气虚寒凝证 qi deficiency mixed cold coagulation pattern

气虚血行不畅，寒邪侵袭，以神疲气短，四肢不温，肢体反应迟钝，冷痛，或挛急不舒，舌淡苔白润等为常见症的证候。

7.4.1.6 气虚气滞证 qi deficiency causing qi stagnation pattern

正气亏虚，运行无力而气机阻滞，以精神不振，呼吸气短，四肢无力，胸腹疼痛，舌淡，脉弦缓等为常见症的证候。

7.4.1.7 气虚血瘀证 qi deficiency causing blood stasis pattern

气虚运血无力，血行瘀滞，可视黏膜晦暗，精神不振，四肢无力，少气不鸣，疼痛，痛处不移，舌质淡紫，或有紫斑，脉沉涩等为常见症的证候。

同义词：气虚血凝证

7.4.2 血虚挟实证 blood deficiency mixed excess pattern

血虚兼有寒痰、风热、瘀血等邪的证候。其症除有血虚表现外，并因实邪不同而各具特征。

7.4.2.1 血虚挟瘀证 blood deficiency mixed stasis pattern

血虚而又有瘀血内阻，可视黏膜淡白，站立不稳，躁动不安，疼痛固定，舌淡紫或有斑点，脉细涩等为常见症的证候。

7.4.2.2 血虚挟痰证 blood deficiency carrying sputum pattern

血虚而又有痰浊内阻，站立不稳，躁动不安，咳吐痰浊，或肌肤不仁、瘿瘤瘰疬，苔腻脉滑等为常见症的证候。

同义词：血虚痰阻证

7.4.2.3 血虚寒凝证 blood deficiency and clod coagulation pattern

血虚而又有寒邪凝滞，血行不畅，以可视黏膜晦暗，站立不稳，唇舌紫暗，四肢不温，苔白脉沉细涩等为常见症的证候。

7.4.2.4 血虚内热证 qi deficiency causing inner heat pattern

血虚而气无依附，浮动而发热，以舌淡脉细、发热为常见症的证候。

7.4.3 阴虚挟实证 yin deficiency mixed excess pattern

阴虚兼有实热、水湿、痰浊、瘀血等邪的证候。其症除有阴虚表现外，又因实邪不同而各具特征。

7.4.3.1 阴虚内热证 yin deficiency causing inner heat pattern

阴液亏虚，虚热内生，以低热不退，口干喜饮，小便短黄，大便干结，舌红少津，脉细数等为常见症的证候。

同义词：虚热证

7.4.3.2 阴虚火旺证 deficiency of yin induces fire hyperactivity pattern

阴液亏虚，虚火亢旺，以躁动不安，口燥，盗汗遗精，小便短黄，大便干结，或咳血、衄血，或舌体、口腔溃疡，舌红少津，脉细数等为常见症的证候。

同义词：虚火证；阴虚火炽证

7.4.3.3 阴虚阳亢证 yin deficiency and yang hyperactivity pattern

阴液亏虚，阳失制约而偏亢，以潮热，站立不稳，躁动，舌红少津，脉细数等为常见症的证候。

同义词：虚阳偏亢证；虚阳偏旺证

7.4.3.4 阴虚血热证 yin deficiency and blood heat pattern

阴液亏虚，热迫血分，以低热不退，躁动不安，口渴喜饮，口干，小便短黄，大便干结，或咳血、衄血、斑疹，舌红苔黄少津，脉细数等为常见症的证候。

7.4.3.5 阴虚外感证 yin deficiency and external affections pattern

阴液亏虚，又感外邪，以发热、微恶寒，低烧不退，躁动不安，脉浮细数等为常见症的证候。

7.4.3.6 阴虚热郁证 yin deficiency and inner heat pattern

阴液亏虚，兼挟郁热，以低烧不退，躁动不安，口干，胸胁疼痛，小便短黄，舌红苔少而干，脉弦细数等为常见症的证候。

同义词：阴虚郁热证

7.4.3.7 阴虚气滞证 yin deficiency with qi stagnation pattern

阴液亏虚，兼气机阻滞，以低烧不退，躁动不安，口燥，胸胁脘腹胀满，肠鸣矢气，或嗳气，舌红，脉弦细等为常见症的证候。

7.4.3.8 阴虚血瘀证 yin deficiency with blood congestion pattern

阴液亏虚，兼挟瘀血，以低烧不退，躁动不安，口燥，午后低热，局部疼痛，或出血挟块、色紫暗，或舌有斑点，脉细涩等为常见症的证候。

同义词：阴虚瘀热证

7.4.3.9 阴虚水停证 yin deficiency mixed body fluids stagnation pattern

阴液亏虚，水饮内停，以肢体或局部水肿，小便短少，口燥，低烧不退，躁动不安，大便干结，舌红苔少，脉细数等为常见症的证候。

7.4.3.10 阴虚湿热证 yin deficiency mixed dampness heat pattern

阴液亏虚，湿热内阻，以潮热低烧不退，躁动不安，小便淋涩，口腻，舌红苔黄腻，脉细滑数等为常见症的证候。

7.4.3.11 阴虚痰热证 yin deficiency mixed sputum heat pattern

阴液亏虚，痰热内阻，以潮热低烧不退，躁动不安，咳嗽，咯痰黄稠，或痰中带血，喜饮，舌红苔黄腻，脉细滑数等为常见症的证候。

7.4.3.12 阴虚痰湿证 yin deficiency mixed sputum dampness blocking pattern

阴液亏虚，痰湿内阻，以低烧不退，躁动不安，痰多黏稠，肢体困重，或生瘰疬，舌质红，苔腻，脉细滑等为常见症的证候。

同义词：阴虚痰阻证；阴虚痰浊证

7.4.3.13 阴虚热盛证 yin deficiency mixed severe heat pattern

阴液亏虚，邪热内盛，以发热躁动不安，口干喜饮，午后热甚，小便短黄，大便干结，或干咳咯血，舌红苔少而干，脉细数等为常见症的证候。

7.4.4 阳虚挟实证 yang deficiency mixed excess pattern

阳虚兼有痰浊、水湿、瘀血、风寒等邪的证候。其症除有阳虚表现外，又因实邪不同而各具特征。

7.4.4.1 阳虚气滞证 yang deficiency mixed qi stagnation pattern

阳虚失温，气机阻滞，以畏寒肢冷，可视黏膜苍白，胸胁、脘腹胀满，肠鸣，大便溏泄，尿清长，舌淡胖，脉沉迟无力等为常见症的证候。

7.4.4.2 阳虚湿困证 yang deficiency causing dampness blocking pattern

阳气亏损，气化失常，水湿内停，以畏寒肢冷，肢体浮肿，小便不利，大便溏泻，食欲减退腹胀，舌淡胖，苔白腻或白滑，脉沉迟而滑等为常见症的证候。

同义词：阳虚湿阻证

7.4.4.3 阳虚饮停证 yang deficiency mixed body fluids stagnation pattern

阳气亏损，饮邪停聚，以畏冷肢凉，或咳嗽吐稀白痰，或心区跳动明显，搏动加快，呕吐清水，或胁肋饱胀，舌淡胖，苔白滑，脉弦滑或弦紧等为常见症的证候。

同义词：寒饮内停证

7.4.4.4 阳虚水泛证 yang deficiency causing water retention pattern

阳气亏损，不能温运、气化水液，以肢体浮肿，小便不利，心区跳动明显，搏动加快，喘促，腹胀濡泄，形寒肢冷，舌淡胖，苔白滑为常见症的证候。

同义词：阳虚水停证

7.4.4.5 阳虚血瘀证 yang deficiency causing blood stasis pattern

阳气亏损，瘀血阻滞，以畏寒肢凉，反应迟钝或痿废不用，或局部固定疼痛，肢体紫斑，出血紫暗夹块，舌淡胖或有斑点，脉沉迟而涩等为常见症的证候。

7.4.4.6 阳虚痰凝证 yang deficiency with sputum coagulation pattern

阳气亏损，痰浊凝滞，以形寒肢冷，嗜睡，痰多，体胖身重，或瘿瘤瘰疬、乳核不散，关节漫肿、僵硬、不热，苔腻脉滑等为常见症的证候。

同义词：阳虚痰阻证

7.4.4.7 阳虚寒凝证 yang deficiency with cold coagulation pattern

阳虚失温，阴寒凝滞，以形寒肢冷，胸胁、脘腹、腰膝冷痛喜温，舌淡胖，苔白滑，脉沉迟等为常见症的证候。

7.4.4.8 阳虚外感证 yang deficiency and external affections pattern

阳气亏损，复感外邪，以恶寒，微发热，汗出恶寒更甚，骨节疼痛，四肢不温，声低，舌淡胖，苔白滑，脉沉迟无力等为常见症的证候。

7.4.5 津亏热结证 body fluids loss and heat accumulating pattern

津液亏虚，热邪内结，以发热口渴，唇舌干燥，小便不利，大便秘结，躁动不宁，舌红苔黄，脉数等为常见症的证候。

7.4.6 正虚邪恋证 vital qi deficiency causing evil detaining pattern

正气亏虚，痰饮、湿热、寒湿、余热等病邪留滞所反映的证候。因正虚与病邪性质的不同而临床表现各异的证候。

同义词：正虚邪留证

7.4.6.1 气虚邪恋证 qi deficiency causing evil detaining pattern

气虚而又有病邪留滞的证候。除有精神不振、四肢无力，气短不鸣，舌淡脉弱等气虚见症外，由于留滞的病邪不同而又兼有各自的特征。

同义词：气虚邪留证

7.4.6.2 气虚毒滞证 qi deficiency causing evil toxin detaining pattern

气虚而又有火毒、热毒、湿毒、虫毒等毒邪留滞的证候。除有精神不振、四肢无力，气短不鸣，舌淡脉弱等气虚症状外，由于留滞的毒邪不同而又兼有各自的特征。

同义词：正虚毒恋证

7.4.6.3 气虚余热证 qi deficiency causing heat detaining pattern

正气亏虚而又有余热留滞，以低热不退，或夜热早凉，神疲体瘦，气短自汗，食欲不振，舌红苔薄黄，脉细数无力等为常见症的证候。

7.4.6.4 正虚毒恋证 vital qi deficiency causing evil toxin detaining pattern

正气亏虚而又有火毒、热毒、湿毒、虫毒等毒邪留滞的证候。因毒邪的性质不同而表现出各自的特征。

7.4.6.5 正虚毒陷证 vital qi deficiency causing evil toxin attacking pattern

正气亏虚，正不胜邪而致毒邪内陷，以昏迷、动风、动血、气息微弱、二便失禁等为常见症的危重证候。

7.4.6.6 阴虚邪恋证 yin deficiency causing evil toxin detaining pattern

阴液亏虚而又有邪毒留滞的证候。除有低热、盗汗、口燥等阴虚见症外，因邪毒性质不同而有各自的特征。

同义词：阴虚毒恋证

7.4.6.7 血虚邪恋证 blood deficiency causing evil toxin detaining pattern

血液亏虚而又有邪毒留滞的证候。除有唇、舌色淡，肢体感觉迟钝，脉细等血虚见症外，由于邪性质不同而有各自的特征。

同义词：血虚毒恋证

7.4.6.8 阳虚邪恋证 yang deficiency causing evil toxin detaining pattern

阳气亏虚而又有邪毒留滞的证候。除有畏寒肢冷喜暖，尿清便溏，舌淡胖，苔白滑等阳虚见症外，由于留滞的邪毒性质不同而有各自的特征。

同义词：阳虚毒恋证

7.4.6.9 阴虚余热证 yin deficiency causing heat detaining pattern

阴液亏虚，余热未尽，以形体消瘦，低热或夜热早凉，皮肤干燥，口渴，舌红苔薄黄，脉细数等为常见症的证候。

7.4.6.10 正虚脓毒证 vital qi deficiency causing sepsis detaining pattern

正气亏虚，脓毒壅滞，以患处肿痛，脓液清稀，疮口久不收敛，神疲气短，烦渴，舌红苔腐等为常见症的证候。

7.4.7 风热血燥证 wind heat causing blood dryness pattern

风热之邪损伤阴血而化燥，形体失其濡养，以皮肤干燥、瘙痒脱屑，目赤干涩，咽喉红肿，口渴，大便干结，舌红苔黄，脉细数等为常见症的证候。

同义词：风热化燥证

7.4.7.1 寒凝阳虚证 cold stagnation causing yang deficiency pattern

阴寒凝滞，损伤阳气，以形寒肢冷，畏寒喜温，精神不振，小便清长，大便清冷，局部冷痛，舌淡苔白滑，脉沉迟等为常见症的证候。

同义词：阴盛阳虚证；阴盛阳衰证

7.4.7.2 寒凝血虚证 cold stagnation and blood deficiency pattern

寒邪凝滞，血液亏虚，以面色苍白紫暗，局部冷痛，形寒肢冷，或挛急不舒，站立不稳，舌淡苔白，脉沉细等为常见症的证候。

7.4.8 暑伤津气证 summer heat injuring body fluids and qi pattern

暑热侵袭，耗气伤津，以身热汗出，口渴多饮，躁动不安，神疲气短，小便短黄，舌红苔黄干，脉浮大无力等为常见症的证候。

7.4.9 邪热伤阴证 evil heat injuring yin pattern

火热、痰热等邪耗损阴津的证候。除有身热，躁动不安，口渴欲饮，舌红，便干尿黄等阴津亏虚等常见症外，因病邪不同而又兼有各自的特征。

7.4.9.1 血热伤阴证 blood heat injuring yin pattern

温热病邪深入血分，耗伤阴液，以身热可视黏膜潮红，口舌干燥，精神不振，低烧不退，躁动不安，脉数无力等为常见症的证候。

同义词：血热阴虚证

7.4.9.2 火热伤阴证 fire heat injuring yin pattern

火热炽盛，损伤阴液，以发热，口渴喜冷饮，大便干结，小便短黄，舌红干，苔黄燥，脉细数等为常见症的证候。

同义词：热盛伤阴证

7.4.9.3 热盛伤津证 severe heat injuring body fluids pattern

火热炽盛，损伤津液，以发热，饮水多，皮肤干瘪，眼眶凹陷，大便干结，小便短黄，舌红干，苔黄燥，脉细数等为常见症的证候。

同义词：热盛耗液证

7.4.9.4 热伤营阴证 heat injuring ying yin pattern

热邪深入营血，损伤营阴，以身热夜甚，躁动不安，口渴欲饮，斑疹隐隐，便结尿黄，舌绛，脉细数等为常见症的证候。

7.4.10 痰热阴虚证 sputum heat causing yin deficiency pattern

痰热内盛，阴液亏虚，以咳喘，吐痰黄稠，或痰黏难咯，发热口渴，低烧不退，躁动不安，便结尿黄，舌红苔黄厚，脉细数等为常见症的证候。

同义词：痰热伤阴证

7.4.11 血瘀风燥证 blood stasis causing wind dryness pattern

瘀血内阻，新血不生，化燥生风，以肌肤甲错，皮肤干涩、瘙痒，肢体感觉迟钝，舌紫暗或有斑点，脉细涩等为常见症的证候。

同义词：血瘀风盛证

7.4.12 实中挟虚证 excess pattern carrying deficiency pattern

实证中兼挟有某些虚弱的症状。

7.4.13 邪陷正脱证 evil toxin attacking while healthy qi exhausting pattern

邪毒内陷，正虚欲脱，以发热口渴，躁动不宁，甚或神昏，可视黏膜苍白，四肢厥冷，冷汗淋漓，脉微欲绝等为常见症的证候。

7.4.14 内闭外脱证 excess evil blocking and healthy qi exhausting pattern

实邪内闭，正虚欲脱，以发热，咳嗽气喘，或腹痛，或二便不通，可视黏膜苍白，四肢厥冷，冷汗淋漓，气息微弱，脉微欲绝等为常见症的证候。

7.5 心系证类

7.5.1 心气虚证 heart qi deficiency pattern

心脏与心神气虚，以心区跳动明显，搏动加快，气短乏力，自汗，使役和运动后尤甚，舌淡苔白，脉虚等为常见症的证候。

同义词：心气亏虚证

7.5.2 心气虚血瘀证 heart qi deficiency causing blood stasis pattern

心气虚弱，运血无力，心脉瘀阻，以心区跳动明显，搏动加快，气短，精神疲倦，可视黏膜紫暗，舌淡紫，脉弱而涩等为常见症的证候。

7.5.3 心气血两虚证 both qi and blood deficiency in heart pattern

气血两虚，心与心神失养，以心区跳动明显，搏动加快，精神不振，身躯摇晃，舌淡，脉弱等为常见症的证候。

7.5.4 心气阴两虚证 both qi and yin deficiency in heart pattern

气阴两虚，心与心神失养，以心区跳动明显，搏动加快，呼吸气短，精神不振，身躯摇晃，舌红少苔，脉弱而数等为常见症的证候。

7.5.5 心阳虚证 yang deficiency of heart pattern

心阳虚衰，温运失司，以心区跳动明显，搏动加快，呼吸喘促，畏冷肢凉，或后肢浮肿，唇舌色暗，苔白，脉弱或结代等为常见症的证候。

同义词：心阳亏虚证

7.5.5.1 心阳暴脱证 sudden collapse of heart yang pattern

阳气突然衰败而欲脱导致心神失守、血行异常且伴亡阳的证候。

7.5.5.2 心阳虚脱证 heart yang deficiency pattern

因病久阳气虚衰，渐致心阳衰败而欲脱，在病重虚衰的基础上出现冷汗淋漓，四肢厥冷，呼吸微弱，心区跳动明显，搏动加快，可视黏膜苍白，脉微欲绝，神志模糊等常见症的证候。

7.5.6 心阳虚血瘀证 heart yang deficiency causing blood stasis pattern

心阳虚衰，运血无力，心脉瘀阻，以心区跳动明显，搏动加快，畏冷肢凉，可视黏膜紫暗，舌淡紫，脉弱而涩或结代等为常见症的证候。

7.5.7 心血虚证 heart blood deficiency pattern

血液亏虚，心与心神失养，以心区跳动明显，搏动加快，躁动不安，身躯摇晃，舌色、结膜苍白，脉细等为常见症的证候。

同义词：心血亏虚证

7.5.8 心阴血虚证 heart yin and blood deficiency pattern

阴虚血亏，心与心神失养，以心区跳动明显，搏动加快，舌红少苔，脉细数等为常见症的证候。

7.5.9 心阴虚火旺证 deficiency of heart yin induces fire hyperactivity pattern

心阴亏虚，虚热内扰，以心区跳动明显，搏动加快，潮热口渴，舌红少津，脉细数等为常见症的证候。

同义词：心阴虚阳亢证

7.5.10 心阴虚血瘀证 heart yin deficiency causing blood stasis pattern

心阴亏虚，瘀阻心脉，以心区跳动明显，搏动加快，运动无力，舌暗红或有斑点，脉细数涩或结代等为常见症的证候。

7.5.11 心阴阳两虚证 heart yin and yang deficiency pattern

心阳心阴均不足，以心区跳动明显，搏动加快，易惊，畏冷肢凉，低烧不退，身躯摇晃，舌暗红，脉结代或弱等为常见症的证候。

7.5.12 心血瘀阻证 heart blood stasis pattern

血行不畅，瘀血阻滞心脉，以心区跳动明显，搏动加快，易惊，唇舌紫暗，脉细涩或结代等为常见症的证候。

同义词：心脉瘀阻证；心血瘀滞证

7.5.12.1 心气滞血瘀证 heart qi stagnation and blood stasis pattern

气机郁滞，血行不畅，瘀阻心脉，以心区跳动明显，搏动加快，易惊，胸胁胀满，唇舌紫暗，脉涩等为常见症的证候。

7.5.12.2 心热血瘀证 heart heat stagnation and blood stasis pattern

热邪内扰，血行不畅，以发热，口渴，心区跳动明显，搏动加快，可视黏膜鲜红，心胸疼痛，胆小易惊，唇舌紫红，脉促或涩等为常见症的证候。

7.5.13 痰阻心脉证 sputum blocking heart channel pattern

痰浊阻痹心脉，血行不畅，以体胖多痰，舌淡紫，苔腻或滑，脉滑等为常见症的证候。

7.5.14 寒滞心脉证 cold evil stagnated in heart pattern

心阳不振，寒邪凝滞，心脉痹阻，以恶寒畏冷，遇寒痛增、得温痛减，苔白，脉沉迟或沉紧等为常见症的证候。

7.5.15 心脉气滞证 heart qi stagnation pattern

心脉气机阻滞，以心胸胀痛，脉弦等为常见症的证候。

同义词：气滞心脉证

7.5.16 饮停心包证 excessive body fluids stagnated in pericardium pattern

饮邪停积心包，阻滞气血运行，以心区跳动明显，搏动加快，气喘，舌淡紫苔白滑，脉沉伏或弱等为常见症的证候。

7.5.17 心火炽盛证 heart fire blazing pattern

火热炽盛，扰乱心神，以高热，大汗，甚或狂乱，气促喘粗，粪干尿少，口渴，舌红，脉洪数等为常见症的证候。

同义词：心经积热证；心火亢盛证

7.5.18 心火上炎证 upward flaming of heart fire pattern

心经火旺，循经燔灼上炎，而致心神不安、口舌生疮的证候。

7.5.19 心热阴虚证 heart heat and yin deficiency pattern

心火炽盛，阴液亏虚，以心区跳动明显，搏动加快，易惊，潮热，口渴，可视黏膜发红，舌红苔黄少津，脉数等为常见症的证候。

7.5.20 热闭心包证 heat blocking pericardium pattern

邪热炽盛而神闭，以发热口渴，神志昏迷，或狂乱，可视黏膜鲜红，呼吸气粗，舌红苔黄，脉滑数等为常见症的证候。

同义词：热陷心包证；热闭心神证

7.5.21 热扰心神证 heat disturbing mind pattern

邪热炽盛，心神被扰，以心区跳动明显，搏动加快，易惊，发热口渴，可视黏膜鲜红，舌红苔黄，脉滑数等为常见症的证候。

同义词：火扰心神证

7.5.22 热入心营证 heat attacking heart ying pattern

邪热陷入心营，以身热，渴不多饮，心区跳动明显，搏动加快，或甚至昏迷，斑疹隐隐，舌绛少苔，脉细数等为常见症的证候。

7.5.23 血热扰神证 blood heat disturbing mind pattern

血分热盛，扰乱神明，以心区跳动明显，搏动加快，易惊，身热夜甚，渴不多饮，斑疹显露，舌色深绛、少津少苔，脉细数等为常见症的证候。

7.5.24 暑热闭神证 summer heat blocking mind pattern

暑热炽盛，内闭心神，以壮热烦渴，突然昏倒，神志不清，四肢厥逆，呼吸气粗，牙关紧闭，脉沉伏等为常见症的证候。

同义词：暑入心包证；暑闭心包证

7.5.25 痰火扰神证 sputum heat disturbing mind pattern

火热痰浊交结，闭扰心神，以发热口渴，可视黏膜发红，呼吸气粗，便秘尿黄，吐痰色黄，或喉间痰鸣，蹬槽越桩，狂躁奔走，咬物伤人，舌红苔黄腻，脉滑数等为常见症的证候。

同义词：痰火闭窍证；痰热扰心证

7.5.26 痰迷心窍证 sputum blocked heart mind pattern

痰浊蒙闭心神，以神识痴呆，行如酒醉，或昏迷嗜睡，口流痰涎或喉中痰鸣，苔腻，脉滑等为常见症的证候。

同义词：痰蒙心神证；痰阻心神证

7.5.27 风痰闭神证 wind sputum blocked mind pattern

肝风挟痰浊上犯，蒙闭心神，以突然昏仆，神志昏迷，喉中痰鸣，苔腻脉滑等为常见症的证候。

同义词：风痰闭窍证

7.5.28 浊毒闭神证 toxin blocked heart apertures pattern

秽浊邪毒蒙闭心神，以感秽浊之邪后，突然神志昏迷，可视黏膜晦暗，脉沉伏等为常见症的证候。

同义词：浊毒闭窍证

7.5.29 气闭神厥证 qi stagnation causing mind syncope pattern

因精神刺激，神气郁闭，以突然昏倒，或神识不清，或牙关紧闭，或四肢抽搐，脉弦或伏等为常见症的证候。

同义词：气郁神闭证

7.5.30 心虚神怯证 heart deficiency and mind timidity pattern

心气亏虚，神气怯弱，以心区跳动明显，搏动加快，胆怯易惊，神疲，脉弱或结等为常见症的证候。

7.5.31 囊虫侵脑证 cysticercus attacking brain pattern

囊虫蕴积于脑，以头痛固定，或发作性昏倒、口吐涎沫，或四肢抽搐等为常见症的证候。

7.5.32 惊恐伤神证 scare caused mental disorder pattern

因卒惊、大恐，损伤神气，以可视黏膜苍白或青紫，心区跳动明显，搏动加快，神识痴呆，甚或昏倒，脉伏或结或动等为常见症的证候。

7.5.33 心神不宁证 nervous pattern

各种原因导致以心区跳动明显，搏动加快，躁动不安，胆怯易惊等为常见症的证候。

7.6 肺系证类

7.6.1 肺气虚证 lung qi deficiency pattern

肺气虚损，以久咳气喘，咳喘无力，动则喘甚，鼻流清涕，畏寒喜暖，易于感冒，容易出汗，日渐消瘦，皮燥毛焦，倦怠肯卧，口色淡白，脉细弱等为常见症的证候。

同义词：肺气亏虚证

7.6.2 肺气阴两虚证 both qi and yin deficiency in lung pattern

肺气虚弱，阴液亏虚，以干咳无力，气短而喘，声低或音哑，低烧不退，躁动不安，脉细无力等为常见症的证候。

7.6.3 肺阳虚证 lung yang deficiency pattern

阳气亏虚，肺失温煦，以咳嗽气喘，畏冷肢凉，吐稀白痰，苔白滑，脉弱等为常见症的证候。

同义词：肺虚寒证

7.6.4 肺阴虚证 lung yin deficiency pattern

肺阴亏虚，虚热内扰，以干咳连声，日轻夜重，或痰中带血，甚则气喘，鼻液黏稠，口干，声音嘶哑，形体消瘦，低热不退，或午后潮热，盗汗，粪球干小，舌红少津，脉细数等为常见症的证候。

同义词：肺虚热证；肺阴亏虚证

7.6.5 肺卫气虚证 lung protective qi deficiency pattern

肺气虚弱，卫表不固，以恶风自汗，时常感冒，气短乏力，舌淡脉弱等为常见症的证候。

同义词：肺虚表疏证；肺卫气不固证

7.6.6 阴虚肺燥证 yin deficiency and lung dryness pattern

阴液亏虚，肺燥失润，以口燥，干咳少痰，鼻燥，少苔少津，脉浮细数等为常见症的证候。

同义词：肺燥津亏证；肺燥阴虚证

7.6.7 肺热炽盛证 lung heat over vigorous pattern

火热炽盛，壅积于肺，以发热口渴，咳嗽，气粗而喘，或胸有压痛、咽痛，鼻气发热，便秘尿黄，舌红苔黄，脉数等为常见症的证候。

同义词：肺实热证；邪热壅肺证

7.6.8 肺热阴虚证 lung heat and yin deficiency pattern

肺热炽盛，阴液亏虚，以发热口渴，咳嗽痰少，气喘，便秘尿黄，舌红苔黄少津，脉数等为常见症的证候。

同义词：阴虚肺热证；肺热津伤证

7.6.9 肺热移肠证 lung heat caused intestinal dysfunction pattern

肺热炽盛，肠失传导，以发热口渴，咳嗽气喘，腹胀便秘，舌红苔黄，脉数或实等为常见症的证候。

7.6.10 风热犯肺证 wind heat attacking the lung pattern

风热邪气侵袭肺卫，而致肺气宣降失常的证候。

7.6.11 风热闭肺证 wind heat blocking lung pattern

风热外侵，肺气郁闭，以发热恶风，咳嗽，气粗而喘，胸压痛，鼻煽，无汗，舌红，脉浮数等为常见症的证候。

7.6.12 肺经风热证 wind heat in lung channel pattern

风热之邪侵袭肺经（系），以发热恶风，鼻塞流涕，咽喉肿痛，咳嗽气喘，脉浮数等为常见症的证候。

7.6.13 肺经郁火证 lung channel fire stagnation pattern

火热郁结肺经，以发热口渴，咽喉鲜红疼痛，咳嗽，吐黄痰，鼻气发热，舌红苔黄，脉数等为常见症的证候。

同义词：肺经郁热证

7.6.14 肺热血瘀证 lung heat and blood stasis pattern

肺热炽盛，血瘀气滞，以发热口渴，咳嗽，痰中夹血，或咯血色暗红，胸部有压痛，舌红苔黄，脉弦数等为常见症的证候。

同义词：肺热瘀滞证

7.6.15 暑伤肺络证 summer-heat injured lung channel pattern

暑热之邪损伤肺络，以发热口渴，咳嗽，咯血色鲜红，舌红苔黄，脉数无力等为常见症的证候。

7.6.16 痰热壅肺证 sputum dampness accumulating in the lung pattern

痰热互结，壅闭于肺，肺失宣降的证候。

同义词：痰火蕴肺证

7.6.17 痰热闭肺证 sputum heat blocking lung pattern

痰热内蕴，阻闭肺气，以发热口渴，胸痛，咳嗽，气粗而喘，或有哮鸣，鼻气发热，舌红苔黄腻，脉滑数等为常见症的证候。

7.6.18 痰浊阻肺证 sputum dampness accumulating in the lung pattern

痰湿壅盛，内阻于肺，肺失宣降的证候。
同义词：痰湿蕴肺证

7.6.19 痰瘀阻肺证 sputum stasis blocking lung pattern

瘀血痰浊蕴阻于肺，以咳嗽气喘，胸部疼痛，痰多或痰中夹血，舌淡紫，苔腻，脉弦滑或弦涩等为常见症的证候。
同义词：瘀痰阻肺证

7.6.20 风寒袭肺证 wind cold evil attacking lung pattern

风寒侵袭，肺气失宣，以恶寒，无汗，咳嗽，气喘，吐白痰，苔白，脉浮紧等为常见症的证候。
同义词：风寒束肺证

7.6.21 寒饮停肺证 cold fluids accumulating in lung pattern

寒饮停聚于肺，肺失肃降，以咳嗽气喘，或哮鸣有声，痰涎稀白，苔白滑，脉弦等为常见症的证候。
同义词：饮邪客肺证；肺寒饮停证

7.6.22 肺郁水停证 lung qi stagnation and body fluids stagnation pattern

肺气郁闭，宣降失常，水饮停聚，以咳嗽，吐清稀痰，头面浮肿，小便不利，苔白滑等为常见症的证候。

7.6.23 肺热饮停证 lung heat and body fluids stagnation pattern

邪热炽盛，水饮停肺，以发热口渴，咳嗽气喘，胸痛，或有哮鸣，舌红苔黄滑，脉弦数等为常见症的证候。
同义词：热饮阻肺证

7.6.24 寒痰阻肺证 cold sputum blocking lung pattern

寒痰停聚于肺，以咳嗽气喘，白痰量多，苔白滑，脉弦紧等为常见症的证候。
同义词：寒痰停肺证

7.6.25 表寒肺热证 exterior cold and lung heat pattern

寒邪外束，肺热内郁，以恶寒发热，口渴，无汗，躁动，咳嗽气喘，苔黄白，脉浮数等为常见症的证候。

7.6.26 燥邪犯肺证 dryness evil attacking lung pattern

秋燥伤津，肺失宣降，以微有寒热，干咳无痰，或痰夹血丝，口渴，舌燥少津，脉浮等为常见症的证候。

同义词：燥邪伤肺证

7.6.27 凉燥袭肺证 cool dryness attacking lung pattern

凉燥犯肺，肺失宣降，以恶寒无汗，干咳少痰，苔白少津，脉浮紧等为常见症的证候。

同义词：燥寒犯肺证

7.6.28 温燥袭肺证 warm dryness attacking lung pattern

燥热之邪侵袭，肺失宣降，以发热，口渴，干咳少痰，痰稠难咯，或咯痰带血，咽喉疼痛，口鼻干燥，舌红，苔薄黄少津，脉浮数等为常见症的证候。

同义词：温燥伤肺证

7.6.29 燥痰结肺证 dryness sputum evil blocking lung pattern

燥痰凝结，停聚于肺，以咳嗽，痰黏成块、难以咯出，胸痛，苔腻，脉弦等为常见症的证候。

同义词：燥痰阻肺证

7.6.30 肺燥郁热证 lung dryness and heat stagnation pattern

郁热内蕴，肺燥津伤，以发热口渴，咳嗽气喘，胸部发热，痰少而黏，舌红苔黄少津，脉弦数等为常见症的证候。

7.6.31 肺燥肠热证 lung dryness and intestine heat pattern

肺燥津亏，肠热腑实，以发热，口渴，咳嗽气喘，大便秘结，腹胀满，舌红苔黄燥，脉沉实或弦数等为常见症的证候。

7.6.32 肺燥肠闭证 lung dryness and intestine blocking pattern

肺燥津亏，腑气闭塞，以咳嗽，口渴，气喘，便秘不通，腹胀满，苔黄燥，脉沉实等为常见症的证候。

7.6.33 瘀阻肺络证 blood stasis blocking lung channel pattern

瘀血内停，阻滞肺络，以胸部压痛，咳嗽，咯血色暗红或成块，舌紫暗或有斑点，脉弦涩等为常见症的证候。

同义词：瘀血停肺证；瘀血乘肺证

7.6.34 热毒闭肺证 heat toxin blocking lung pattern

火热毒邪炽盛，阻闭肺气，以发热肢厥，口渴，咳嗽，气粗而喘，鼻翼煽动，鼻气发热，舌红苔黄，脉数等为常见症的证候。

同义词：火毒闭肺证

7.6.35 虫毒犯肺证 parasite toxin attacking lung pattern

虫毒外侵，上袭于肺，以微有寒热，咳嗽，或咯痰带血丝等为常见症的证候。

同义词：虫毒侵肺证

7.6.36 气郁伤肺证 qi stagnation caused lung injuring pattern

气机郁滞，肺失宣降，以精神沉郁，咳嗽等为常见症的证候。

7.7 脾系证类

7.7.1 脾气虚证 spleen qi deficiency pattern

气虚脾失健运，以草料迟细，体瘦毛焦，倦怠喜卧，肚腹虚胀，肢体浮肿，尿短，粪稀，口色淡黄，舌苔白，脉缓弱等为常见症的证候。

同义词：脾气亏虚证

7.7.2 脾气下陷证 sinking of spleen qi pattern

脾气虚弱，中气下陷，以久泻不止，脱肛或子宫脱或阴道脱，并伴有体瘦毛焦，倦怠肯卧、多卧少立，草料迟细，腹胀，口色淡白，苔白，脉虚等为常见症的证候。

同义词：脾虚气陷证；中气下陷证

7.7.3 脾气郁结证 spleen qi stagnation pattern

脾失健运，气机郁结，以食欲减退，腹胀，便溏，脉弦等为常见症的证候。

7.7.4 脾不统血证 spleen failing to control the blood pattern

脾气虚弱，不能统摄血液、血溢脉外的证候。

同义词：脾不摄血证

7.7.5 脾阳虚证 spleen yang deficiency pattern

脾阳虚衰，失于温运，以腹胀，食欲减退，腹痛喜温、喜按，畏冷肢凉，大便稀溏，或四肢水肿，舌淡苔白润，脉沉迟无力等为常见症的证候。

同义词：脾阳虚衰证；脾虚寒证

7.7.6 脾阴虚证 spleen yin deficiency pattern

阴液亏虚，脾失健运，以食欲不振，腹胀便结，体瘦倦怠，涎少唇干，低热，舌红少苔，脉细数等为常见症的证候。

同义词：脾阴亏虚证

7.7.7 脾虚营亏证 spleen ying deficiency pattern

脾失健运，营气亏虚，以食欲减退，腹胀，大便稀溏，形体消瘦，疲倦乏力，舌淡脉弱等为常见症的证候。

7.7.8 脾虚血亏证 spleen deficiency and blood depletion pattern

脾气虚弱，生血不足，以食欲减退，腹胀，便溏，四肢无力，精神不振，舌淡，脉细无力等为常见症的证候。

7.7.9 脾虚血燥证 spleen deficiency caused blood dryness pattern

脾气虚弱，血亏失濡，以食欲减退，腹胀，大便干结，舌淡脉细无力等为常见症的证候。

7.7.10 脾虚气滞证 spleen deficiency caused qi stagnation pattern

脾气虚弱，气机阻滞，以食欲减退，腹胀，便溏，肠鸣矢气，精神不振、四肢无力，脉弦等为常见症的证候。

7.7.11 脾虚水泛证 spleen deficiency caused water retention pattern

脾气虚弱，运化失职，水液内停，以食欲减退，腹胀，便溏，面浮肢肿，或有腹水，精神不振、四肢无力，可视黏膜苍白，舌淡胖，苔白滑，脉濡或弱等为常见症的证候。

同义词：脾气虚水泛证；脾气虚水停证

7.7.12 脾阳虚水泛证 spleen yang deficiency caused water retention pattern

脾阳虚衰，温运失职，水液内停，以食欲减退，腹胀，便溏，畏冷肢凉，面浮肢肿，或有腹水，可视黏膜苍白，舌淡胖，苔白滑，脉濡或弱等为常见症的证候。

同义词：脾阳虚水停证

7.7.13 脾虚湿困证 spleen deficiency with dampness pattern

脾气虚衰，运化失司而致湿浊内停、虚实夹杂的证候。

同义词：脾虚挟湿证；脾虚湿泛证

7.7.14 脾虚湿热证 spleen deficiency and dampness heat stagnation pattern

脾气虚弱，湿热内蕴，以食欲减退，腹胀，便溏，身热不扬，身体困重，舌红胖，苔黄滑，脉滑数等为常见症的证候。

7.7.15 脾虚痰湿证 spleen deficiency and sputum dampness stagnation pattern

脾气虚弱，痰湿内蕴，以食欲减退，腹胀，便溏，体肥困重，精神不振，嗜睡，舌淡胖，苔白腻，脉濡缓等为常见症的证候。

7.7.16 脾虚食积证 spleen deficiency caused food stagnating pattern

脾失健运，食积胃肠，以肚腹虚胀，经常腹泻，嗳腐吐酸，腹泻，大便腐臭，口淡苔腻等为常见症的证候。

同义词：脾虚夹食证

7.7.17 脾虚虫积证 spleen deficiency and parasite stagnation pattern

脾气虚弱，虫积肠道，以食欲减退，腹胀，便溏，体瘦疲乏，大便排虫，舌淡苔白等为常见症的证候。

7.7.18 湿热蕴脾证 damp heat affecting the spleen pattern

脾失健运，水湿停滞，湿蕴化热，湿热郁蒸的证候。

同义词：脾经湿热证；湿热困脾证

7.7.19 脾经热毒证 heat toxin in spleen channel pattern

火热邪蕴结脾经，以发热口渴，口腔或口唇赤烂疼痛，便秘尿黄，舌红苔黄，脉数等为常见症的证候。

7.7.20 寒湿困脾证 cold dampness affecting the spleen pattern

寒湿内盛，困阻脾阳，运化失职的证候。

同义词：湿困脾阳证；太阴寒湿证

7.7.21 思伤脾气证 worry injuring spleen qi pattern

思虑太过，脾气郁滞，以神情呆滞，不思饮食，胸胁脘腹作胀，大便黏滞，脉弦等为常见症的证候。

同义词：思虑伤脾证

7.7.22 胃气虚证 stomach qi deficiency pattern

胃气虚弱，纳运失司，以胃部胀满、喜按，食欲不振，或食后痛缓，精神不振，舌淡嫩，脉弱等为常见症的证候。

同义词：胃气亏证

7.7.23 胃气阴两虚证 both stomach qi and yin deficiency pattern

胃气虚弱，胃阴亏虚，以胃部胀满、喜按，饥不欲食，或干呕呃逆，口微渴，便结，脉弱等为常见症的证候。

同义词：胃气阴两亏证

7.7.24 胃气虚血瘀证 stomach qi deficiency and blood stasis pattern

胃气虚弱，瘀血阻滞于胃，以胃部胀满、疼痛、拒按，食欲不振，精神不振，舌淡或有斑点，脉细涩等为常见症的证候。

7.7.25 胃气上逆证 ascending of stomach qi pattern

胃的通降功能障碍，胃气下降不及，反而上逆的证候。

7.7.25.1 胃热气逆证 stomach heat and qi upward inversion pattern

火热之邪侵袭，胃失和降而上逆，以呕吐或呃逆、嗳气，胃脘疼痛，口渴，舌红苔黄，脉数等为常见症的证候。

同义词：胃火气逆证

7.7.25.2 胃寒气逆证 stomach cold and qi upward inversion pattern

寒冷侵袭，胃气失于和降而上逆，以呕吐或呃逆、嗳气，胃脘疼痛，苔白，脉弦紧等为常见症的证候。

7.7.25.3 胃滞气逆证 stomach stagnation and qi upward inversion pattern

外邪侵袭或采食不当，胃气失于和降而上逆，以呕吐或呃逆、嗳气，精神沉郁，肚腹胀满，脉弦等为常见症的证候。

7.7.26 胃阳虚证 stomach yang deficiency pattern

阳气虚衰，胃失温煦，食欲减退，脘痞，形寒肢凉，舌淡苔白，脉沉迟无力等为常见症的证候。

同义词：胃虚寒证；胃阳亏虚证

7.7.26.1 胃阳虚气滞证 stomach yang deficiency and qi stagnation pattern

阳气虚衰，气机阻滞，胃失温煦和降，以肚腹胀满，冷痛喜温，呃逆嗳气，食欲减退，便溏，形寒肢凉，舌淡苔白，脉沉迟无力等为常见症的证候。

7.7.26.2 胃阳虚血瘀证 stomach yang deficiency and blood stasis pattern

阳气虚衰，瘀血阻滞于胃，以胃脘疼痛、拒按，食欲减退，脘痞，形寒肢凉，舌淡紫或有斑点，苔白，脉沉迟而涩等为常见症的证候。

7.7.27 胃气滞血瘀证 stomach qi stagnation and blood stasis pattern

气机不畅，瘀血阻滞于胃，以胃脘胀满，疼痛拒按，呕恶呃逆，舌紫或有瘀斑瘀点，脉涩等为常见症的证候。

7.7.28 胃阴虚证 stomach yin deficiency pattern

阴液亏虚，胃失濡润、和降，以体瘦毛焦，皮肤松弛，食欲减退，口干舌燥，粪球干小，尿少色浓，口色红，苔少或无苔，脉细数等为常见症的证候。

同义词：胃虚热证；胃阴亏虚证

7.7.29 胃阴虚气滞证 stomach yin deficiency and qi stagnation pattern

胃阴亏虚，气机阻滞，以口燥，食欲减退，胃脘胀满，或干呕呃逆，便结，舌红少津，脉弦细等为常见症的证候。

7.7.30 胃阴虚血瘀证 stomach yin deficiency and blood stasis pattern

胃阴亏虚，瘀血阻滞于胃，以口燥，饥不欲食，胃脘压痛，便结，舌暗红或有斑点，脉细涩等为常见症的证候。

7.7.31 胃火证 stomach heat pattern

火热炽盛，壅滞于胃，以耳鼻温热，草料迟细，粪球干小而尿少，口干舌燥，口渴贪饮，口腔腐臭，齿龈肿痛，口色鲜红，舌有黄苔，脉洪数等为常见症的证候。

同义词：胃实热证；胃火炽盛证

7.7.32 胃热气滞证 stomach heat and qi stagnation pattern

胃热炽盛，气机阻滞，以肚腹胀满、拒按，嗳气，口臭，便结，舌红苔黄，脉弦数等为常见症的证候。

7.7.33 胃热津伤证 stomach heat injuring body fluids pattern

胃热炽盛，津液亏损，以腹痛，口渴欲饮，或消谷善饥，大便干结，舌红苔黄少津，脉数等为常见症的证候。

同义词：胃热津亏证

7.7.34 胃热阴虚证 stomach heat and yin deficiency pattern

胃热炽盛，阴液亏损，以腹痛，口渴，低烧不退，躁动不安，便结，舌红少苔少津，脉细数等为常见症的证候。

7.7.35 胃燥津伤证 stomach dryness injuring body fluids pattern

津液耗损，胃失濡润，以饥不欲食，口渴，便结，舌干少津等为常见症的证候。

同义词：胃燥津亏证

7.7.36 寒邪犯胃证 cold evil attacking stomach pattern

寒邪侵袭胃腑，胃失和降，以胃脘冷痛，痛势急剧，喜温，肢冷，苔白，脉弦紧等为常见症的证候。

同义词：痰饮停胃证

7.7.37 寒饮停胃证 cold water accumulating in stomach pattern

寒性水液输布障碍，停积于胃，以胃脘痞胀，胃中有振水声，呕吐清水稀涎，苔白滑，脉弦等为常见症的证候。

7.7.38 瘀阻胃络证 blood stasis blocking stomach channel pattern

瘀血阻滞胃络，以胃脘疼痛、拒按，或胃脘部触及包块，或呕血，色暗成块，舌有斑点，脉弦涩等为常见症的证候。

同义词：胃脘瘀血证

7.7.39 寒滞肠道证 cold stagnated intestine pattern

寒邪侵袭肠道，传化失常，以腹部冷痛、痛势急剧、喜温，腹泻清稀，肢冷，苔白，脉弦紧等为常见症的证候。

7.7.40 肠道湿热证 intestine dampness heat pattern

湿热内蕴，阻滞肠道，以腹胀腹痛，暴注下泻，或下痢脓血，或腹泻、粪质黏稠腥臭，肛门发热，身热口渴，尿短黄，舌红苔黄腻，脉滑数等为常见症的证候。

同义词：大肠湿热证

7.7.41 肠道实热证 intestine excess heat pattern

肠道热盛，腑气不通，以发热，腹痛起卧，泻痢腥臭，甚则脓血混杂，口干舌燥，口渴贪饮，尿液短赤，口色红黄，舌苔黄腻或黄干，脉滑数等为常见症的证候。

7.7.41.1 小肠实热证 excess heat in the small intestine pattern

心火炽盛，循经移热小肠，小肠热盛，泌别清浊功能失调的证候。

7.7.41.2 肠热阴虚证 intestine heat and yin deficiency pattern

热侵肠道，阴津亏虚，传化失常，以口渴，潮热，大便干结难下，唇舌、皮肤干燥，小便短黄，苔少而干，脉细数等为常见症的证候。

7.7.41.3 肠热气滞证 intestine heat and qi stagnation pattern

热侵肠道，气机阻滞，传导失常，以发热口渴，腹胀便结，或腹泻，矢气频作，肛门坠胀，舌红苔黄，脉弦数等为常见症的证候。

7.7.42 大肠热结证 heat tangling in the large intestine pattern

大肠邪热炽盛，气机阻滞，腑气不通，燥屎内结而迫津下泄的证候。

同义词：肠热腑实证；热结肠燥证

7.7.43 热毒蕴肠证 heat toxin accumulating in intestine pattern

毒邪蕴结肠道，以腹痛腹胀，便秘或腹泻，或便脓血腥臭等为常见症的证候。

7.7.44 肠道津亏证 intestine dryness and fluids exhausting pattern

津液亏损，肠失濡润，以粪球干小而硬，或粪便秘结干燥，努责难下，舌干少津，脉弦涩等为常见症的证候。

同义词：大肠津亏证；肠燥津亏证

7.7.45 血虚肠燥证 blood deficiency and intestine dryness pattern

血液亏虚，肠失濡润，以大便干结、努责难下，或有便血，舌淡，脉细涩等为常见症的证候。

同义词：血虚肠结症

7.7.46 阴虚肠燥证 yin deficiency and intestine dryness pattern

阴液亏虚，肠失濡润，以大便干结、努责难下，粪球干小而硬，口鼻、皮肤干燥，舌红少津，脉细数涩等为常见症的证候。

同义词：肠燥阴虚证

7.7.47 血热肠燥证 blood heat and intestine dryness pattern

血分热盛，耗伤阴液，肠道失濡，以发热口渴，大便干燥、秘结，甚或便血，舌红绛少津，脉细数等为常见症的证候。

7.7.48 湿阻肠道证 dampness blocking intestine pattern

湿邪蕴阻肠道，传化失常，以腹胀，大便濡泻，便质黏垢而腥臭，苔白滑，脉濡缓等为常见症的证候。

同义词：湿滞肠道证；湿蕴肠道证

7.7.49 肠道寒湿证 intestine cold dampness pattern

寒湿之邪蕴阻肠道，以腹胀疼痛、喜温，腹泻便质清稀，形寒肢凉，苔白滑，脉弦紧或濡缓等为常见症的证候。

7.7.50 肠道气滞证 qi stagnation in intestine pattern

肠道气机阻滞，以腹部胀满疼痛，肠鸣矢气，脉弦等为常见症的证候。

同义词：小肠气滞证

7.7.51 虫积肠道证 parasite accumulating in intestine pattern

蛔虫等寄生肠道，耗吸营养，阻滞气机，以腹部阵痛，痛时有块，贪食而瘦，或大便排虫等为常见症的证候。

同义词：虫积小肠证

7.7.52 虫扰魄门证 parasite gathering at anus pattern

蛲虫侵扰肛门，以肛门瘙痒，发现米粒样虫体等为常见症的证候。

7.7.53 风伤肠络证 wind injuring intestine channel pattern

风邪伤及肠道络脉，以大便下血，大便中夹泡沫等为常见症的证候。

同义词：肠风伤络证

7.7.54 肛门热毒证 heat toxin at anus pattern

热毒蕴聚肛门，以肛门红肿疼痛，甚至化脓、溃烂、流脓血，或直肠翻出、紫暗糜烂，舌暗红，苔黄，脉数等为常见症的证候。

7.7.55 肛门湿热证 anus dampness heat pattern

湿热之邪蕴结肛门，以肛门起丘疹、瘙痒、湿烂，舌红苔黄，脉滑数等为常见症的证候。

7.7.56 气血瘀滞肛门证 qi and blood stagnating at anus pattern

肛门处气血瘀滞，以肛门发生肿块，或肿胀、紫暗，脉弦涩等为常见症的证候。

7.7.57 脾胃气虚证 spleen and stomach qi deficiency pattern

脾胃气虚，中焦失运，以食欲不振，腹胀，大便溏薄，神疲，肢体倦怠，舌淡脉弱等为常见症的证候。

7.7.58 脾胃阴虚证 yin deficiency of the spleen and stomach pattern

脾胃阴液亏损，失于濡润，受纳运化功能减退，虚热内扰的证候。

同义词：中焦虚热证

7.7.59 脾胃阳虚证 spleen and stomach yang deficiency pattern

脾胃阳气虚衰，失于温运，以腹胀，食欲减退，脘腹冷痛喜温、喜按，形寒肢凉，大便稀溏，舌淡苔白润，脉沉迟无力等为常见症的证候。

同义词：中焦虚寒证

7.7.60 脾胃阳虚气滞证 spleen and stomach yang deficiency causing qi stagnation pattern

脾胃阳气虚衰，气机阻滞，以食欲减退、肚腹冷痛、喜温、嗳气肠鸣，形寒肢凉，大便稀溏，舌淡苔白润，脉沉弦等为常见症的证候。

同义词：中焦虚寒气滞证

7.7.61 脾胃气阴两虚证 deficiency of qi and yin in spleen and stomach pattern

脾胃气虚，阴液亏损，以肚腹胀满疼痛，食欲不振，或干呕呃逆，口微渴，便结，脉弱等为常见症的证候。

同义词：脾胃气阴两亏证

7.7.62 脾胃实热证 spleen and stomach excess heat pattern

火热炽盛，壅滞脾胃，以胃脘痛、喜冷、发热口渴，腹胀满，便秘，或口臭、口腔赤烂疼痛、牙龈肿痛、齿衄，舌红苔黄，脉数等为常见症的证候。

同义词：中焦积热证；中焦热盛证

7.7.63 脾胃湿热证 damp heat in the spleen and stomach pattern

湿热内蕴中焦，阻碍脾胃气机，纳运失司，升降失常的证候。

同义词：湿热中阻证；中焦湿热证

7.7.64 湿困脾胃证 dampness blocking spleen and stomach pattern

湿浊内盛，困阻中焦，以脘腹，口腻纳呆，欲呕，口淡不渴，腹痛便溏，或身目发黄而晦暗，舌淡胖，苔白腻，脉濡缓等为常见症的证候。

7.7.65 胃热脾虚证 stomach heat and spleen deficiency pattern

胃有郁热，脾气虚弱，以烦热口渴，食欲不振，腹痛，腹胀，便溏或结，消瘦疲乏等为常见症的证候。

7.7.66 脾胃不和证 spleen and stomach qi disharmonyp pattern

气机阻滞，脾胃失健，以食欲不振，腹胀，便溏，嗳气肠鸣，脉弦等为常见症的证候。

同义词：中焦气滞证

7.7.67 胃肠湿热证 stomach and intestine dampness heat pattern

湿热内蕴，阻滞胃肠，以脘腹痞胀，纳呆，便溏，或下痢脓血，或呕吐、腹泻如注，发热口渴，舌红苔黄腻，脉滑数等为常见症的证候。

7.7.68 胃肠实热证 stomach and intestine excess heat pattern

火热炽盛，壅滞胃肠，以胃脘痛、喜冷、发热、口渴、口臭、腹胀作痛、大便秘结、腐臭、小便短黄、舌红苔黄、脉数等为常见症的证候。

同义词：胃肠积热证

7.7.69 食滞胃肠证 stomach and intestine food stagnation pattern

饮食停滞胃肠，以脘腹痞胀疼痛，厌食，呕吐馊食，肠鸣矢气，泻下，便臭如败卵，苔厚腻，脉滑或沉实等为常见症的证候。

同义词：食积胃肠证

7.7.70 食滞胃热证 food stagnation in stomach and intestine pattern

食滞胃肠，胃热壅盛，以胃脘痞胀，或呕酸馊苦水，腹泻，便质腐臭如败卵，舌红苔厚腻，脉滑数等为常见症的证候。

7.7.71 胃肠气滞证 stomach and intestine qi stagnation pattern

胃肠气机阻滞，以胃脘或腹部痞胀、胀痛，嗳气、肠鸣、矢气，脉弦等为常见症的证候。

同义词：气滞胃肠证

7.7.72 瘀滞胃肠证 blood stasis blocking stomach and intestine pattern

瘀血阻滞胃肠，以胃脘、腹部疼痛、拒按，或触及包块，或呕血、便血色暗成块，舌有斑点，脉弦涩等为常见症的证候。

同义词：肠胃瘀滞证；肠胃血瘀证

7.7.73 寒滞胃肠证 cold evil attacking stomach and intestine pattern

寒邪侵袭胃肠，阻滞气机，以胃脘、腹部冷痛，痛势急剧、喜温，呕吐，腹泻清稀，肢冷，苔白，脉弦紧等为常见症的证候。

7.7.74 痰湿中阻证 sputum dampness blocking zhongjiao pattern

痰湿内蕴，阻滞胃肠，以纳呆，欲呕，脘腹痞胀，大便清稀，舌淡胖，苔白腻，脉濡缓等为常见症的证候。

同义词：痰浊中阻证；痰饮中阻证

7.7.75 肠道瘀滞证 intestinal stasis pattern

肠道瘀血内阻，气机郁滞，以腹部疼痛，胀满肠鸣，大便黏滞，便血或色黑，脉涩或弦等为常见症的证候。

7.8 肝系证类

7.8.1 肝阴虚证 liver yin pattern

阴液亏虚，肝失濡润，以两目干涩，视力障碍，或胁肋疼痛，低烧不退或午后低热，躁动不安，舌红少苔，脉细数等为常见症的证候。

同义词：肝虚热证；肝阴亏虚证

7.8.2 肝血虚证 liver blood pattern

血液亏虚，肝失濡养，以视力障碍，或夜盲，或肢体反应迟钝，爪甲、舌色淡，脉细等为常见症的证候。

同义词：肝血亏虚证

7.8.3 肝气虚证 deficiency of liver qi pattern

气虚肝失疏泄，以两胁胀满，精神沉郁，疲乏气短，舌淡脉弱等为常见症的证候。

同义词：肝气亏虚证

7.8.4 肝阳虚证 deficiency of liver yang pattern

阳气虚弱，肝失条达，以两胁胀满，形寒肢凉，苔白润，脉沉迟无力等为常见症的证候。

同义词：肝虚寒证；肝阳亏虚证

7.8.5 肝阳上亢证 hyperactivity of the liver yang pattern

肝肾阴虚，阴不制阳，阳亢于上，导致本虚标实、上盛下虚的证候。

同义词：肝阳上扰证；肝阳亢盛证

7.8.6 肝阳暴亢证 sudden hyperactivity of the liver yang pattern

肝阳亢盛，迫扰于上，以阵发头痛，目赤，急躁暴怒，站立不稳，口干，舌红苔黄，脉弦数等为常见症的证候。

同义词：肝风暴亢证

7.8.7 肝阴虚阳亢证 deficiency of liver yin and hyperactivity of liver yang pattern

阴液亏虚，肝阳偏亢，以腰痛，肢体感觉迟钝，低烧不退，躁动不安，易怒，口干，舌红少苔，脉细数等为常见症的证候。

同义词：阴虚肝旺证

7.8.8 肝郁证 stagnation of liver qi pattern

肝失疏泄，气机郁滞，以精神沉郁，胸胁，或少腹疼痛，乳房胀满，脉弦等为常见症的证候。

同义词：肝气郁结证；肝气滞证

7.8.9 肝郁血虚证 stagnation of liver qi and blood deficiency pattern

血液亏虚，肝气郁滞，以两胁作胀，精神沉郁，舌淡紫，脉弦细等为常见症的证候。

同义词：血虚肝郁证

7.8.10 肝郁血瘀证 stagnation of liver qi and blood stasis pattern

肝气郁结，血瘀于肝，以两胁胀满或疼痛，或胁下、少腹有肿块，精神沉郁，舌紫暗或有斑点，脉弦涩等为常见症的证候。

同义词：肝血瘀滞证；肝瘀气滞证

7.8.11 肝郁阴虚证 stagnation of liver qi and yin deficiency pattern

肝气郁结，肝阴亏虚，以两胁胀满或有触痛，低烧不退，躁动不安，口干，精神沉郁，舌暗红少苔，脉弦细等为常见症的证候。

同义词：阴虚肝郁证

7.8.12 肝瘀化热证 liver stasis transforming heat pattern

肝经血瘀，淤久化热，以两胁疼痛，或胁下有痞块、拒按，口干，舌紫暗或有斑点，苔黄，脉弦涩等为常见症的证候。

7.8.13 肝瘀证 blood stasis pattern

瘀血阻滞肝络，以胁肋疼痛、拒按，或胁下包块，舌紫暗或有斑点，脉弦涩等为常见症的证候。

同义词：肝经血瘀证；瘀滞肝络证

7.8.14 肝虚血瘀证 liver deficiency and blood stasis pattern

血行瘀滞，肝血亏虚，以两胁疼痛，舌淡紫，脉细涩等为常见症的证候。

同义词：血虚肝瘀证；肝瘀血虚证

7.8.15 肝阴虚血瘀证 liver yin deficiency and blood stasis pattern

阴液亏虚，肝经血瘀，以低烧不退，躁动不安，胁肋发热、疼痛，口干，便结，舌紫暗或有斑点，苔少，脉细数涩等为常见症的证候。

7.8.16 肝气虚血瘀证 liver qi deficiency and blood stasis pattern

肝气虚弱，血行瘀滞，以两胁胀满，偶有疼痛，或胁下有痞块，胸闷，舌淡紫，脉弱而涩等为常见症的证候。

7.8.17 肝郁化火证 liver depression transforming into fire pattern

肝气郁滞，郁热内蕴，以两胁胀痛、发热，躁动易怒，口干，舌红苔黄，脉弦数等为常见症的证候。

同义词：肝经郁热证；肝气化火证

7.8.18 肝郁血热证 liver qi stagnation and blood heat pattern

肝气郁滞，血热内扰，以两胁胀满疼痛、发热，躁动易怒，口干，舌绛少苔，脉弦数等为常见症的证候。

7.8.19 肝郁痰火证 liver qi stagnation and sputum heat pattern

肝气郁滞，痰热内蕴，以两胁胀痛、发热，躁动易怒，身体摇晃，吐黄痰，舌红苔黄腻，脉弦数等为常见症的证候。

同义词：痰热郁肝证；肝郁痰热证

7.8.20 肝瘀痰结证 liver qi stagnant and sputum stasis pattern

瘀痰蕴结于肝，以胁下痞块，胀满疼痛拒按，舌紫或有斑点，苔腻，脉弦涩等为常见症的证候。

同义词：肝瘀痰阻证

7.8.21 肝火炽盛证 liver fire blazing pattern

火热炽盛，内扰于肝，以胁肋疼痛，口干，或呕吐苦水，急躁不安，目赤，便秘尿黄，舌红苔黄，脉弦数等为常见症的证候。

同义词：肝火旺盛证；肝火充盛证

7.8.22 肝火上炎证 upward flaming of liver heat pattern

肝火炽盛，气火上冲，循经上攻头目，气血壅盛络脉的证候。

7.8.23 肝经火旺证 fire hyperactivity in liver channel pattern

火热炽盛，壅滞肝经，以胁肋发热，或头目赤肿胀，口干，舌红苔黄，脉弦数等为常见症的证候。

同义词：肝经火盛证

7.8.24 肝经风热证 wind heat affecting the liver channel pattern

风热邪气，侵袭肝经，循经侵扰头目的证候。

7.8.25 肝经湿热证 damp heat affecting the liver channel pattern

湿热邪气，蕴结于肝，肝气郁滞，循经下注的证候。

7.8.26 肝郁湿热证 stagnation of liver qi and dampness heat pattern

湿热内蕴，肝气郁滞，以两胁胀痛，胁下痞块，或身目发黄，口渴，舌红苔黄腻，脉滑数等为常见症的证候。

同义词：肝滞湿热证

7.8.27 热毒淤肝证 heat toxicity and liver stasis pattern

火热毒邪，淤滞于肝，以胁胀疼痛，或胁下有肿块，壮热，口渴，身目深黄，甚至昏迷，舌红苔黄，脉弦数等为常见症的证候。

同义词：热毒瘀肝证

7.8.28 肝热气滞证 liver heat and qi stasis pattern

邪热内蕴，肝气郁滞，以两胁胀满疼痛、发热，口干，舌红苔黄，脉弦数等为常见症的证候。

7.8.29 肝热血瘀证 liver heat and blood stasis pattern

邪热内蕴，肝血瘀滞，以两胁疼痛、发热，口干，舌暗红苔黄，脉涩而数等为常见症的证候。

7.8.30 肝热阴虚证 liver heat and yin deficiency pattern

邪热内蕴，肝阴亏虚，以两胁发热，低烧不退，躁动不安，躁动易怒，口干，舌红少苔，脉弦细数等为常见症的证候。

7.8.31 寒滞肝脉证 cold stagnating in the liver vessel pattern

寒邪侵袭，凝滞肝经，以少腹冷痛，或阴器硬肿发凉，遇寒痛增，得温痛缓，后肢运步困难，呕吐清涎，苔白，脉弦紧等为常见症的证候。

同义词：肝经实寒证；肝寒证

7.8.32 肝风内动证 internal stirring of liver wind pattern

肝之阴阳气血失调，体内阳气亢逆变动而为风的证候。

7.8.32.1 肝阳化风证 liver yang transforming into wind pattern

肝阳上亢，肝风内动，以神昏似醉，站立不稳，时欲倒地或头向左或向右盘旋不停，偏头直颈，歪唇斜眼，肢体感觉迟钝，拘挛抽搐，舌质红，脉弦数等为常见症的证候。

7.8.32.2 肝热动风证 liver heat stirring wind pattern

邪热炽盛，热极动风，以高热口渴，四肢痉挛抽搐，项强，甚则角弓反张，撞墙，圆圈运动，舌质红绛，脉弦数等为常见症的证候。

同义词：热动肝风证

7.8.32.3 肝阴虚动风证 liver yin deficiency and stirring wind pattern
肝阴亏损，虚风内动，以形体消瘦，肢体抽搐或震颤，四肢蠕动，午后潮热，低烧不退，躁动不安，体瘦，口干，舌红少苔少津，脉细数等为常见症的证候。

7.8.32.4 肝血虚动风证 liver blood deficiency and stirring wind pattern
肝血亏虚，虚风内动，以站立不稳，蹄壳干枯皱裂，肢体感觉迟钝，震颤，四肢拘挛抽搐，口色淡白，脉细等为常见症的证候。

7.8.33 胆气虚证 gallbladder qi deficiency pattern
胆气亏虚，心神不宁，以胆怯易惊，恐惧，舌淡脉弱等为常见症的证候。
同义词：胆虚气怯证；胆气亏虚证

7.8.34 胆郁痰扰证 depressed gallbladder with harassing sputum pattern
痰浊内扰，胆郁失宣，以躁动不宁，胆怯易惊，恶心欲呕，口吐痰涎，苔白腻，脉弦缓等为常见症的证候。

7.8.35 胆热痰扰证 bile heat and sputum disturbance pattern
痰热内扰，胆失疏泄，以躁动不宁，胆怯易惊，胸胁胀满，舌红苔黄腻，脉弦数等为常见症的证候。
同义同：胆经痰火证

7.8.36 虫扰胆膈证 parasite disturbing biliary and diaphragmatic pattern
蛔虫内扰，上窜胆膈，以阵发性腹部剧痛，呕吐苦水或吐蛔虫，脉弦等为常见症的证候。
同义词：虫扰胆腑证

7.8.37 肝胆湿热证 liver gallbladder dampness heat pattern
湿热内蕴，肝胆疏泄失常，以黄疸鲜明如橘色，尿液短赤或黄而浑浊，母畜带下黄臭，外阴瘙痒，公畜睾丸肿胀热痛，阴囊湿疹，舌苔黄腻，脉弦数等为常见症的证候。

7.8.38 肝胆火旺证 fire hyperactivity in liver and gallbladder pattern
火热炽盛，内扰肝胆，以胁肋胀痛，急躁多怒，口干，听力减退，舌红苔黄，脉弦数等为常见症的证候。
同义词：肝胆郁热证；肝胆火热证

7.8.39 胆经郁热证 stagnant heat of gallbladder channel pattern
情志内郁化热，或邪居少阳、热郁胆经、上扰心神的证候。

7.8.40 肝胆寒湿证 dampness and cold in liver and gallbladder pattern

寒湿内侵，肝胆疏泄不利，以黄疸晦暗如烟熏，食欲减退便溏，舌苔滑腻，脉沉迟为常见症的证候。

7.8.41 肝胆瘀滞证 liver gallbladder stasis pattern

肝胆气滞血瘀，以胁肋胀痛、固定不移，或胁下有肿块、拒按，舌紫暗或有斑点，脉弦涩等为常见症的证候。

7.8.42 肝胆气滞证 liver gallbladder qi stagnation pattern

肝胆气机郁滞，以胁肋、乳房及少腹触痛，精神沉郁，脉弦等为常见症的证候。

7.9 肾系证类

7.9.1 肾气虚证 kidney qi deficiency pattern

肾气虚弱，固摄无权，以腰胯无力，性欲减退，舌淡苔白，脉弱等为常见症的证候。
同义词：肾气亏虚证

7.9.2 肾气不固证 kidney qi insecurity pattern

肾气亏虚，固摄无权，以小便频数而清，余沥不尽，遗尿，小便失禁或大便失禁，公畜遗精，早泄，母畜胎动易滑，腰胯无力，脉弱等为常见症的证候。
同义词：下元不固证；肾气虚证

7.9.3 肾虚水泛证 water retention due to kidney deficiency pattern

肾阳虚衰，气化失常，体内水液代谢障碍，水湿泛溢的证候。
同义词：肾虚水停证

7.9.3.1 肾气虚水泛证 kidney qi deficiency and water retention pattern

肾气亏虚，气化无权，水液泛溢，以水肿后肢为甚，尿少，腰胯无力，舌淡胖，苔白滑，脉弱等为常见证候。
同义词：肾气虚水停证

7.9.3.2 肾阳虚水泛证 water retention due to deficiency of kidney yang pattern

肾阳虚衰，气化无权，水液泛滥，以畏冷肢凉，水肿，腹胀，小便短少，舌淡胖，苔白滑，脉沉迟等为常见症的证候。
同义词：肾阳虚水停证

7.9.4 肾阳虚证 kidney yang deficiency pattern

肾阳亏虚，机体失却温煦，以畏寒肢冷，腰胯无力，小便清长，夜尿多，舌淡苔白，脉弱等为常见症的证候。

同义词：元阳亏虚证；命门火衰证

7.9.5 肾阴虚证 kidney yin deficiency pattern

肾阴亏损，虚热内扰，以腰胯无力，公畜遗精，母畜不孕，低烧不退，躁动不安，午后潮热，舌红少苔，脉细数等为常见症的证候。

同义词：元阴亏虚证；肾水亏虚证

7.9.6 肾阴虚火旺 kidney yin deficiency and fire hyperactivity pattern

肾阴亏虚，虚热内扰，以潮热，低烧不退，躁动不安，公畜举阳遗精或精少不育，母畜不孕，尿黄，舌红苔黄少津，脉细数等为常见症的证候。

同义词：相火偏旺证；肾阴虚内热证

7.9.7 肾精亏虚证 kidney essence depletion pattern

肾精亏损，以幼畜生长发育迟缓，成年公畜生殖机能减退，牙齿松动等为常见症的证候。

同义词：肾气亏虚证

7.9.8 肾虚髓亏证 kidney deficiency and marrow depletion pattern

肾精亏虚，精髓不足，以生长发育迟缓，或骨折久不愈合，或骨痿等为常见症的证候。

7.9.9 肾阴阳两虚证 consumptive disease with deficiency of both kidney yin and kidney yang pattern

肾阴阳俱虚，以畏冷肢凉，低烧不退，躁动不安，腰胯无力，遗精早泄等为常见症的证候。

7.9.10 肾虚寒湿证 cold dampness due to kidney deficiency pattern

肾经阳气亏虚，寒湿浸著，以腰膝冷痛，活动受限，畏冷肢凉，苔白腻，脉濡缓等为常见症的证候。

同义词：肾经寒湿证

7.9.10.1 肾虚寒凝证 kidney deficiency cold stasis pattern

肾阳亏虚，寒邪凝滞，以畏冷肢凉，后肢尤甚，腰膝冷痛，小便清长，苔白脉沉迟等为常见症的证候。

同义词：阳虚肾寒证

7.9.10.2 肾虚寒痰证 kidney deficiency cold sputum pattern

肾阳亏虚，寒痰凝滞，以畏寒肢凉，腰胯无力，或于深处触及柔韧肿块、不红不痛，或溃后流脓气腥，苔白腻，脉沉迟或弦滑等为常见症的证候。

7.9.10.3 肾虚血瘀证 kidney deficiency and blood stasis pattern

肾虚而瘀血阻滞于肾，以腰胯无力、腰脊拒按，舌淡紫，脉细涩等为常见症的证候。

7.9.11 湿热蕴肾证 damp heat brewing in the kidney pattern

湿热之邪壅滞于肾，以腰部胀痛，小便疼痛，血尿，或排脓尿，发热口渴，舌红苔黄腻，脉滑数等为常见症的证候。

同义词：肾经湿热证

7.9.12 脓毒蕴肾证 septic kidney pattern

脓毒蕴积于肾，以腰部肿痛，排脓尿，小便疼痛，苔腻脉滑等为常见症的证候。

同义词：肾经脓毒证

7.9.13 膀胱湿热证 damp heat in the urinary bladder pattern

湿热下注膀胱，导致膀胱气化不利的证候。

7.9.13.1 膀胱湿热气滞证 damp heat in the urinary bladder and qi stasis pattern

湿热蕴结膀胱，气机阻滞，以小便频急、疼痛，小腹胀满疼痛，舌红苔黄腻，脉滑数等为常见症的证候。

7.9.13.2 膀胱湿热血瘀证 damp heat in the urinary bladder and blood stasis pattern

湿热蕴结膀胱，瘀血阻滞，以小便频急、疼痛，血尿，小腹疼痛，舌红有斑点，苔黄腻，脉滑数等为常见症的证候。

7.9.14 膀胱蕴热证 bladder heat amassment pattern

邪热蕴积膀胱，以小腹胀，小便疼痛，发热口渴，舌红苔黄，脉数有力等为常见症的证候。

同义词：热积膀胱证；膀胱实热证

7.9.15 膀胱蓄水证 bladder water amassment pattern

膀胱气化失司，水蓄膀胱，以小腹膨大、胀急作痛，小便不利等为常见症的证候。

7.9.16 膀胱蓄血证 bladder blood amassment pattern

小腹受损，或邪热内侵，血蓄膀胱，以小腹胀满疼痛，小便不畅、疼痛，舌紫或有斑点，脉弦涩等为常见症的证候。

7.9.17 膀胱虚寒证 bladder cold deficiency pattern

肾阳虚衰，虚寒内生，膀胱气化功能减弱，贮尿排尿功能失常的证候。

7.9.18 寒凝胞宫证 coagulated cold in womb pattern

寒邪凝滞胞宫,以小腹冷痛,喜温,或带下清稀色白,苔白,脉沉紧等为常见症的证候。

同义词:胞宫寒滞证;寒滞胞宫证

7.9.19 痰凝胞宫证 coagulated sputum in womb pattern

痰湿阻滞胞宫,以带下色白量多,或不孕,肥胖乏力,舌淡,苔白腻,脉滑或濡缓等为常见症的证候。

同义词:痰湿凝结胞宫证

7.9.20 瘀阻胞宫证 static blood blocking in womb pattern

瘀血阻滞胞宫,以小腹疼痛、拒按,或有肿块,尿血,舌紫暗或有斑点,脉弦涩等为常见症的证候。

同义词:瘀滞胞脉证

7.9.21 胞宫虚寒证 deficient cold in uterus pattern

阳气亏虚,胞宫失却温煦,以畏寒肢凉,小腹喜温喜按,或带下清稀,或不孕,或流产,可视黏膜苍白,舌淡苔白等为常见症的证候。

同义词:胞宫阳虚证

7.9.22 胞宫湿热证 dampness heat in uterus pattern

湿热侵袭,蕴结胞宫,以带下量多、色黄、黏稠秽臭,阴部瘙痒、糜烂,舌红苔黄腻,脉滑数等为常见症的证候。

同义词:湿热蕴胞证

7.9.23 胞宫血热证 uterine blood heat pattern

热邪蕴积胞宫,以小腹疼痛,带下量多、色鲜红或黄稠气臭,舌红苔黄,脉数等为常见症的证候。

同义词:胞宫积热证

7.9.24 冲任失调证 disharmony of the thoroughfare and controlling vessels pattern

冲任二脉功能失调,以发情不规律,小腹胀满疼痛等为常见症的证候。

同义词:冲任不调证

7.9.25 冲任不固证 insecurity of the thoroughfare and controlling vessels pattern

冲任二脉不能固摄,甚或崩漏,或滑胎流产等为常见症的证候。

7.9.26 冲任瘀阻证 thoroughfare and controlling vessel stasis obstruction pattern

瘀血阻滞冲任二脉，以少腹胀痛、拒按，舌紫暗或有斑点，脉弦涩等为常见症的证候。

同义词：瘀阻冲任证；瘀滞冲任证

7.9.27 湿热阻滞精室证 dampness heat blocks essence chambe pattern

湿热侵袭，蕴结于精室，以会阴部发热胀满，遗精，或精中夹脓液，阴部瘙痒、糜烂，舌红苔黄腻，脉滑数等为常见症的证候。

7.9.28 痰湿阻滞精室证 sputum dampness blocking essence chamber pattern

痰湿之邪阻滞精室，以精稀，阳痿，性欲低下，肥胖，四肢无力，舌淡，苔白腻，脉滑或濡缓，小腹胀满疼痛等为常见症的证候。

同义词：痰阻精室证；痰湿凝结精室证

7.9.29 瘀血阻滞精室证 blood stasis and sperm blocking essence chamber pattern

瘀血阻滞精室，以会阴部固定疼痛、拒按，或有肿块，或精少，或阳痿，或射精疼痛，或为血精，舌紫暗或有斑点，脉弦涩等为常见症的证候。

同义词：精室瘀阻证；瘀阻精室证

7.9.30 瘀浊阻滞精室证 stasis turbidity blocking essence chambe pattern

瘀血痰浊等邪阻滞精室，以会阴部肿胀、疼痛、拒按，小便不畅甚至尿闭，或便后有脓血、白浊脂液经前阴流出等为常见症的证候。

同义词：精室瘀浊证

7.9.31 惊恐伤肾证 fright and fear damage the kidney pattern

大惊大恐暴伤肾气，以惊慌不定，阳痿，滑精，或二便失禁等为常见症的证候。

同义词：恐伤肾气证

7.10 脏腑兼证类

7.10.1 心肾阴虚证 heart and kidney yin deficiency pattern

心与肾的阴液亏虚，以心区跳动明显，搏动加快或精神沉郁，腰胯无力，舌红少苔，脉细数无力等为常见症的证候。

同义词：心肾虚热证

7.10.2 心肾不交证 disharmony between the heart and kidney pattern

因久病伤阴，或劳损过度致使肾水亏虚于下，不能上济于心，心火亢于上，不能下交于肾；或因外感热病，致使心阴耗损，心阳亢盛，心火不能下交于肾，造成心肾水火不相既济而致虚火亢动、烦躁不安的证候。

同义词：心肾阴虚阳亢证；心肾阴虚火旺证

7.10.3 心肾阳虚证 yang deficiency of the heart and kidney pattern

心肾阳气亏虚，虚寒内生，失于温煦，功能减退，气化失常，血行无力的证候。

同义词：心肾虚寒证

7.10.4 水气凌心证 water retention affecting the heart pattern

肾阳亏虚，气化无力，水液泛溢，上凌于心，抑遏心阳的证候。

同义同：肾水凌心证；心肾阳虚水泛证

7.10.5 心肾气虚证 heart kidney qi deficiency pattern

心肾两脏气虚，以心区跳动明显，搏动加快，气短，腰胯无力，夜尿多，小便不尽，舌淡脉弱等为常见症的证候。

7.10.6 心肾气阴两虚证 both heart and kidney deficiency of qi and yin pattern

心与肾的气阴亏虚，以心区跳动明显，搏动加快，腰胯无力，精神不振，四肢无力，脉细无力等为常见症的证候。

7.10.7 心肾阴阳两虚证 both heart and kidney deficiency of yin and yang pattern

心与肾的阴液、阳气均虚，以畏冷肢凉，低烧不退，心区跳动明显，搏动加快，躁动不安易惊，腰胯无力，脉弱等为常见症的证候。

同义词：心肾阴阳亏虚证

7.10.8 心肺气虚证 qi deficiency of the heart and lung pattern

心气与肺气俱虚，心动失常，血行无力，并发肺失宣降、气机不畅的证候。

7.10.9 心肺阴虚证 heart and lung yin deficiency pattern

心与肺的阴液亏虚，以心区跳动明显，搏动加快，咳嗽，低烧不退，躁动不安，午后潮热，舌红少苔，脉细数等为常见症的证候。

7.10.10 心肺气阴两虚证 both heart and lung deficiency of yin and qi pattern

心与肺的气阴亏虚，以心区跳动明显，搏动加快，咳嗽，气短而喘，低烧不退，躁动不安，自汗或精神不振，四肢无力，脉弱等为常见症的证候。

7.10.11 心肺阴虚血瘀证 heart and lung yin deficiency ang blood stasis pattern

心与肺的阴液亏虚，瘀血内阻，以心区跳动明显，搏动加快，咳嗽，低烧不退，躁动不安，午后潮热，或痰中带血，舌红少苔，脉细数涩等为常见症的证候。

7.10.12 心肺阳虚证 heart and lung yang deficiency pattern

心与肺的阳气亏虚，以心区跳动明显，搏动加快，咳嗽，畏冷肢凉，吐稀白痰，舌淡紫，脉弱等为常见症的证候。

7.10.13 心肺热盛证 heart and lung exuberant heat pattern

火热炽盛，内扰心肺，以咳嗽气喘，吐痰黄稠，甚至昏迷，舌红苔黄，脉数有力等为常见症的证候。

同义词：热炽心肺证；心肺实热证；心肺火旺证

7.10.14 心肾火热证 heart and kidney fiery pattern

火热炽盛，内扰心肾，以发热口渴，甚至精神不振、昏迷，尿黄发热，腰痛，舌红苔黄，脉数等为常见症的证候。

7.10.15 心脾两虚证 deficiency of the heart and spleen pattern

心脾气血不足，心动失常，烦躁不安，并发脾失健运的证候。

7.10.16 心脾气虚证 heart and spleen qi deficiency pattern

心脾气虚，以心区跳动明显，搏动加快，神疲，食欲减退，腹胀，便溏，舌淡脉弱等为常见症的证候。

7.10.17 心脾阳虚证 heart and spleen yang deficiency pattern

心脾阳气亏虚，失却温运，以形寒肢凉，心区跳动明显，搏动加快，神疲，食欲减退，腹胀，便溏，舌淡紫，苔白滑，脉弱等为常见症的证候。

同义词：心脾虚寒证

7.10.18 心脾气血两虚证 heart and spleen qi deficiency pattern

心血与脾气亏虚，以心区跳动明显，搏动加快，神疲，食欲减退，腹胀，便溏，舌淡脉弱等为常见症的证候。

7.10.19 心脾积热证 heart and spleen heat accumulation pattern

邪热壅滞心脾，以发热口渴，口舌生疮、溃烂、疼痛，舌红苔黄，脉数有力等为常见症的证候。

同义词：心脾实热证

7.10.20 心肝火旺证 fire hyperactivity of the heart and liver pattern

心肝邪热炽盛，循经上炎，躁扰神明，甚至灼伤脉络、迫血妄行的证候。

同义词：心肝实热证；心肝热盛证

7.10.21 心肝血瘀证 heart and liver blood stasis pattern

瘀血阻滞心肝，以胸胁疼痛，心区跳动明显，搏动加快，舌紫或有斑点，脉弦涩等为常见症的证候。

7.10.22 心肝血虚证 blood deficiency of the heart and liver pattern

血液亏虚，心血虚少，肝血亦亏，神志、头目、筋脉、爪甲、血海失养的证候。

7.10.23 心肝阴虚证 heart liver yin deficiency pattern

心肝阴液亏虚，虚热内扰，以心区跳动明显，搏动加快，易惊，低烧不退，躁动不安，视力减退，舌红少苔，脉细数等为常见症的证候。

7.10.24 心肝血虚挟瘀证 heart and liver blood deficiency and stasis pattern

血虚挟瘀，心肝失养，以心区跳动明显，搏动加快，站立不稳，视力障碍，胸胁疼痛，母畜不孕，舌淡或有斑点，脉细涩等为常见症的证候。

7.10.25 心肝气血两虚证 heart and liver deficiency of qi and blood pattern

气血亏虚，心肝失养，以心区跳动明显，搏动加快，视力障碍，两胁疼痛，舌、爪甲色淡白，脉细无力等为常见症的证候。

同义词：心肝气血亏虚证

7.10.26 心肝气虚血瘀证 heart and liver both qi deficiency and blood stasis pattern

心肝两脏气虚，瘀血内阻，以心区跳动明显，搏动加快，神疲，胸胁或头部疼痛，脉细涩等为常见症的证候。

7.10.27 肝肾亏虚证 deficiency of kidney essence pattern

肝肾精血阴液亏虚，以两胁痛疼，腰胯无力，性欲减退，不孕不育及产蛋量减少等为常见症的证候。

同义词：精血亏虚证；精血亏损证

7.10.28 肝肾阴虚证 liver and kidney yin deficiency pattern

肝肾阴液亏虚，虚热内扰，以站立不稳，两眼干涩，夜盲内障，腰胯软弱，公畜可见举阳滑精，母畜发情周期不正常，低热不退，午后潮热，口色红，舌无苔，脉细数等为常见症的证候。

同义词：肝肾虚火证

7.10.29 肝肾阴虚阳亢证 liver and kidney yin deficiency and hyperactive yang pattern

肝肾阴液亏虚,虚阳偏亢,以站立不稳,急躁不安,腰膝疼痛,遗精,舌红少苔,脉弦细数等为常见症的证候。

同义词：水不涵木证；肾虚肝旺证

7.10.30 肝脾两虚证 liver spleen deficiency pattern

肝脾两脏虚证,以胁胀,食欲减退,腹胀,便溏等为常见症的证候。

7.10.31 肝脾气血两虚证 liver and spleen deficiency of qi and blood pattern

肝脾气血均亏虚,以肢体感觉迟钝,带下量少,食欲减退,腹胀,便溏,舌淡脉弱等为常见症的证候。

7.10.32 肝脾气阴两虚证 liver and spleen deficiency of qi and yin pattern

肝脾两脏气阴亏虚,以胁胀疼痛,气短乏力,食欲减退,腹胀,低烧不退,躁动不安,自汗或脉弱等为常见症的证候。

7.10.33 肝郁脾虚证 liver stagnant and spleen deficiency pattern

肝气郁结,疏泄失常,脾失健运,以躁动不安,草料迟细,粪便稀薄,肠鸣矢气,腹痛泄泻,泄必痛,苔白,脉弦等为常见症的证候。

同义词：肝木乘土证；肝滞脾虚证

7.10.34 土壅侮木证 effulgent liver and weak spleen pattern

脾失健运,气滞于中,肝失疏泄,以精神沉郁,草料迟细,便溏,肠鸣矢气,腹痛欲泄,泄后痛减,口红干,苔腻,脉弦数为常见症的证候。

同义词：土壅木郁证

7.10.35 肝脾湿热证 liver and spleen dampness heat pattern

湿热中阻,肝失疏泄,脾失健运,以腹胀,或黄疸,大便黏滞,舌红苔黄腻,脉弦滑数等为常见症的证候。

同义词：中焦湿热证

7.10.36 肝脾气滞证 liver and spleen qi stagnation pattern

肝脾气机阻滞,以胁胀胁痛,腹胀肠鸣,腹泻,脉弦等为常见症的证候。

7.10.37 肝脾血瘀证 liver and spleen blood stasis pattern

瘀血阻滞肝脾,以胁下肿块,或疼痛拒按,舌暗或有斑点,脉弦涩等为常见症的证候。

同义词：肝脾瘀滞证

7.10.38 肝热脾虚证 liver heat and spleen deficiency pattern

肝经热盛，脾气亏虚，以胁肋热痛，躁动不安，口干，腹胀，食欲减退，便溏，脉弦数等为常见症的证候。

7.10.39 肝胃不和证 liver and stomach disharmony pattern

肝气郁滞，横逆犯胃，胃失和降，以胃脘、胁肋胀满疼痛，嗳气、呃逆，精神沉郁，食欲不振，苔薄黄，脉弦等为常见症的证候。

同义词：肝气犯胃证；肝胃不调证

7.10.40 肝胃热盛证 liver and stomach heat filled pattern

邪热炽盛，肝胃火旺，以发热口渴，躁动不安，胁痛，胃脘灼痛，舌红苔黄，脉数有力等为常见症的证候。

同义词：肝胃积热证

7.10.41 肝火犯胃证 liver fire attacking the stomach pattern

肝火炽盛，横逆犯胃，胃失和降，以胁肋、胃脘发热作痛，口干，呕吐，便结尿黄，舌红苔黄，脉弦数等为常见症的证候。

7.10.42 肝胃气滞证 liver and stomach qi stagnation pattern

肝胃气机阻滞，以胁肋、胃脘胀满，嗳气、呃逆，脉弦等为常见症的证候。

7.10.43 肝胃气滞血瘀证 liver and stomach qi stagnation and blood stasis pattern

瘀血内阻，肝胃气滞，以胁肋、脘腹胀满，或于腹部或胁下触及肿块、拒按，嗳气，脉弦涩等为常见症的证候。

同义词：肝胃瘀滞证

7.10.44 肝胃气虚血瘀证 liver and stomach qi deficiency and blood stasis pattern

肝胃气虚，瘀血阻滞，以精神不振、四肢无力，食欲减退，胁肋、脘腹胀满，或触及肿块，舌淡紫或有斑点，脉涩无力等为常见症的证候。

7.10.45 肝胃气滞阴虚证 liver and stomach qi stagnation and yin deficiency pattern

肝胃阴液亏虚，气机阻滞，以胁肋、脘腹胀满，口干，潮热，便结尿黄，脉弦细等为常见症的证候。

同义词：肝胃阴虚气滞证

7.10.46 肝胃阴虚证 yin deficiency of liver and stomach pattern

阴液亏虚，肝胃失和，以口干，便结尿黄，舌红少津，脉弦细数等为常见症的证候。

7.10.47 肝胃阴虚血瘀证 liver and stomach yin deficiency and blood stasis pattern

肝胃阴液亏虚，瘀血阻滞，以胁肋触及肿块，口干，便结，尿黄，舌红少苔，或有斑点，脉弦细涩等为常见症的证候。

7.10.48 肝胃虚寒证 liver and stomach cold deficiency pattern

阳气亏虚，肝胃不和，以胁胀脘痞，脘腹冷痛、喜按，食欲减退，舌淡，脉沉迟等为常见症的证候。

7.10.49 肝火犯肺证 liver fire attacking the lung pattern

肝火炽盛，上逆犯肺，肺失清肃，或肺络受伤的证候。
同义词：木火刑金证

7.10.50 肝肺风热证 liver lung wind heat pattern

风热之邪，侵袭肝肺，以发热口渴，咳嗽，目赤肿痛，舌红苔薄黄，脉弦数等为常见症的证候。

7.10.51 肝肺热盛证 liver and lung heat flourishing pattern

邪热炽盛，肝肺火旺，以发热口渴，咳嗽气喘，舌红苔黄，脉弦数等为常见症的证候。
同义词：肝肺实热证；肝肺实火证

7.10.52 肝郁肾虚证 liver stagnant and kidney deficiency pattern

肝气郁滞，肾气亏虚，以胁胀作痛，腰胯无力，精神沉郁等为常见症的证候。

7.10.53 脾肺两虚证 deficiency of both spleen and lung pattern

脾肺气虚，以咳嗽声低，气短而喘，吐痰清稀，食欲减退，腹胀，便溏，舌淡苔白滑，脉弱等为常见症的证候。
同义词：土不生金证；脾肺气虚证

7.10.54 脾肺气阴两虚证 spleen and lung both qi and yin deficiency pattern

肺与脾的气阴亏虚，以咳嗽气短，乏力，食欲减退，腹胀，口干，低烧不退，躁动不安，自汗或脉弱等为常见症的证候。

7.10.55 脾肾阳虚证 yang deficiency of the spleen and kidney pattern

脾肾阳气虚衰，温煦气化无力，健运失职，气化无权，水谷不化，水液泛滥的证候。

同义词：脾肾虚寒证

7.10.56 脾肾气虚证　spleen and kidney qi deficiency pattern

脾肾两脏气虚，以神疲气短，食欲减退，腹胀，便溏或久泄，腰胯无力，腰痛，舌淡脉弱等为常见症的证候。

7.10.57 脾肾两虚证　deficiency of both spleen and kidney pattern

脾肾亏虚，以食欲减退，腹胀，便溏，腰胯无力、腰痛等为常见症的证候。

同义词：脾肾亏虚证

7.10.57.1 脾肾气虚水停证　spleen and kidney qi deficiency and water stasis pattern

脾肾气虚，水液内停，以神疲气短，食欲减退，腹胀，腰胯无力，小便不利，肢体浮肿，舌淡苔白，脉弱等为常见症的证候。

7.10.57.2 脾肾阳虚水停证　spleen kidney yang deficiency and water stasis pattern

脾肾阳气亏虚，水液气化失常，以形寒肢冷，腰膝腹部冷痛，小便不利，肢体浮肿，后肢尤甚，舌淡胖，苔白滑，脉沉迟无力等为常见症的证候。

7.10.58 肺肾阴虚证　yin deficiency of the lung and kidney pattern

因久咳耗伤肺阴，进而累及肾阴，或由于肾阴亏损，不能滋养肺阴，加虚火内生而致肺失清肃的证候。

7.10.59 肺肾气虚证　qi deficiency of the lung and kidney pattern

肺肾之气俱虚，肺失肃降，肾不纳气的证候。

同义词：肾不纳气证；肾失摄纳证

7.10.60 肺肾阳虚证　yang deficiency of lung and kidney pattern

肾阳虚衰，水液泛溢，上射于肺，以畏寒肢冷，咳嗽气喘，吐多量清稀痰，后肢水肿，尿少，舌淡胖，苔白滑，脉弱等为常见症的证候。

同义词：水寒射肺证

7.10.61 肺胃风热证　lung and stomach wind heat pattern

风热之邪，侵袭肺胃，以发热，口渴，咳嗽，斑疹，舌红苔薄黄，脉浮数等为常见症的证候。

7.10.62 肺胃火热证　intense lung stomach fire pattern

邪热炽盛，侵及肺胃，以发热，口渴，汗多，咳嗽气喘，便秘尿黄，舌红苔黄，脉滑数等为常见症的证候。

同义词：肺胃热盛证

7.10.63 肺胃阴虚证 yin deficiency of lung and stomach pattern

肺胃阴液亏虚，以饮水多，干咳少痰，善食易饥，舌红少津，脉细数等为常见症的证候。

同义词：肺胃津亏证

7.10.64 毒陷心肝证 toxin attacking heart and liver pattern

邪毒内陷心肝，以发热，甚至昏迷，肢体抽搐，可视黏膜紫暗，舌暗红，脉沉细数等为常见症的证候。

7.10.65 肺虚肠脱证 lung deficiency and intestinal withdrawal pattern

肺气亏虚，后阴不固，以咳嗽，气短而喘，大便失禁，或久泄不止，或粪随咳出，或脱肛，舌淡脉弱等为常见症的证候。

7.10.66 脾虚肠脱证 spleen deficiency and intestinal withdrawal pattern

脾气亏虚，中气下陷，肛门松弛，以食欲减退，腹胀，气短而喘，大便失禁，或久泄不止，或脱肛，舌淡脉弱等为常见症的证候。

7.10.67 肾虚肠脱证 kidney deficiency and intestinal withdrawal pattern

肾气亏虚，后阴不固，以腰膝疼痛，气短而喘，大便失禁，或久泄不止，或粪随矢气出，或脱肛，舌淡脉弱等为常见症的证候。

7.10.68 肝肠气滞证 liver intestinal qi stagnation pattern

肝气不舒，肠道气滞，以胁肋、腹部胀满，肠鸣矢气，腹泻，脉弦等为常见症的证候。

7.11 卫表肌肤证类

7.11.1 邪袭卫表证 evil attack of wei-defence surface pattern

风热、风寒、风温、温热等邪侵袭卫表，以新起发热恶风寒，脉浮苔薄等为常见症的证候。

同义词：卫表证

7.11.2 风袭表疏证 wind evil attacking exterior pattern

风邪侵袭肤表，腠理不固，以发热，恶风，自汗，或鼻鸣干呕，舌淡红，苔薄白，脉浮缓等为常见症的证候。

同义词：风邪袭表证；表虚证

7.11.3 风寒束表证 wind cold fettering the exterior pattern

风寒之邪侵袭肤表，腠理闭塞，以发热，无汗，或鼻塞流清涕，气喘，苔薄白，脉浮紧等为常见症的证候。

同义词：表实寒证；风寒表证

7.11.4 风热犯表证 wind heat attacking exterior pattern

风热之邪侵袭肤表，以发热较重，恶风(寒)较轻，伴有少食，咳嗽，流鼻涕，口津微干，舌尖红，脉浮数等为常见症的证候。

同义词：风热表证；风热犯卫证

7.11.5 风湿袭表证 wind dampness attacking exterior pattern

风湿之邪侵袭肤表，阻遏卫气，以全身胀满困重，恶寒发热，汗而热不解，口不渴，苔白滑，脉濡缓等为常见症的证候。

同义词：湿郁卫分证；表湿证

7.11.6 暑湿袭表证 exterior attacked by summer heat dampness pattern

暑季因暑湿之邪侵袭肤表，卫气失调，以发热，微恶风寒，无汗或有汗，口渴，舌红苔黄，脉濡数等为常见症的证候。

7.11.7 外燥袭表证 external dryness attacking the exterior pattern

风燥外邪侵袭肤表，卫气失调，以微有发热，恶风寒，无汗，鼻燥，口渴，脉浮等为常见症的证候。

同义词：风燥袭表证

7.11.8 风毒犯表证 wind and poison attacking the exterior pattern

风毒之邪侵袭肤表，以突起皮肤瘙痒，丘疹或风团肿块，脉浮等为常见症的证候。

7.11.9 温毒袭表证 wind poison attacking the exterior pattern

火热毒邪侵袭卫表，以发热微恶寒，舌红苔黄白，脉浮数等为常见症的证候。

同义词：热毒袭表证；火毒袭表证

7.11.10 风湿蕴肤证 warm and toxic attacking skin pattern

风湿邪毒蕴结皮肤，以皮肤瘙痒、溃烂、出疹或流水等为常见症的证候。

7.11.11 风毒蕴肤证 wind poison attacking skin pattern

风毒侵袭皮肤，以皮肤赤红、瘙痒、疼痛，出疹或为肿块，甚或赤烂脱皮等为常见症的证候。

7.11.12 湿毒蕴结肌肤证 dampness and poison accumulate to the skin pattern

湿浊毒邪蕴结肌肤，以皮肉生疮，或湿疹、瘙痒、糜烂、流水等为常见症的证候。

同义词：肌肤湿毒证

7.11.13 热毒蕴结肌肤证 heat toxicity in the skin pattern

火热毒邪蕴结皮肤肌肉，以皮肉生疮疖痈疡，红肿灼痛，溃烂流脓，或皮肤赤红、发热、糜烂，口渴便秘，舌红、苔黄、脉数等为常见症的证候。

同义词：肌肤热毒证；火毒蕴结肌肤证

7.11.14 虫毒蕴肤证 parasite poison smolder at skin pattern

虫毒蕴结皮肤，以皮肤出疹、瘙痒、糜烂、疼痛等为常见症的证候。

同义词：虫毒袭肤证

7.11.15 寒湿蕴肤证 cold dampness smolder at skin pattern

寒湿之邪蕴结皮肤，以皮肤感觉迟钝、僵硬、肿胀、刺激感觉不灵，触之发凉等为常见症的证候。

7.11.16 湿热组结肌肤证 damp heat smolder at skin pattern

湿热之邪蕴结肌肤，以皮肤起水疱，或皮肤潮红，肌肤肿胀、糜烂、流水、结痂、浸淫成片，瘙痒，苔黄腻等为常见症的证候。

同义词：肌肤湿热证

7.11.17 湿痰蕴结肌肤证 dampness sputum smolder at skin pattern

痰湿之邪蕴结肌肤，以皮肉间起柔韧肿块、圆滑、无压痛，或肌肤肿硬，刺激感觉不灵，或身体肥胖，或肌肤漫肿疼痛、溃后流稀水，舌淡胖、苔白滑等为常见症的证候。

同义词：肌肤痰湿证

7.11.18 风热郁滞肌肤证 wind heat stagnation skin pattern

风热之邪外袭而蕴结肌肤，以突起风团、丘疹、瘙痒、发热，舌红、苔黄、脉数等为常见症的证候。

同义词：风热蕴肤证；肌肤风热证

7.11.19 虫毒侵袭肌肤证 parasite poison attacking skin pattern

虫毒侵袭，蕴结肌肤，以皮肤出疹、瘙痒，肌肤溃烂、疼痛等为常见症的证候。

同义词：肌肤虫毒证

7.11.20 风水证 wind and water retention with each other pattern

外感风邪，肺卫失宣，水湿浸淫肌肤，以眼睑、头面、四肢水肿，或起风团肿块，微恶风寒，小便短少，脉浮等为常见症的证候。

同义词：风袭水泛证；风水相搏证

7.11.21 表闭水停证 skin is closed and the water stasis pattern

外邪束表，肤表闭塞，水道不利，水湿内停，以恶风寒，无汗，小便不利，肢体浮肿，脉浮等为常见症的证候。

7.11.22 瘀滞肌肤证 skin stasis pattern

瘀血阻滞肌肤，以皮肤干燥、粗糙，或为血丝红缕，或为紫暗斑块，肤痒不适，或见肌肤甲错，或肌肤感觉迟钝，脉浮细涩等为常见症的证候。

同义同：肌肤瘀滞证

7.11.23 寒凝血涩肌肤证 cold coagulation astringent skin pattern

寒邪凝滞，气血瘀结于肌肤，以肢端厥冷，甚至冷痛，肤色紫暗，或为皲裂、冻疮，舌淡紫，脉沉细等为常见症的证候。

7.11.24 肌肤失养证 skin dysplasia pattern

血液及阴精亏少，肌肤失养，以皮肤干燥粗糙、瘙痒或感觉迟钝、脱屑、开裂，甚至肌肤甲错，舌淡脉细等为常见症的证候。

7.12 头面官窍证类

7.12.1 实邪犯头证 excess evil attacking the head pattern

风寒、湿热、疫毒、痰浊、瘀血等侵犯头部的证候。其症除头痛等之外，因不同的实邪而各具特征。

7.12.1.1 风寒犯头证 wind cold attacking head pattern

风寒之邪侵犯头部，连及项背，以恶寒遇风则痛增，苔薄白，脉浮紧等为常见症的证候。

7.12.1.2 风热犯头证 wind heat attacking head pattern

风热之邪侵犯头部，以挠头，烦躁不安，用头撞墙，发热或恶风，口渴，可视黏膜发红，舌尖边红，苔薄黄，脉浮数等为常见症的证候。

7.12.1.3 风湿犯头证 wind dampness attacking head pattern

风湿之邪侵犯头部，以挠头，烦躁不安，用头撞墙，微恶风寒，肢体困重，苔白滑，脉濡等为常见症的证候。

7.12.1.4 风痰上扰证 wind sputum attacking head pattern

肝风挟痰上扰于头，以挠头，烦躁不安，用头撞墙，可视黏膜发红，舌红苔黄腻，脉弦滑等为常见症的证候。

同义词：风痰上攻证

7.12.1.5 痰浊犯头证 turbidness of sputum attacking the head pattern

痰浊上蒙清窍，以不识主人，不听呼唤，视物模糊，精神沉郁，呕痰涎，苔白腻，脉滑或弦滑等为常见症的证候。

同义词：痰湿犯头证

7.12.1.6 肝阳上扰证 wind yang harassing the upper body pattern

肝阳有余，上扰头面清窍，以目赤，挠头，用头撞墙，烦躁不安，舌红脉弦等为常见症的证候。

同义词：风阳上扰证

7.12.1.7 瘀血犯头证 blood stasis and head offense pattern

外伤等致瘀血于头部脉络，以头痛经久不愈、固定不移，舌紫或有斑点，脉弦涩等为常见症的证候。

7.12.1.8 瘀热犯头证 stasis heat attacking the head pattern

热邪与瘀血上犯头部，以发热，挠头，烦躁不安，用头撞墙，甚至昏迷，面色暗红，舌绛紫，脉数而涩等为常见症的证候。

同义词：瘀热上蒙证

7.12.1.9 热毒壅聚头面证 heat toxicity stagnated in head and face pattern

火热疫毒壅结头面部及其官窍，以头部红肿发热，或颜面生疔疖，烦热口渴，便结尿黄，舌红脉数，甚则壮热，甚至昏迷等为常见症的证候。

7.12.1.10 肝火犯头证 liver fire attacking the head pattern

肝火上炎，侵扰头部，以挠头，烦躁不安，用头撞墙，目赤，急躁，舌红脉弦数等为常见症的证候。

7.12.2 实邪犯目证 excess evil attacking the eyes pattern

风热、痰湿、瘀血等侵犯于目的证候。除有目痛等症之外，并因不同的实邪而各具特征。

7.12.2.1 风热犯目证 wind heat attacking the eyes pattern

风热之邪上犯于目，以发热恶风，两目红赤疼痛、眵多、流泪，脉浮数等为常见症的证候。

同义词：风火犯目证；风火攻目证

7.12.2.2 风湿凌目证 wind dampness attacking eye pattern

风湿之邪上犯于目，以胞睑肿胀、色赤、瘙痒，流泪等为常见症的证候。

7.12.2.3 外伤目络证 exogenous injury attacking eye channel pattern

外力损伤目络，以胞睑肿痛色紫暗，或睛色赤或紫暗，或眼底出血，目痛等为常见症的证候。

7.12.2.4 虫积化疳证 parasite accumulation and chancre pattern

蛔虫等耗损营气，以气血亏虚，目失所养，眼球干涩无光泽等为常见症的证候。

7.12.3 气轮证 qi ring pattern

风热、湿热、瘀血、热毒等侵犯白睛，或正虚气轮失养所见证候。

7.12.3.1 气轮风热证 heat wind in qi ring pattern

风热之邪外袭白睛，以白睛红赤、发热、多眵等为常见症的证候。

同义词：风热外袭白睛证

7.12.3.2 气轮湿热证 dampness heat in qi ring pattern

湿热之邪侵及白睛，以白睛红赤、微肿、眵白黏稠或白睛发黄等为常见症的证候。

同义词：湿热郁结白睛证

7.12.3.3 气轮血瘀证 blood stasis in qi ring pattern

郁热、外伤，瘀血阻滞气轮脉络，以白睛赤脉紫胀、结节隆起，疼痛拒按，白睛积血等为常见症的证候。

同义词：气轮白睛证

7.12.3.4 气轮热毒证 heat toxin in qi ring pattern

火热疫毒炽盛，损伤气轮脉络，以白睛焮红、肿胀，或出血等为常见症的证候。

7.12.3.5 气轮阴虚证 qi ring yin deficiency pattern

燥热伤阴，或肺失肃降，津液不能上布，以白睛干燥失泽、干涩，目赤等为常见症的证候。

7.12.4 血轮证 blood ring pattern

实热、虚火等侵犯内眼角与外眼角所见之证。

7.12.4.1 血轮实热证 excess heat pattern in blood ring pattern

心火上炎，以内眦部红肿疼痛，脉粗大，挠头，烦躁不安，用头撞墙，口干，尿黄，舌红苔黄，脉数等为常见症的证候。

7.12.4.2 血轮虚热证 heat deficiency of blood ring pattern

心阴不足，虚火上炎，以两眦部赤脉微红，瘙痒，或小眦赤脉显露，口燥，舌红少苔，脉细数等为常见症的证候。

7.12.5 肉轮证 flesh ring pattern

风热、湿热、瘀血、热毒、痰湿等侵犯内眼角与外眼角，或正虚肉轮失养所见证候。

7.12.5.1 肉轮血瘀证 blood stasis pattern flesh ring pattern

外伤胞睑，或邪热阻络，胞睑气血瘀滞，以胞睑肿胀青紫，睑硬疼痛，或胞睑内发红、脉紫胀，生红肉等为常见症的证候。

7.12.5.2 肉轮风热证 heat wind in flesh ring pattern

风热之邪侵袭胞睑，以胞睑肿胀、丘疹、刺痒疼痛等为常见症的证候。

同义词：胞睑风热外袭证

7.12.5.3 肉轮热毒证 heat toxin in flesh ring pattern

火热毒邪蕴结于胞睑脉络，以胞睑红赤，肉腐化脓等为常见症的证候。

7.12.5.4 肉轮湿热证 dampness heat pattern in flesh ring pattern

湿热缊结胞睑，以胞睑红肿、疼痛、赤烂浸淫，舌红，苔白腻，脉濡缓等为常见症的证候。

同义词：胞睑湿热侵淫证

7.12.5.5 肉轮痰湿证 sputum dampness flesh ring pattern

痰湿阻于胞睑脉络，以胞睑有肿块等为常见症的证候。

7.12.5.6 肉轮气虚证 flesh ring qi deficiency pattern

脾虚气弱，运化失司，水湿上泛，以上睑下垂，或胞睑虚肿等为常见症的证候。

7.12.5.7 肉轮血虚证 blood deficiency in flesh ring pattern

血虚胞睑失养，风胜化燥，以胞睑皮肤干燥皲裂、脱屑、瘙痒等为常见症的证候。

7.12.6 风轮证 wind ring pattern

风热、热毒、湿热等侵犯黑睛，或正虚黑睛失养所见证候。

7.12.6.1 风轮风热证 wind heat in wind ring pattern

风热外邪侵袭风轮，以黑睛起星翳，目赤疼痛，畏光流泪，鼻塞等为常见症的证候。

7.12.6.2 风轮湿热证 dampness heat in wind ring pattern

湿热侵目，或肝胆湿热上攻，以抱轮红赤，黑睛混浊、色红或白，或凝脂色白、表面粗糙，缠绵不愈，苔黄腻，脉弦数等为常见症的证候。

7.12.6.3 风轮热毒证 heat toxicity in wind ring pattern

热毒结聚，灼伤风轮，以黑睛生翳陷下，黄液上冲，目赤，苔黄，脉弦数等为常见症的证候。

7.12.6.4 风轮阴虚证 yin deficiency of wind ring pattern

邪热伤阴，或肝肾阴亏，风轮失于滋养，以黑睛失却光泽，起细小星翳，干涩，畏光眨目，舌红苔少，脉细数等为常见症的证候。

7.12.7 水轮证 water ring pattern

实热、痰火、痰湿、瘀血、热毒等侵犯瞳仁，或正虚瞳仁失养所见证候。

7.12.7.1 水轮实热证 excess heat in water ring pattern

火热上炎水轮，以视力障碍，舌红苔黄，脉数等为常见症的证候。

7.12.7.2 水轮痰火证 sputum fire in water ring pattern

痰火与风邪相搏于目，以瞳孔散大，眼压升高，抱轮红赤，甚则呕吐，舌红苔黄腻，脉滑数等为常见症的证候。

7.12.7.3 水轮阴亏证 yin deficiency of water ring pattern

阴液亏损，不能上荣水轮，以两目干涩，视力下降，腰胯无力，舌红少苔，脉细数等为常见症的证候。

7.12.7.4 水轮气虚证 qi deficiency water ring pattern

气虚不升，不能上荣目窍，以视物不能持久，舌淡苔薄白，脉弱等为常见症的证候。

7.12.7.5 水轮气虚血亏证 pattern qi deficiency with blood stasis of water ring pattern

气血亏虚，不能濡养瞳仁，以瞳孔散大，眼压升高，结膜苍白，舌淡，脉弱等为常见症的证候。

7.12.7.6 水轮气虚血瘀证 pattern qi deficiency with blood stasis of water ring pattern

气虚推运无力，瘀血阻滞瞳仁，以突然目盲不见，舌紫暗，脉细涩等为常见症的证候。

7.12.7.7 水轮气滞血瘀证 qi stagnation and blood stasis of water ring pattern

气滞而血行不畅，瘀阻瞳仁，以暴盲，舌暗红，脉弦涩等为常见症的证候。

7.12.7.8 水轮水湿停聚证 water dampness stagnate at water ring pattern

阳气亏虚，水液上泛，内停瞳仁，以视力异常，舌淡胖，苔白滑，脉濡缓等为常见症的证候。

7.12.7.9 水轮痰瘀互结证 sputum and static blood tangling at water ring pattern

湿痰与瘀血蕴结于瞳仁，以视力减退，舌暗红，苔腻，脉弦涩等为常见症的证候。

7.12.7.10 水轮火邪伤络证 fire evil injuring channels of water ring pattern

火热上炎,灼伤脉络,迫血溢于瞳仁,以云雾移睛,视力下降,舌红苔黄,脉数等为常见症的证候。

7.12.7.11 水轮血脉痹阻证 blood obstruction water ring pattern

瞳仁血脉阻塞不通,以视力下降,或视物异常,或暴盲,舌暗红,脉弦涩等为常见症的证候。

7.12.7.12 水轮络痹精亏证 bi jing deficiency in channels of water ring pattern

瞳仁脉络痹阻,精血亏虚,不能升运于目,以视力下降,视野缩小,眼底色晦暗、脉络细窄等为常见症的证候。

7.12.8 耳窍证 ear pattern

肝火、风热、湿热、瘀血、热毒等侵犯耳窍,或正虚耳窍失养所见证候。

7.12.8.1 肝火燔耳证 liver fire attacking ear pattern

肝火内炽,上燔耳窍,以耳窍疼痛,目赤,躁烦不安,鼓膜充血或穿孔,或耳道流脓、流血,舌红苔黄,脉弦数等为常见症的证候。

7.12.8.2 毒火犯耳证 toxic fire attacking ear pattern

邪毒外袭,火毒上攻耳窍,以耳部剧痛,鼓膜充血或穿孔流脓,或外耳道生疮疖,舌红苔黄,脉数有力等为常见症的证候。

同义词:热毒犯耳证

7.12.8.3 风热犯耳证 wind heat attacking ear pattern

风热之邪外犯,壅滞耳窍,以耳内分泌物、排泄物堵塞,发热微恶风寒,鼓膜充血或内陷,苔薄黄,脉浮数等为常见症的证候。

7.12.8.4 湿热犯耳证 dampness heat attacking ear pattern

湿热之邪侵袭耳窍,以耳道或耳廓红肿疼痛、糜烂、渗液、结痂,或耳内流脓黄稠,或耳胀,苔黄腻,脉滑数等为常见症的证候。

7.12.8.5 痰湿泛耳证 sputum dampness attacking ear pattern

痰湿内停,上泛耳窍,以耳内流脓量多,听力障碍,透过鼓膜见有液平面,或耳廓局部肿胀、皮色小变,苔滑腻,脉弦滑等为常见症的证候。

7.12.8.6 邪恋耳窍证 evil stop in ear pattern

正气不足,邪毒留恋于耳,以耳内微痛或不适,或有听力障碍,鼓膜混浊内陷或穿孔,耳道流脓,经久不愈等为常见症的证候。

7.12.8.7 气滞耳窍证 qi stagnation in ear pattern

肝气郁结，气机不利，气滞耳窍，以卒然耳窍失聪，或耳内堵塞，精神沉郁，脉弦等为常见症的证候。

7.12.8.8 血瘀耳窍证 blood stasis in ear pattern

血行受阻，瘀滞耳窍，以听力障碍，或耳内生赘生物，舌紫暗或有斑点，脉涩等为常见症的证候。

7.12.8.9 阴虚耳窍失濡证 yin deficiency and loss of osmosis from ear pattern

阴液亏虚，耳窍失濡，以听力障碍，站立不稳或鼓膜穿孔，耳内流脓，口燥，潮热，脉细数等为常见症的证候。

7.12.8.10 阳虚耳窍失煦证 yang deficiency ear loss of heat pattern

阳气亏虚，耳窍失于温煦，以听力障碍，或中耳积液，或鼓膜穿孔、流脓清稀，形寒肢凉，舌淡，脉沉细等为常见症的证候。

7.12.8.11 气虚耳窍失充证 qi deficiency ear insufficiency pattern

正气亏虚，耳窍失于充养，以四肢无力，气短不鸣，听力障碍，或鼓膜穿孔，脓液清稀，舌淡脉弱等为常见症的证候。

7.12.8.12 血虚耳窍失养证 blood deficiency of ear pattern

血液亏少，耳窍失养，以听力障碍，站立不稳，舌淡，脉细等为常见症的证候。

7.12.8.13 血虚耳燥证 blood deficiency dryness pattern

血虚风燥，耳廓肌肤失养，以耳道、耳廓及其周围皮肤增厚、粗糙、皲裂、脱屑，舌淡，脉细等为常见症的证候。

7.12.9 鼻窍证 nasal pattern

风寒、风热、痰湿、湿热、燥火、瘀血、热毒等侵犯鼻窍，或正虚鼻窍失养所见证候。

7.12.9.1 风寒袭鼻证 wind cold attacking nose pattern

风寒之邪侵袭鼻窍，以鼻塞、流清涕，或鼻痒、喷嚏，鼻甲肿大、黏膜色淡，分泌物清稀，恶风寒，苔薄白，脉浮紧等为常见症的证候。

7.12.9.2 风热犯鼻证 wind heat attacking nose pattern

风热之邪侵袭鼻窍，以鼻塞，流涕，发热微恶风寒，口微渴，鼻甲肿大，黏膜充血，分泌物黏稠，苔薄黄，脉浮数等为常见症的证候。

7.12.9.3 湿壅鼻窍证 dampness attacking nose pattern

湿浊壅塞鼻窍，以鼻塞，鼻涕量多，鼻甲肿胀，苔白腻，脉濡或滑等为常见症的证候。

7.12.9.4 痰聚鼻窍证 sputum gathering nasal orifice
痰浊停聚，凝滞鼻窍，以鼻塞，浊涕量多，或鼻窍生息肉，鼻甲肿胀，苔腻脉滑等为常见症的证候。

7.12.9.5 痰热犯鼻证 sputum heat attacking nose pattern
火热与痰浊搏结，阻遏鼻窦，以鼻流腥臭浊涕、量多色黄，鼻塞，鼻黏膜红肿，舌红苔黄腻，脉滑数等为常见症的证候。

7.12.9.6 湿热蒸鼻证 dampness heat steaming nose pattern
湿热内蕴，熏蒸鼻窍，以外鼻、鼻前庭及鼻窍肌肤潮红、糜烂，或黄水浸淫、渗液，或鼻甲充血肿大，鼻涕浓稠量多，舌红苔黄腻，脉滑数等为常见症的证候。

7.12.9.7 燥伤鼻窍证 dryness attacking nose pattern
气候干燥，耗伤津液，鼻失濡润，鼻窍不利，鼻孔干燥，鼻内黏膜干燥少津，或鼻涕胶结而成痂皮，或鼻窍皮肤皲裂、衄血等为常见症的证候。

7.12.9.8 火毒犯鼻证 nasal fire poison pattern
火热毒邪侵袭鼻窍，以外鼻及鼻前庭红肿疼痛，或生疔疖、溃烂，或鼻衄量多势剧，发热口渴，舌红苔黄，脉数等为常见症的证候。

7.12.9.9 肺热熏鼻证 lung heat disturbing nose pattern
肺热炽盛，上灼鼻窍，以鼻前孔红肿疼痛，鼻干涕稠，鼻黏膜充血肿大或糜烂，发热口渴，舌红苔黄，脉数等为常见症的证候。

7.12.9.10 气虚鼻窍失充证 qi deficiency nose insufficiency pattern
正气亏虚，鼻窍失于充养，以四肢无力，气短不鸣，鼻塞，清涕自流，喷嚏时作，自汗恶风，鼻甲肿胀、色淡白，舌淡脉弱等为常见症的证候。

7.12.9.11 血虚鼻窍失养证 blood deficiency and loss of nose pattern
血虚不能上荣，鼻窍失养，以鼻干，鼻内黏膜萎缩、鼻甲缩小、鼻腔宽大，站立不稳，舌淡，脉细等为常见症的证候。

7.12.9.12 阴虚鼻窍失濡证 yin deficiency nose loss pattern
阴液亏虚，鼻窍失濡，以鼻腔宽大、发热不适，鼻黏膜干燥，鼻甲萎缩，涕少结痂或带血，或时有鼻衄，口燥，潮热，舌干苔燥，脉细数等为常见症的证候。

7.12.9.13 阳虚鼻窍失煦证 yang deficiency nose loss of warm pattern
阳气亏虚，鼻窍失于温煦，以鼻塞难通，遇冷尤甚，流清涕，喷嚏时作，鼻黏膜肿胀、色淡，形寒肢凉，面白，舌淡，脉弱等为常见症的证候。

7.12.10 咽喉证 pharyngeal pattern

风寒、风热、痰湿、湿热、火毒、瘀血等侵犯咽喉,或正虚咽喉失养所见证候。

7.12.10.1 风寒袭咽证 wind cold assailing the pharyngeal pattern

风寒之邪侵袭咽喉,以咽喉疼痛不适,恶寒微发热,无汗,苔薄白,脉浮紧等为常见症的证候。

同义词:风寒袭喉证

7.12.10.2 风热侵咽证 wind heat assailing the pharyngeal pattern

风热之邪侵袭咽喉,以咽喉红肿疼痛,或喉核充血肿大,或声音不利、声嘶,发热微恶风寒,口微渴,苔薄黄,脉浮数等为常见症的证候。

同义词:风热侵喉证

7.12.10.3 痰湿凝阻咽喉证 sputum dampness coagulation blocking throat pattern

痰浊湿邪内蕴,凝滞咽喉,以咽部肿胀,异物感,声音不扬或声嘶,或声带肿胀、生息肉,呼吸不利,痰涎增多,苔腻脉濡等为常见症的证候。

7.12.10.4 气滞痰凝咽喉证 qi stagnation and coagulated sputum in throat pattern

气机阻滞,痰浊凝聚咽喉,以精神沉郁,咽部黏膜肿胀,苔腻,脉弦滑等为常见症的证候。

7.12.10.5 痰毒壅喉证 sputum poison obstructing throat pattern

痰火邪毒壅塞咽喉,以喉部痰涎壅盛,痰鸣如拽锯,呼吸不利,或局部充血肿胀,舌红苔腻,脉滑数等为常见症的证候。

7.12.10.6 湿热蒸喉证 dampness heat steam throat pattern

湿热蕴结,熏蒸咽喉,以咽喉红肿疼痛,声音不扬或嘶哑,咽喉充血肿胀,痰黄不易咯出,舌红苔黄腻,脉滑数等为常见症的证候。

7.12.10.7 热毒攻喉证 heat toxicity attacking throat pattern

热毒上攻咽喉,以咽喉红肿疼痛,吞咽困难,甚至溃烂、化脓,口气臭秽,壮热口渴,舌红苔黄,脉数有力等为常见症的证候。

同义词:胃火燔咽证;火毒攻喉证

7.12.10.8 气滞声带证 qi stagnation of vocal cords pattern

气机郁滞,声带不利,以声音不扬、声嘶或低微,喉部不适,脉弦等为常见症的证候。

7.12.10.9 瘀阻声带证 stasis of vocal cords pattern

瘀血阻滞声带,以声音嘶哑,鸣叫困难,声带暗红、边缘增厚,或有小结、息肉,舌有斑点，脉弦涩等为常见症的证候。

同义词：血痹阻声带证

7.12.10.10 瘀阻咽喉证 static blood stagnated in throat pattern

瘀血阻滞咽喉，以咽喉部疼痛、异物感，吞咽不利，咽喉黏膜暗红，或有赘生物，舌有斑点，脉弦涩等为常见症的证候。

同义词：血痹阻咽喉证

7.12.10.11 气虚咽喉失充证 qi deficiency and throat insufficiency pattern

正气亏虚，咽喉失于充养，以气短，声音低微，咽喉黏膜色淡，自汗恶风，舌淡脉弱等为常见症的证候。

7.12.10.12 阴虚咽喉失濡证 loss of moistening of throat due to yin deficiency pattern

阴液亏虚，咽喉失养，以咽喉发热微痛，声音嘶哑，或有异物感，咽喉微红或潮红，或局部糜烂，舌红少津，脉细数等为常见症的证候。

7.12.11 齿龈证 gum pattern

风火、湿热、瘀血等侵犯齿龈，或正虚齿龈失养所见证候。

7.12.11.1 风火犯齿证 wind fire attacking teeth pattern

风火热毒侵犯牙齿，以齿龈红肿，张口不便，咀嚼痛甚，饮冷痛减，舌红苔薄黄，脉浮数等为常见症的证候。

同义词：风热犯齿证

7.12.11.2 胃火燔龈证 stomach fire flaring gum pattern

胃火内炽，上灼牙龈，以龈肉红肿疼痛，齿缝间渗血渗脓，口渴口臭，便秘，舌红苔黄，脉数等为常见症的证候。

7.12.11.3 胃火燔齿证 stomach fire burnt tooth pattern

胃火内炽，上灼牙体，以齿缝间渗血渗脓，口渴口臭，便秘，舌红苔黄，脉数等为常见症的证候。

7.12.11.4 湿热蒸齿证 dampness heat steaming tooth pattern

湿热内蕴，上蒸牙齿，以牙痛或牙齿被蛀蚀成洞，或齿龈红肿，口气臭秽，舌红苔黄腻，脉滑数等为常见症的证候。

7.12.11.5 火毒犯齿证 tooth fire poison offender pattern

火毒炽盛，燔灼牙齿，以咀嚼痛，发热口渴，舌红苔黄，脉数等为常见症的证候。

7.12.11.6 火毒犯龈证 gingival fire poison pattern

火毒炽盛，燔灼齿龈，以龈肉红肿疼痛，遇热痛甚，发热口渴，舌红苔黄，脉数等为常见症的证候。

7.12.11.7 血瘀齿龈证 blood stasis gingival pattern

瘀血内阻，齿龈瘀滞，以齿龈疼痛，龈肉紫暗或有赘生物，或龈肉腐溃流出污黑血水，舌暗，脉涩等为常见症的证候。

7.12.11.8 阴虚齿燥证 teeth dryness due to yin deficiency pattern

阴液亏虚，牙齿失于濡养，以牙齿干燥枯槁，牙齿疏豁松动，咀嚼无力，潮热，舌红少津，脉细数等为常见症的证候。

7.12.11.9 虚火灼龈证 deficient fire flaring gum pattern

阴液亏虚，龈肉失濡，虚火灼龈，以龈肉干燥萎缩、潮红，齿根宣露，齿牙枯槁、疏豁松动，低烧不退，躁动不安，潮热，舌红少津，脉细数等为常见症的证候。

7.12.11.10 血虚龈肉失养证 blood deficiency gingival meat loss pattern

血液亏虚，龈肉失于濡养，以龈肉淡白或萎缩，齿根宣露，齿牙松动，咀嚼无力，舌淡，脉细等为常见症的证候。

7.12.11.11 血虚齿槽失养证 blood deficiency alveolar dysplasia pattern

血液亏虚，齿槽失于濡养，以齿槽骨腐溃，齿牙松动，咀嚼无力，舌淡，脉细等为常见症的证候。

7.12.11.12 气虚齿动证 qi deficiency and tooth movement pattern

正气亏虚，齿龈失充，以牙齿浮动，咀嚼无力，气短不鸣，脉弱等为常见症的证候。

7.12.12 口唇证 mouth and lips pattern

风热、火毒、湿热等侵犯口唇，或正虚嘴唇失养所见证候。

7.12.12.1 风邪犯唇证 wind evil attacking mouth and lips pattern

风邪侵袭口唇或口腔，以口唇潮红肿，或口腔微红，恶风发热，苔薄黄，脉浮数等为常见症的证候。

同义词：风邪犯口证

7.12.12.2 毒火攻唇证 poison fire attack lip pattern

火毒炽盛，燔灼口唇，以嘴唇红肿剧痛，或局部溃烂、臭秽流脓，壮热口渴，舌红苔黄，脉数等为常见症的证候。

同义词：毒火攻口证

7.12.12.3 湿热蒸口证 damp heat attacking mouth and lips pattern

湿热内蕴，上蒸口唇，以口腔黏膜红肿疼痛，或肌膜腐溃、溢脓臭秽，或口唇红肿糜烂、黄水浸淫，口角红赤皲裂，舌红苔黄腻，脉濡数等为常见症的证候。

同义词：湿热蒸唇证

7.12.12.4 虚火灼口证 deficiency fire attacking mouth and lips pattern

虚火上炎，熏灼口腔或口唇，以口腔干燥，黏膜潮红少津，或局部腐溃久不愈合，或口唇潮红干燥、皲裂、脱屑，舌红少苔，脉细数等为常见症的证候。

同义词：虚火灼唇证

7.12.12.5 血虚唇燥证 blood deficiency and lip dryness pattern

血液亏虚，口唇失养，以口唇肌膜干燥少津、皲裂、瘙痒、脱屑，或嘴唇黏膜有黄白色小溃疡，反复发作，唇舌淡白，脉细等为常见症的证候。

同义词：血虚口燥证

7.12.12.6 湿热蒸舌证 dampness heat steaming tongue pattern

湿热内蕴，熏蒸舌体，以舌体红肿疼痛，或局部溃烂流脓，苔黄腻，脉濡数等为常见症的证候。

7.12.12.7 热毒攻舌证 heat toxicity attacking tongue pattern

火热邪毒炽盛，攻犯舌体，以舌体红肿疼痛，或舌体局部红肿高突、疼痛，舌体活动不灵，发热口渴，脉数有力等为常见症的证候。

同义词：火毒攻舌证

7.12.12.8 血瘀舌下证 sublingual blood stasis pattern

瘀血阻于舌下，以舌下赘生紫暗色肿物，舌体颜色紫暗或有斑点、运动不灵等为常见症的证候。

7.12.13 邪犯清窍证 evil crime clearing the orifice pattern

燥热、痰湿、瘀血等侵犯清窍所见的证候。

7.12.13.1 燥干清窍证 dryness affecting the clear orifices pattern

气候干燥，津液耗损，清窍失濡，以口燥，两眼干涩、少泪，少涕，少津，甚至衄血等为常见症的证候。

7.12.13.2 瘀阻清窍证 stasis blocking clear orifice pattern

瘀血阻滞头面清窍，以反应迟钝或五官疼痛，或五官溢血，舌暗红，脉弦涩等为常见症的证候。

7.13 经脉筋骨证类

7.13.1 风中经络证 channel hit by wind pattern

风邪侵袭经络筋脉，以肌肤感觉迟钝、瘙痒，或突起，口眼歪斜等为常见症的证候。

同义词：风邪袭络证

7.13.1.1 风痰入络证 wind sputum attacking channel pattern

肝风挟痰阻闭经络，以肢体感觉迟钝不仁，甚或瘫痪不遂，或肌肤感觉迟钝瘙痒，口角流涎，苔腻等为常见症的证候。

同义词：风痰阻络证

7.13.1.2 风热中络证 wind heat in channels pattern

风热之邪中于经络，以患处感觉迟钝、发热、瘙痒、色赤等为常见症的证候。

同义词：风热阻络证

7.13.1.3 寒滞经脉证 cold stagnation in channels pattern

寒邪凝滞经脉，血行不畅，以肢体冷痛、拘急或感觉迟钝，肤色紫暗或苍白，苔白，脉弦紧等为常见症的证候。

同义词：风寒袭络证；风寒阻络证

7.13.1.4 风毒入络证 fire poison entering channels pattern

风热邪毒窜入脉络，以肢体迅速出现线条状、感觉迟钝、疼痛、斑点，或见出血症状等为常见症的证候。

同义词：风毒入脉证

7.13.1.5 火毒窜络证 dampness heat blocking channels pattern

风热火毒窜入经络，以发热口渴，肢体患处发热、疼痛，或色赤瘙痒等为常见症的证候。

同义词：热毒窜络证；热毒入络证

7.13.1.6 湿热阻络证 dampness heat blocking channel pattern

湿热之邪阻滞经脉，以发热口不甚渴，患处糜烂、瘙痒，苔黄腻，脉滑数等为常见症的证候。

7.13.1.7 寒湿阻络证 cold dampness blocking channels pattern

寒湿之邪阻滞经络，以肢体肿胀，形寒肢凉，苔白滑等为常见症的证候。

同义词：寒湿入络证

7.13.1.8 痰湿阻络证 sputum dampness blocking channels pattern

痰浊湿邪阻痹经络，以肢体或关节等处肿胀，或皮肤肿硬、感觉迟钝、瘙痒，苔白腻等为常见症的证候。

同义词：痰湿阻痹证

7.13.1.9 瘀热入络证 stagnant heat attacking channels pattern

邪热与瘀血阻结于脉络，以低热，患处发热疼痛、色赤，舌绛或紫，脉细涩数等为常见症的证候。

同义词：血热伤络证；热蕴络瘀证

7.13.1.10 瘀血阻络证 static blood blocking channels pattern

瘀血阻于经络，以患处固定疼痛，或见紫斑、肿块，或见出血色暗，舌紫或有斑点，脉涩等为常见症的证候。

同义词：瘀阻脉络证

7.13.1.11 虫湿壅络证 obstruction parasite dampness pattern

虫毒湿邪蕴结经络，以患处肢体或肌肤肿胀、瘙痒或疼痛，起丘疹或流水，或有柔韧结节等为常见症的证候。

7.13.2 经气不利证 adverse qi pattern

风寒湿热、瘀血等邪阻滞，使经气不利，或正虚经络失养而经气不利，以肢体感觉迟钝，活动不利，感觉异常等为常见症的证候。

7.13.2.1 风胜行痹证 wind prevails over the marching bi pattern

风寒湿邪阻滞筋骨关节而以风邪为主，以肢体关节游走疼痛为常见症的证候。

7.13.2.2 寒胜痛痹证 cold beats pain and bi pattern

风寒湿邪阻滞筋骨关节而以寒邪为主，以肢体关节固定冷痛为常见症的证候。

7.13.2.3 湿胜着痹证 dampness conquers bi pattern

风寒湿邪阻滞筋骨关节而以湿邪为主，以肢体关节沉重疼痛、肿胀等为常见症的证候。

7.13.2.4 热邪阻痹证 heat induced obstruction of bi pattern

风湿热邪阻滞筋骨关节而以热邪为主，肢体关节出现肿胀、发热、疼痛等为常见症的证候。

7.13.2.5 湿热阻痹证 dampness heat blocking bi pattern

湿热蕴阻于筋骨关节，以身热，肢体关节肿胀，发热疼痛，舌红苔黄腻，脉滑数等为常见症的证候。

7.13.2.6 风寒湿凝滞筋骨证 wind cold and dampness stagnation of muscles and bones pattern

风寒湿邪阻滞筋骨关节，以肢体关节游走性疼痛等为常见症的证候。

同义词：风寒湿阻证

7.13.2.7 寒湿犯腰证 cold and dampness attacking waist pattern

寒湿之邪侵及腰部，以腰脊冷痛，遇寒痛甚为常见症的证候。

7.13.2.8 痰湿犯腰证 sputum dampness causes lumbar pattern

痰浊湿邪侵及腰部，以腰脊感觉迟钝、疼痛、转侧不利，苔白腻等为常见症的证候。

7.13.2.9 湿热犯腰证 damp heat attacking the waist pattern

湿热之邪侵及腰部，以腰脊疼痛，发热口渴，苔黄腻，脉滑数等为常见症的证候。

7.13.2.10 瘀血犯腰证 blood stasis and waist offense pattern

瘀血阻滞于腰部，以腰脊固定疼痛、拒按等为常见症的证候。

7.13.2.11 伤损筋骨证 muscle and bone injury pattern

外伤导致筋骨损伤，以患处肿胀、疼痛、活动障碍等为常见症的证候。

同义词：筋伤骨断证

7.13.2.12 瘀滞筋骨证 stasis of muscles and bones pattern

外伤或病久，瘀血阻滞筋骨，以筋骨固定疼痛、拒按，活动障碍，或关节局部肿硬变形、皮色紫暗等为常见症的证候。

7.13.2.13 络伤出血证 bleeding due to collateral injury pattern

外伤而损伤脉络，以患处出血，或局部紫暗斑块、疼痛等为常见症的证候。

同义词：外伤络损证

7.13.2.14 外伤瘀滞证 traumatic stasis pattern

外伤导致气血瘀滞，以患处出现紫暗斑块、疼痛拒按等为常见症的证候。

7.13.2.15 痰湿流注证 sputum damp obstruction pattern

痰浊湿邪流窜于经脉、筋骨等处，以肢体深处触及柔韧肿块、疼痛，或触及脓液，苔腻脉滑等为常见症的证候。

同义词：痰湿流注筋骨证

7.14 六经病证

7.14.1 太阳病证 taiyang disease pattern

外邪袭体，正邪相争于表，以致肌表营卫失调的病证。多见于外感病的初期，在八纲辨证中属表证。根据感受外邪的不同和机体体质的差异，太阳病证分为太阳伤寒和太阳中风。

7.14.1.1 太阳经证 taiyang channel pattern

风寒之邪侵犯肌肤，正邪抗争，营卫失和，以恶寒、脉浮、头项强痛为常见症的证候。

7.14.1.1.1 太阳中风证 taiyang wind attack pattern

以风邪为主的风寒之邪侵犯太阳经脉所致的表虚证。证见恶风，发热，汗出，脉浮缓。

7.14.1.1.2 太阳伤寒证 taiyang cold attack pattern

以寒邪为主的风寒之邪侵犯太阳经脉所致的表实证。证见恶寒、发热、关节疼痛、跛行、无汗、咳嗽、气喘、脉浮紧。

7.14.2 太阳腑证 taiyang-fu organ pattern

太阳经证不解，病邪内传其腑所表现的证候。

7.14.2.1 太阳蓄水证 taiyang water amassment pattern

太阳经证不解，邪与水结，膀胱气化不利，水液停蓄，以发热恶寒，小便不利，少腹满，口渴但饮水不多，脉浮或浮数等为常见症的证候。

7.14.2.2 太阳蓄血证 taiyang disease with stagnated blood pattern

太阳经证不解，邪热内传，与血相结于少腹，以少腹硬满而痛，甚则发狂，小便自利，舌质紫或有瘀斑，大便色黑，脉沉涩等为常见症的证候。

7.14.3 阳明病证 yangming disease pattern

伤寒病发展过程中，阳热亢盛，胃肠燥热所表现的证候。

7.14.4 阳明经证 yangming channel pattern

热邪弥漫全身，充斥阳明之经，里热炽盛，而肠道尚无燥粪内结的实热证。证见身热汗出、口渴贪饮、呼吸喘粗、口色红燥、舌苔黄干、脉象洪大等。

7.14.5 阳明腑证 yangming-fu organ pattern

邪热传里与肠中糟粕相搏，致使燥粪内结的里实证。证见食欲大减或废绝、恶热喜凉、身热、呈日晡潮热、汗出、口鼻干燥，腹部胀痛拒按，粪便燥结，甚或闭而不通、舌苔黄燥、脉沉实有力等。

7.14.6 少阳病证 shaoyang disease pattern

邪犯少阳胆腑，枢机不运，经气不利，以寒热往来，食欲减少，精神沉郁，呕吐，苔白滑等为常见症的证候。

7.14.7 太阴病证 taiyin disease pattern

脾阳虚衰，寒湿内生，以食欲不振，腹满呕吐，腹痛，腹泻，喜温喜按，口不渴，舌淡苔白，脉迟缓等为常见症的证候。

7.14.8 少阴病证 shaoyin disease pattern

病邪深入，畜体正气衰微，心肾功能活动减退的全身性虚弱证。少阴经内连心肾，因此病邪内犯少阴，则必累及心肾，以致畜体阴阳俱损，少阴病多为疾病过程中的危重阶段。可由他经传来，亦可因体质素虚，外邪直中少阴所致。

注：临床上，少阴病证有寒化和热化两种证候。

7.14.8.1 少阴寒化证　shaoyin cold transformation pattern

少阴病过程中比较多见的一种证候，多为阳气不足、病邪入内、从阴化寒所致，呈现出全身性的虚寒证候。证见恶寒、嗜睡、立少喜卧、耳鼻发凉、四肢厥冷、体温偏低、脉沉细。

7.14.8.2 少阴热化证　shaoyin heat transformation pattern

少阴阴虚阳亢，从阳化热的证候。证见口燥、咽痛、烦躁不安、舌红绛、脉细数。

7.14.9 厥阴病证　jueyin disease pattern

外感病发展过程中，阳气衰微，阴寒盛极，或阳气被外邪所郁，出现以四肢厥冷为主要证候的病证。厥阴病是外感疾病发展的最后阶段，病变的表现极为错综复杂。

7.14.9.1 厥阴寒化证　jueyin cold reversal pattern

邪从厥阴，里寒化生，阴虚而阳气衰微，以四肢厥冷，口色青，体温低下，脉微欲绝等为常见症的证候。

7.14.9.2 厥阴热化证　jueyin heat reversal pattern

阴液亏耗，阳热极盛，阳郁不能外达，以四肢厥冷继而发热，或先发热，目赤，口渴烦躁，舌红，苔黄，脉滑数等为常见症的证候。

7.15 其他证类

7.15.1 邪扰胸膈证　chest diaphragm evil disturbance pattern

邪气阻扰胸膈，以胸膈胀满等为常见的症候。

7.15.1.1 热扰胸膈证　chest and diaphragm disturbed by heat pattern

邪热扰于胸膈，以胸中烦热、躁扰不宁，口渴，咳嗽吐黄痰，舌红苔黄，脉数等为常见症的证候。

7.15.1.2 热实结胸证　heat excess chest bind pattern

热邪阻结胸膈，以发热口渴，咳嗽气喘，便秘尿黄，舌红苔黄，脉数等为常见症的证候。

7.15.1.3 寒实结胸证　chest binding pattern with cold fluid pattern

寒邪阻结胸膈，以咳嗽，胸膈痞闷，吐稀白痰，苔白，脉弦紧等为常见症的证候。

7.15.1.4 痰热结胸证　chest binding pattern with heat sputum pattern

痰浊热邪结于胸膈，以胸腹胀满疼痛，咳嗽吐黄痰，或脘部硬满、按之则痛，舌红苔黄腻，脉滑数等为常见症的证候。

7.15.1.5 痰气阻膈证 spittoon obstructing diaphragm pattern

痰浊阻于胸膈，气机郁滞，以进食梗塞，呃逆，呕吐痰涎，苔腻或滑，脉弦等为常见症的证候。

7.15.1.6 痰瘀阻膈证 sputum stasis blocking diaphragm pattern

瘀血与痰浊阻于胸膈，以胸膈痞闷疼痛，进食梗塞，呕吐痰涎，舌紫或有斑点，苔腻脉弦涩等为常见症的证候。

7.15.1.7 饮停胸胁证 fluid retained in chest and hypochondrium pattern

水饮停积胸腔，以胸廓饱满，胸部胀痛，咳嗽气喘，苔白滑，脉弦滑等为常见症的证候。

7.15.1.8 瘀阻胸胁证 blood stasis in chest and stress pattern

瘀血阻于胸胁部脉络，以胸胁部固定疼痛、拒按，或唇紫，舌暗或有斑点，脉弦涩等为常见症的证候。

同义词：胸络不和证；瘀滞胸胁证

7.15.1.9 瘀血阻膈证 static blood obstructing the network vessels pattern

瘀血阻于胸膈，以胸膈或上脘部固定疼痛，或进食梗塞，舌紫或有斑点，脉弦涩等为常见症的证候。

同义词：膈下瘀阻证

7.15.2 湿热弥漫三焦证 diffusive dampness heat in sanjiao pattern

湿热弥漫全身，累及上中下三焦，以身热不扬，渴不多饮，咳嗽，腹胀呕恶，便溏，小便短涩，舌红苔黄腻，脉濡数或滑数等为常见症的证候。

同义词：三焦湿热证

7.15.3 上焦湿热证 damp heat in the upper jiao pattern

湿热侵袭上焦，困遏卫阳，肺失宣降的证候。

7.15.4 中焦湿热证 damp heat in middle jiao pattern

湿热病邪传入中焦，困阻脾胃，脾失健运的证候。

7.15.4.1 中焦实热证 excessive heat in middle jiao pattern

中焦邪热炽盛，燥实内结，以发热口渴，脘腹胀满，大便秘结，尿短黄，舌红苔黄燥，脉数有力等为常见症的证候。

7.15.4.2 中寒虫扰证 moderate cold parasite disturbance pattern

中焦虚寒，蛔虫不得安宁，向上窜扰，以腹部冷痛，大便稀溏，脘部阵发疼痛，烦热口渴等为常见症的证候。

7.15.5 下焦湿热证 damp heat in lower jiao pattern

湿热流注下焦大肠、膀胱，湿聚热蕴，气机阻滞，气化失司，腑气不利的证候。

7.15.5.1 湿热下注证 dampness heat diffusing downward pattern

湿热之邪向下侵及肠道、膀胱、子宫、阴部、后肢等处，以小便频急淋漓涩痛，或大便腥臭溏烂，或带下黄臭，或阴部湿疹、瘙痒，或后肢生疮、溃烂流水等为常见症的证候。

7.15.5.2 瘟毒下注证 scourge toxin pouring downward pattern

温热毒邪向下流窜，以睾丸肿痛，或疔疮等走窜而见下部脓肿，或后肢溃烂灼痛等为常见症的证候。

7.15.5.3 瘀阻下焦证 dampness stasis the lower jiao pattern

瘀血阻滞于肠道、膀胱、子宫等处，以小腹疼痛或胀满、拒按，或可触及包块，或二便疼痛带血，舌紫暗或有斑点，脉弦涩等为常见症的证候。

同义词：下焦瘀滞证

7.15.6 邪犯少腹证 hypoabdominal of evil offenders pattern

湿热、瘀血、热毒等邪侵及少腹，以少腹疼痛等为常见症的证候。

7.15.6.1 少腹血瘀证 hypoabdominal blood stasis pattern

瘀血阻滞于少腹，以少腹胀满、拒按，或可触及包块，或伴二便不调，舌紫暗或有斑点，脉弦涩等为常见症的证候。

7.15.6.2 少腹气滞证 hypoabdominal qi stagnation pattern

邪阻少腹，气机不畅，以少腹胀满，或母畜不孕，或有痞块聚散无常，脉弦等为常见症的证候。

7.15.6.3 少腹热滞证 hypoabdominal heat stagnation pattern

热邪壅滞少腹，以少腹发热胀满，口渴，舌红苔黄，脉弦数等为常见症的证候。

7.15.6.4 少腹湿热阻滞证 hypoabdominal dampness heat block pattern

湿热蕴结，阻滞于少腹，以少腹发热胀满，带下色黄腥臭，大便臭秽，舌红苔黄腻，脉滑数等为常见症的证候。

7.15.6.5 少腹瘀滞证 low abdominal stasis pattern

瘀血阻于少腹，气机不利，以少腹胀满拒按，或触及质硬包块，发情周期不稳定或不孕，脉弦涩等为常见症的证候。

7.15.6.6 少腹瘀热证 low abdominal stasis and heat pattern

瘀血与热邪互结于少腹，以发热口渴，少腹满硬疼痛、发热拒按，或触及包块，发情周期不稳定或不孕，舌暗红，苔黄等为常见症的证候。

7.15.7 邪入少阳证 evil entering shaoyang pattern

外邪侵袭，由表入里的过渡阶段，郁阻少阳胆腑，以寒热往来，胸胁胀满，目眩等为常见症的证候。

7.15.8 胎毒蕴热证 accumulated heat due to fetal toxicity pattern

幼畜因在母体时染受邪热毒气所致，以皮肤赤烂、脱皮，或口疮、目赤烂，或大便腥臭夹脓血等为常见症的证候。

7.15.9 营卫不和证 patterns of disharmony between ying and wei-defence pattern

卫弱营强，或卫强营弱，以身微热或微恶风寒，时有汗出，脉缓等为常见症的证候。

7.15.10 表寒里热证 patterns of exterior cold and interior heat pattern

寒邪外束，郁热于内，以恶寒，发热，无汗，气喘，口渴，舌红苔黄白，脉浮数等为常见症的证候。

7.15.11 表热里寒证 superficies heat and interior cold pattern

内有阳气不足，而外感风热之邪，以发热微恶风寒，尿清，大便稀溏，舌淡胖，苔薄黄，脉浮等为常见症的证候。

7.15.12 表里俱寒证 cold in both superficies and interior pattern

寒邪外侵，表里同时受邪，以恶寒肢冷，无汗，或咳喘吐白痰，或脘腹冷痛、吐泻清稀等为常见症的证候。

同义词：表里实寒证

7.15.13 表里俱热证 heat in both exterior and interior pattern

风热、瘟毒等邪侵袭，表里俱热，以发热微恶风寒，气喘，口渴，便秘，尿黄，舌红苔薄黄，脉浮数等为常见症的证候。

同义词：表里实热证

7.15.14 上盛下虚证 upper body exuberance and lower body deficiency pattern

肝肾亏虚，气血痰热等上壅所致的证候。

7.15.15 上热下寒证 upper heat and lower cold pattern

上部有热，下部有寒的证候。

7.15.16 上寒下热证 upper cold and lower heat pattern

上部有寒，下部有热的证候。

第八章 治 法

8.1 内治法

8.1.1 汗法

8.1.1.1 辛温解表 releasing superficies with pungent-warm

用性味辛温的方药，发散风寒，解除表证的治疗方法。
同义词：发汗解表

8.1.1.2 辛凉解表 releasing superficies with pungent-cool

用性味辛凉的方药，疏散风热，解除表证的治疗方法。
同义词：疏散风热

8.1.1.3 辛凉清热 clearing heat with pungent-cool

用性味辛凉的方药，解表清热，适用于卫气同病证的治疗方法。

8.1.1.4 疏邪解表 dispersing pathogenic factors and releasing the external

用祛散表邪的方药，适用于邪气在表的治疗方法。
同义词：疏邪透表

8.1.1.5 疏风透疹 dispelling wind and promoting eruption

用疏散风邪的方药，达到透疹目的的治疗方法。适用于风疹等疾病。

8.1.1.6 疏表通经 dispersing the external and unblocking the meridians

通过使用方药，达到疏散表邪、舒畅经气目的的治疗方法。适用于外邪束表，经气不利的治疗方法。

8.1.1.7 宣肺解表 ventilating lung and releasing superficies

用发散表邪，宣发肺气的方药，适用于邪在肺卫所致风寒袭肺证、风寒闭肺证的治疗方法。

8.1.1.8 调和营卫 regulating nutrient and defensing

用解散风邪、收敛益阴的方药，以治卫阳、收敛益阴，使营卫恢复正常协调状态，适用于营卫不和证的治疗方法。

8.1.1.9 祛湿解表 removing dampness and releasing superficies

用解散表邪、芳香化湿的方药，适用于风湿袭表证的治疗方法。

8.1.1.10 理气解表 regulating qi and releasing superficies

发散解表药与调理气机药并用，适用于外有表邪、内有气滞证候的治疗方法。

8.1.1.11 扶正解表 strengthening resistance and relieving exterior syndrome

发散解表药与补养气血药并用，适用于气虚外感证、血虚外感证、阴虚外感证、阳虚外感证等的治疗方法。

8.1.2 吐法

8.1.2.1 涌吐痰涎 inducing vomit of phlegm and drool

催吐而达到祛除痰涎、毒物或积滞作用，适用于痰浊壅盛而病位偏上、毒物或积滞在胃的病症的治疗方法。

8.1.2.2 涌吐风痰 inducing vomit of wind-phlegm

催吐而达到祛除风痰作用，适用于风痰壅盛证的治疗方法。

8.1.2.3 涌吐痰食 inducing vomit of phlegm and retained food

催吐而达到祛除痰食作用，适用于痰涎食滞互结证的治疗方法。

8.1.2.4 涌吐宿食 inducing vomit of retained food

催吐而达到祛除宿食作用，适用于食积证的治疗方法。

8.1.2.5 开关涌吐 resuscitation through vomiting

催吐而达到开窍通闭作用，适用于痰浊壅盛所致神昏等的治疗方法。

同义词：涌吐开关

8.1.3 下法

8.1.3.1 清热攻下 clearing heat and purgation

清热药配伍攻下药，适用于大肠热结证的治疗方法。

同义词：苦寒通下

8.1.3.2 泻结行滞 dispersing mass and expelling retention

攻下药配伍行气药，适用于热盛气滞腑实证的治疗方法。

8.1.3.3 温下寒积 warming and purging accumulated cold

通里攻下药配伍温阳散寒药，适用于阳虚寒凝便秘的治疗方法。

8.1.3.4 润燥通便 moistening dryness for relaxing bowels

增液润燥药配伍润下药，适用于肠燥津亏证、阴虚肠燥证的治疗方法。

同义词：润肠通便

8.1.3.5 益气通下 replenishing qi to relax the bowels
通过补气而达到通便作用，适用于气虚所致便秘的治疗方法。
同义词：益气通便

8.1.3.6 润肠泄热 moistening the intestines and purging heat
滋阴润燥药配伍清热攻下药，适用于热结肠燥证的治疗方法。

8.1.3.7 软坚润燥 softening hardness and moistening dryness
软化坚硬、滋润燥结，适用于津亏燥结病证的治疗方法。

8.1.3.8 峻下逐水 removing water retention by purgation
通过攻下以消除水饮，适用于水饮内停的治疗方法。
同义词：泻下逐水

8.1.4 和法

8.1.4.1 和解少阳 harmonizing and releasing shaoyang
运用具有和解表里作用的方药，适用于邪在少阳半表半里证的治疗方法。

8.1.4.2 调和肝脾 harmonizing the liver and spleen
疏肝健脾、调理气机，使肝脾协调，适用于肝郁脾虚证、肝旺脾虚证等的治疗方法。
同义词：调理肝脾

8.1.4.3 疏肝理脾 soothing the liver and regulating the spleen
调理肝脾气机，使肝脾协调，适用于肝脾不和证的治疗方法。
同义词：疏肝和脾

8.1.4.4 疏肝健脾 soothing the liver and fortifying the spleen
调理肝气、健运脾气而使肝脾协调，适用于肝郁脾虚证的治疗方法。
同义词：疏肝补脾

8.1.4.5 抑肝扶脾 suppressing the liver and reinforcing the spleen
泻肝理气、健运脾气而使肝脾协调，适用于肝旺脾虚证的治疗方法。
同义词：补脾泻肝

8.1.4.6 疏肝和胃 soothing the liver and harmonizing the stomach
调理肝胃气机而使肝胃调和，适用于肝胃气滞、肝胃不和的治疗方法。

8.1.4.7 调理肠胃 regulating the intestines and stomach
调理肠胃气机而使肠胃和健，适用于胃肠气滞证的治疗方法。

同义词：调和肠胃

8.1.4.8 调理脾胃 regulating the spleen and stomach

调理脾胃气机而使脾胃和健，适用于脾胃不和证的治疗方法。

同义词：调和脾胃

8.1.4.9 和中缓急 harmonizing the middle and relaxing convulsion

运用具有缓急止痛和中作用的方药，适用于中焦气机不和而挛急所致疼痛的治疗方法。

8.1.4.10 调和气血 regulating and harmonizing qi and blood

运用有理气活血作用的方药，适用于气血不和的治疗方法。

同义词：理气和血

8.1.4.11 调气和营 regulating qi and harmonizing the nutrient

运用具有理气和营作用的方药，适用于营气不和的治疗方法。

8.1.4.12 平调寒热 mildly regulating cold and heat

运用具有清热、祛寒作用而性味较为平和的方药，适用于阴阳寒热不调的治疗方法。

8.1.4.13 分消走泄 separating elimination through urination and defecation

清利小便，导泻大便，使病邪分别通过二便排出的治疗方法。

8.1.4.14 分消上下 separating elimination from the upper and lower

涌吐、祛痰等使病邪从上排出或清利二便使病邪从下排出，以达到病愈目的的治疗方法。

8.1.4.15 表里分消 separating elimination from the external and internal

发汗解表和清泄里实，使病邪从表里分消而达到愈病目的的治疗方法。

8.1.4.16 调理冲任 regulating the thoroughfare vessel and conception vessel

运用具有调理冲任气血作用的方药，适用于冲任不调的治疗方法。

同义词：调摄冲任

8.1.5 温法

8.1.5.1 回阳救逆 reviving yang for resuscitation

用大补阳气的方药治疗阴寒内盛危重症的方法。

同义词：扶阳救逆

8.1.5.2 温中散寒 warming zhongjiao for dispelling cold

温补脾胃阳气以散寒和中，适用于脾胃阳虚证的治疗方法。

同义词：温中祛寒

8.1.5.3 温经散寒 warming jingmai for dispelling cold

运用具有温阳散寒通经作用的方药，适用于寒滞经脉证的治疗方法。

同义词：温经祛寒

8.1.5.4 散寒止痛 eliminating cold for stopping pain

温散寒邪而达到止痛目的的治疗方法，适用于寒邪凝滞的治疗方法。

8.1.5.5 温阳散寒 warming yang for dispelling cold

用具有温补阳气、祛散寒邪作用的方药，适用于寒凝阳虚证、阳虚内寒证的治疗方法。

8.1.5.6 温肺散寒 warming lung for dispelling cold

用具有温阳补肺散寒作用的方药，适用于肺阳虚寒凝所见证候的治疗方法。

8.1.5.7 温肾散寒 warming kidney for dispelling cold

用具有温阳补肾散寒作用的方药，适用于肾阳虚寒凝所见证候的治疗方法。

8.1.5.8 暖肝散寒 warming liver for dispelling cold

用具有温阳行气、散寒止痛作用的方药，适用于寒滞肝脉证的治疗方法。

8.1.5.9 温胃散寒 warming stomach for dispelling cold

用具有温阳散寒和胃作用的方药，适用于寒邪犯胃证的治疗方法。

8.1.5.10 温通小肠 warming and dredging small intestine

用具有温阳散寒行气作用的方药，适用于寒滞肠道证的治疗方法。

8.1.5.11 温经活血 warming channel and activating blood circulation

用具有温阳通经、活血化瘀作用的方药，适用于寒凝经脉、血行不畅所致病证的治疗方法。

8.1.5.12 温经止血 warming channel for stopping bleeding

用具有温经散寒止血作用的方药，适用于寒邪所致出血病证的治疗方法。

8.1.6 清法

8.1.6.1 清热泻火 clearing heat and purging fire

运用性寒味苦的方药，清除火热之邪，适用于火热炽盛证的治疗方法。

8.1.6.2 清热解毒 clearing heat and removing toxin

运用具有寒凉解毒作用的方药，适用于火（热）毒证、火毒流窜证、火毒入络证的治疗方法。

同义词：泻火解毒

8.1.6.3 清热凉血 clearing heat and cooling the blood
运用具有凉血清热作用的方药，适用于血分证、血热炽盛证的治疗方法。

同义词：清气凉血

8.1.6.4 清营凉血 clearing the nutrient and cooling the blood
运用具有清营凉血作用的方药，适用于热入营血证的治疗方法。

8.1.6.5 清宣郁热 clearing and diffusing stagnant heat
运用具有清热泻火、理气解郁除烦作用的方药，适用于气郁化火证、热扰心神证的治疗方法。

同义词：解郁泻火

8.1.6.6 清虚热 clearing and purging deficiency-heat
运用具有滋阴清热作用的方药，适用于阴虚内热证的治疗方法。

同义词：清虚火

8.1.6.7 清里热 clearing and purging internal heat
运用具有清泻内脏邪热作用的方药，适用于脏腑实热证的治疗方法。

8.1.6.8 清心泻火 clearing the heart and purging fire
运用具有清心泻火作用的方药，适用于心火炽盛证的治疗方法。

同义词：清心泻热

8.1.6.9 清心导赤 clearing the heart and downbearing heat
运用具有清心泻火、导热下行作用的方药，适用于心火炽盛证、小肠实热证（心移热膀胱证）的治疗方法。

同义词：清泻火腑

8.1.6.10 清心解毒 clearing the heart and removing toxin
运用具有清热泻火解毒作用的方药，适用于热毒扰心的治疗方法。

8.1.6.11 清心凉营 clearing the heart and cooling the nutrient
运用具有清心泻火、凉营血作用的方药，适用于热入心营证的治疗方法。

8.1.6.12 清心凉血 clearing the heart and cooling the blood
运用具有清心泻火、凉血作用的方药，适用于血热扰神证的治疗方法。

同义词：凉血清心

8.1.6.13 清心养阴 clearing the heart and nourishing yin
运用具有清心泻火滋阴作用的方药，适用于心热阴虚证、心阴虚火旺证的治疗方法。

8.1.6.14 清心安神 clearing the heart and tranquilizing the spirit

运用具有清心泻火宁神作用的方药，适用于热扰心神证的治疗方法。

8.1.6.15 清热泻肺 clearing heat and purging the lung

运用具有清泻肺热作用的方药，适用于肺热炽盛证的治疗方法。

8.1.6.16 清胃泻热 clearing the stomach and purging heat

运用具有清胃泻火作用的方药，适用于胃火炽盛证的治疗方法。

同义词：清泻胃热

8.1.6.17 清脾泻热 clearing the spleen and purging heat

运用具有清泻脾胃火热作用的方药，适用于脾胃实热证的治疗方法。

同义词：清脾泻火

8.1.6.18 清肝泻火 clearing the liver and purging fire

运用具有清泻肝经火热作用的方药，适用于肝经火旺证、肝火炽盛证、肝火上炎证的治疗方法。

同义词：清肝泻热

8.1.6.19 清泻胆热 clearing and purging gallbladder heat

运用具有清泻胆经火热作用的方药，适用于胆经郁热证的治疗方法。

同义词：清泻胆火

8.1.6.20 清泻肝胆 clearing the liver and gallbladder

运用具有清泻肝胆火热作用的方药，适用于肝胆火旺证的治疗方法。

8.1.6.21 清心泻脾 clearing the heart and purging the spleen

运用具有清心泻脾火热作用的方药，适用于心脾积热证的治疗方法。

8.1.6.22 清心泻肝 clearing the heart and purging the liver

运用具有清泻心肝火热作用的方药，适用于心肝火旺证的治疗方法。

8.1.6.23 清心泻肺 clearing the heart and purging the lung

运用具有清泻心肺火热作用的方药，适用于心肺热盛证的治疗方法。

8.1.6.24 清心泻肾 clearing the heart and purging the kidney

运用具有清心肾火热作用的方药，适用于心肾火热证的治疗方法。

8.1.6.25 泻肝清肺 purging the liver and clearing the lung

运用具有清泻肝肺火热作用的方药，适用于肝火犯肺证、肝肺热盛证的治疗方法。

8.1.6.26 泻肝清胃 purging the liver and clearing the stomach

运用具有清泻肝胃火热作用的方药，适用于肝火犯胃证、肝胃热盛证的治疗方法。

8.1.6.27 清泻肺胃　clearing and purging the lung and stomach
运用具有清泻肺胃火热作用的方药，适用于肺胃热盛证的治疗方法。

8.1.6.28 清泻膈热　clearing and purging diaphragm heat
运用具有清热凉膈作用的方药，适用于热扰胸膈证的治疗方法。
同义词：清热凉膈

8.1.6.29 清泻肠热　clearing and purging intestinal heat
运用具有清泻肠道火热作用的方药，适用于肠道实热证的治疗方法。

8.1.6.30 清泻相火　clearing and purging ministerial fire
运用具有清热泻相火作用的方药，适用于相火偏旺证的治疗方法。

8.1.6.31 清热通淋　clearing heat to relieve stranguria
通过清泻膀胱火热达到通淋目的，适用于热淋、膀胱蓄热证的治疗方法。
同义词：泻火通淋

8.1.6.32 清热安胎　clearing heat and preventing abortion
通过清泻火热达到安胎目的，适用于热扰胞宫所致胎动不安的治疗方法。

8.1.6.33 清热生津　clearing heat and producing fluid
味甘性凉之清热药与生津药并用，适用于火热伤津证、阴虚内热证的治疗方法。

8.1.6.34 清热除蒸　clearing heat and eliminating steaming
运用具有清热降火、透热除蒸作用的方药，适用于虚火内伏所致骨蒸发热的治疗方法。

8.1.6.35 清热解暑　clearing summer heat
用清热药或解暑药，适用于外感暑热证的治疗方法。
同义词：祛暑清热

8.1.6.36 祛暑解表　dispelling summer and reliving the exterior
清热解暑药与解表药并用，适用于暑湿袭表证的治疗方法。
同义词：透表清暑

8.1.6.37 清暑解毒　clearing summer and removing toxin
运用具有清暑化湿解毒作用的方药，适用于暑湿蕴毒证的治疗方法。

8.1.6.38 清暑利湿　clearing summer and promoting diuresis
清热解暑药与化湿药并用，适用于暑湿内蕴证的治疗方法。
同义词：清化暑湿

8.1.6.39 清暑益气 clearing summer and replenishing qi

清热解暑药与补气生津药并用，适用于暑伤津气证的治疗方法。

8.1.6.40 清心涤暑 clearing away heart-heat and dispelling summer

运用具有清心解暑作用的方药，适用于暑热闭神证的治疗方法。

8.1.6.41 清暑宣肺 clearing summer to ventilate the stagnated lung qi

清热祛暑以宣降肺气，适用于暑伤肺络证所致烦渴、咳嗽、咯血等的治疗方法。

8.1.7 消法

8.1.7.1 芳香化湿 resolving dampness with aromatics

用芳香化浊辟秽之品，以祛除秽浊之邪，适用于秽浊湿邪侵袭的治疗方法。
同义词：辟秽泄浊

8.1.7.2 利湿化浊 promoting diuresis and resolving turbidity

通利小便以化湿泄浊，适用于湿邪阻滞的治疗方法。
同义词：祛湿化浊

8.1.7.3 宣散湿邪 diffusing and dissipating pathogenic dampness

用芳香辛散之品，宣化湿浊，适用于湿郁卫表证的治疗方法。
同义词：宣散湿浊

8.1.7.4 行气化湿 moving qi and resolving dampness

理气行滞药与芳香化湿药并用，适用于湿阻气滞证的治疗方法。
同义词：理气化湿

8.1.7.5 化湿和营 resolving dampness and harmonizing the nutrient

具有芳香化湿、调和营卫作用，适用于湿郁肌表所致证候的治疗方法。
同义词：燥湿和营

8.1.7.6 化湿和中 resolving dampness and harmonizing the middle

祛湿化浊，以健脾和胃，适用于湿困脾胃证、脾虚湿困证的治疗方法。
同义词：化湿运脾

8.1.7.7 健脾化湿 fortifying the spleen and resolving dampness

补益脾气，使脾气旺而能祛湿化浊，适用于脾虚湿困证的治疗方法。
同义词：扶脾祛湿

8.1.7.8 淡渗利湿 promoting diuresis with bland drugs

淡渗利湿以祛除湿邪的治疗方法。
同义词：淡渗分利

8.1.7.9 宽中利湿 loosening the middle and promoting diuresis
祛湿化浊，以健脾和胃，适用于湿困脾胃证的治疗方法。
同义词：宽中化湿

8.1.7.10 清心利湿 clearing the heart and promoting diuresis
清心泻火药与淡渗利尿药并用，导心火下泄，适用于心火炽盛证的治疗方法。
同义词：清心利水

8.1.7.11 清胆利湿 clearing the gallbladder and promoting diuresis
清胆热药与利湿药并用，适用于湿热阻滞胆腑所致病症的治疗方法。

8.1.7.12 燥湿行气 drying dampness and moving qi
理气行滞药与燥湿药并用，适用于气滞湿阻证的治疗方法。
同义词：燥湿行滞

8.1.7.13 燥湿和胃 drying dampness and harmonizing the stomach
用辛燥方药燥湿而宽中和胃，适用于湿阻中焦的治疗方法。
同义词：燥湿和中

8.1.7.14 祛风燥湿 dispelling wind and drying dampness
运用具有祛风燥湿作用的方药，适用于风湿侵袭所致病症的治疗方法。
同义词：祛风化湿

8.1.7.15 散寒除湿 dissipating cold and eliminating dampness
运用具有辛温祛寒燥湿作用的方药，适用于寒湿阻滞证的治疗方法。
同义词：散寒燥湿

8.1.7.16 清热祛湿 clearing heat and eliminating dampness
清热药与祛湿药并用，适用于湿热蕴结证的治疗方法。
同义词：清热除湿

8.1.7.17 清热利湿 clearing heat and promoting diuresis
清热药与利湿药并用，适用于湿热蕴结证的治疗方法。

8.1.7.18 清热化湿 clearing away heat and resolving dampness
清热药与化湿药并用，适用于湿热蕴结证的治疗方法。

8.1.7.19 清泻湿热 clearing heat and purging dampness
分利二便以清热祛湿，适用于湿热蕴结证的治疗方法。
同义词：化湿泻热

8.1.7.20 清利三焦 clearing and disinhibiting the triple energizer

清热药与祛湿药并用，适用于湿热弥漫三焦证的治疗方法。

同义词：清化湿热

8.1.7.21 清热燥湿解毒 clearing heat and drying dampness to remove toxin

清热化湿以解毒，适用于湿热毒蕴证的治疗方法。

同义词：泄热化湿解毒

8.1.7.22 利水消肿 disinhibiting water and alleviating edema

运用具有利水作用的方药或其他疗法，达到消除水肿目的的治疗方法，适用于水肿病症。

8.1.7.23 渗湿利水 draining dampness and disinhibiting water

渗利水湿，通利小便，适用于水湿内停证的治疗方法。

同义词：利水除湿

8.1.7.24 健脾利水 fortifying the spleen and disinhibiting water

补脾益气药与利水渗湿药并用，适用于脾虚水泛证的治疗方法。

同义词：温脾制水

8.1.7.25 温肾利水 warming the kidney and disinhibiting water

温补肾阳以利水，适用于肾阳虚水泛证的治疗方法。

同义词：温肾行水

8.1.7.26 宣肺利水 diffusing the lung and disinhibiting water

宣通肺气，上窍开而下窍泄，适用于肺失宣降所致皮水、风水等的治疗方法。

同义词：泻肺行水

8.1.7.27 疏风利水 dispersing wind and disinhibiting water

祛风解表以利水消肿，适用于风水等的治疗方法。

同义词：祛风利水

8.1.7.28 解表利水 releasing the external and disinhibiting water

发汗解表以利水消肿，适用于风水等的治疗方法。

同义词：发汗行水

8.1.7.29 散寒利水 dissipating cold and disinhibiting water

温阳散寒药与化气行水药并用，适用于阳虚水泛证、阳虚湿困证的治疗方法。

8.1.7.30 化气利水 transforming qi and disinhibiting water

温阳助气化，以利水消肿，适用于气滞水停证或气化失常所致水肿的治疗方法。

同义词：化气行水

8.1.7.31 除湿通络 eliminating dampness and unblocking the collaterals

温阳祛湿药与活血通络药并用，适用于寒湿阻络证的治疗方法。

同义词：化湿通络

8.1.7.32 消食化滞 resolving food stagnation

运用具有消食行滞作用的方药，适用于食积病证的治疗方法。

同义词：消食导滞

8.1.7.33 消食和胃 removing food stagnation and regulating the stomach

消食导滞以宽中和胃，适用于食滞胃肠证的治疗方法。

同义词：消食和中

8.1.7.34 消食止呕 removing food stagnation and stopping vomiting

消食导滞以降逆止呕，适用于食滞胃肠所致呕吐的治疗方法。

8.1.7.35 消食止痛 removing food stagnation and relieving pain

消食导滞以和胃止痛，适用于食滞胃肠所致脘腹疼痛的治疗方法。

8.1.7.36 消食解毒 removing food stagnation and removing toxin

运用具有消食导滞解毒作用的方药，适用于食毒证的治疗方法。

8.1.7.37 清热导滞 clearing heat and removing food stagnation

消食导滞药与清热药并用，适用于食滞胃热证、食滞胃肠证的治疗方法。

同义词：泻热导滞

8.1.7.38 消痞化积 relieving oppression and masses

运用具有行滞消痞化积作用的方药，适用于食积气滞所致痞胀等病证的治疗方法。

8.1.7.39 泻热消痞 clearing heat and dissolving distension

运用具有清热泻下、消食行滞作用的方药，适用于食积热滞所致痞满胀痛的治疗方法。

8.1.7.40 和胃消痞 regulating the stomach and dissolving distension

运用具有理气行滞、宽中和胃作用的方药，适用于肝脾胃肠气机不和所致痞满胀痛的治疗方法。

同义词：和中消痞

8.1.7.41 软坚散结 softening and resolving hard mass

运用具有行气活血、软坚散结等作用的方药，适用于气血瘀滞等所致瘿瘤、肿块、癥积等的治疗方法。

同义词：软坚消瘿

8.1.8 补法

8.1.8.1 补气 reinforcing qi
运用具有补益正气作用的方药，适用于气虚证的治疗方法。

8.1.8.2 补益心气 reinforcing the heart qi
运用具有补气养心作用的方药，适用于心气虚证的治疗方法。
同义词：补心益气

8.1.8.3 补益肺气 reinforcing the lung qi
运用具有补气益肺作用的方药，适用于肺气虚证的治疗方法。
同义词：补肺益气

8.1.8.4 补益中气 reinforcing and replenishing middle qi
运用具有补气健脾和胃作用的方药，适用于脾胃气虚证的治疗方法。
同义词：补脾益胃

8.1.8.5 补益肝气 reinforcing the liver qi
运用具有补气养肝作用的方药，适用于肝气虚证的方法。
同义词：补肝益气

8.1.8.6 补益肾气 reinforcing the kidney qi
运用具有补肾益气作用的方药，适用于肾气虚证的治疗方法。
同义词：补肾益气

8.1.8.7 补益心肺 reinforcing heart and lung
运用具有补益心肺之气作用的方药，适用于心肺气虚证的治疗方法。
同义词：补心益肺

8.1.8.8 补益脾肺 reinforcing spleen and lung
运用具有补益脾肺作用的方药，脾肺之气旺，适用于脾肺两虚证的治疗方法。
同义词：补脾益肺

8.1.8.9 补益脾肾 reinforcing spleen and kidney
运用具有补益脾肾之气作用的方药，适用于脾肾气虚证的治疗方法。
同义词：补脾益肾

8.1.8.10 补气固脱 reinforcing qi and securing collapse
运用具有大补元气作用的方药，适用于气脱证的治疗方法。
同义词：益气固脱

8.1.8.11 补血 reinforcing blood

运用具有补养血液作用的方药，适用于血虚证的治疗方法。

8.1.8.12 补血养心 reinforcing the blood and nourishing the heart

运用具有补血养心作用的方药，适用于心血虚证的治疗方法。

8.1.8.13 补血养肝 reinforcing the blood and nourishing the liver

运用具有补血养肝作用的方药，适用于肝血虚证的治疗方法。

8.1.8.14 补血固脱 reinforcing the blood and securing collapse

运用具有止血、大补气血等作用的方药，以补血固脱，适用于血脱证的治疗方法。

8.1.8.15 养血润肠 nourishing the blood to moisten the intestines

补血达到润肠通便的目的，适用于血虚肠燥证的治疗方法。
同义词：养血通便

8.1.8.16 养血调血 nourishing the blood and activating the blood

补血药与活血药并用，适用于血虚且运行不畅所致病症的治疗方法。
同义词：养血和血

8.1.8.17 养血安胎 nourishing the blood and preventing abortion

补血达到安胎的目的，适用于血虚所致胎动不安的治疗方法。

8.1.8.18 养血止血 nourishing the blood and stopping bleeding

补血药与止血药并用，适用于出血而血虚病症的治疗方法。

8.1.8.19 补血养神 reinforcing the blood and nourishing the spirit

运用具有补血养心安神作用的方药，适用于心血亏虚、心神失养所致病症的治疗方法。
同义词：补心安神

8.1.8.20 滋阴 enriching yin

用味甘性凉之方药，滋补阴液，适用于阴虚证的治疗方法。
同义词：甘凉滋阴

8.1.8.21 滋补心阴 enriching and reinforcing heart yin

运用具有滋阴养心作用的方药，适用于心阴虚证的治疗方法。
同义词：滋养心阴

8.1.8.22 滋补肺阴 enriching and reinforcing lung yin

运用具有滋阴补肺作用的方药，适用于肺阴虚证的治疗方法。
同义词：滋养肺阴

8.1.8.23 滋阴益胃 enriching yin and reinforcing stomach

运用具有滋阴生津、益气养胃作用的方药，适用于胃津气亏虚证、胃气阴两虚证、胃燥津伤证的治疗方法。

同义词：养胃生津

8.1.8.24 滋补脾阴 enriching and reinforcing the spleen and stomach

运用具有滋补脾阴作用的方药，适用于脾阴虚证的治疗方法。

同义词：滋养脾阴

8.1.8.25 滋补肝阴 enriching and reinforcing liver yin

运用具有滋阴养肝作用的方药，适用于肝阴虚证的治疗方法。

同义词：滋养肝阴

8.1.8.26 滋阴柔肝 enriching yin and softening the liver

运用具有滋阴养肝、柔肝作用的方药。适用于肝阴亏虚、肝失柔润所致病症的治疗方法。

同义词：养阴柔肝

8.1.8.27 滋阴疏肝 enriching yin and soothing the liver

滋阴药与疏肝理气药并用，适用于肝郁阴虚证的治疗方法。

同义词：养阴舒肝

8.1.8.28 滋阴平肝 enriching yin and calming the liver

运用具有滋阴养肝、平肝潜阳作用的方药，适用于肝阴虚阳亢证的治疗方法。

同义词：养阴平肝

8.1.8.29 滋补肾阴 enriching and reinforcing kidney yin

运用具有滋阴补肾作用的方药，适用于肾阴虚证的治疗方法。

同义词：滋养肾阴

8.1.8.30 滋肾纳气 enriching the kidney to receive qi

运用具有滋肾阴、补肾气作用的方药，适用于肾之气阴亏虚、摄纳无权所致病症的治疗方法。

8.1.8.31 滋补心肺 enriching and reinforcing the heart and lung

运用具有滋阴养心补肺作用的方药，适用于心肺阴虚证的治疗方法。

同义词：滋养心肺

8.1.8.32 滋阴清热 enriching yin and clearing heat

滋阴药与清热药并用，适用于阴虚内热证的治疗方法。

8.1.8.33 滋阴凉营 enriching yin and cooling the nutrient
滋阴药与清热凉营药并用，适用于热伤营阴证的治疗方法。

8.1.8.34 滋阴凉血 enriching yin and cooling the blood
滋阴药与清热凉血药并用适用于血热伤阴证、阴虚血热证的治疗方法。

8.1.8.35 滋阴生津 enriching yin and producing fluid
运用具有滋阴生津增液作用的方药，适用于津液亏虚证、阴虚津亏证的治疗方法。
同义词：养阴生津

8.1.8.36 滋阴止渴 enriching yin to quench thirst
运用具有滋阴生津以止口渴作用的方药，适用于阴虚津亏证的治疗方法。

8.1.8.37 滋阴降火 enriching yin and downbearing fire
滋阴药与清热降火药并用，适用于阴虚火旺证的治疗方法。

8.1.8.38 滋阴潜阳 enriching yin and subduing yang
滋阴药与重镇潜阳药并用，适用于阴虚阳亢证、阴虚阳浮证的治疗方法。
同义词：育阴潜阳

8.1.8.39 滋阴止咳 enriching yin and suppressing cough
滋阴润肺而止咳，适用于阴虚内热所致咳嗽的治疗方法。

8.1.8.40 滋阴止汗 enriching yin and checking sweating
滋阴清热而止汗，适用于阴虚内热所致盗汗的治疗方法。

8.1.8.41 滋阴通便 enriching yin and relaxing bowels
滋阴润肠而通便，适用于阴虚肠燥证的治疗方法。
同义词：滋阴润肠

8.1.8.42 补阳益气 reinforcing yang and replenishing qi
补阳药与补气药并用，适用于阳气亏虚证的治疗方法。
同义词：温阳益气

8.1.8.43 补阳 reinforcing yang
运用具有温补阳气作用的方药，适用于阳虚证的治疗方法。
同义词：壮阳

8.1.8.44 温补心阳 warming and reinforcing heart yang
运用具有温阳补心作用的方药，适用于心阳虚证的治疗方法。

8.1.8.45 温补肺阳 warming and reinforcing lung yang
运用具有温阳补肺作用的方药，适用于肺阳虚证的治疗方法。

8.1.8.46 温补脾阳 warming and reinforcing spleen yang

运用具有温阳补脾作用的方药，适用于脾阳虚证的治疗方法。

同义词：温运脾阳

8.1.8.47 温补胃阳 warming and reinforcing stomach yang

运用具有温阳健胃作用的方药，适用于胃阳虚证的治疗方法。

同义词：温中和胃

8.1.8.48 温补肝阳 warming and reinforcing liver yang

运用具有温阳补肝作用的方药，适用于肝阳虚证的治疗方法。

8.1.8.49 温补肾阳 warming and reinforcing kidney yang

运用具有温阳补肾作用的方药，适用于肾阳虚证的治疗方法。

同义词：补肾壮阳

8.1.8.50 温补命火 warming and reinforcing life fire

运用具有温肾阳补命火作用的方药，适用于命门火衰证的治疗方法。

8.1.8.51 温补纳气 warming and reinforcing receive qi

运用具有温阳益气补肾作用的方药，适用于肾阳虚而摄纳无权的治疗方法。

8.1.8.52 温补下元 warming and reinforcing the lower primordium

运用具有滋补强壮、填精益血作用的方药，以温补肾阳，适用于肾阳虚衰证的治疗方法。

同义词：温补元阳

8.1.8.53 补肾止泻 reinforcing the kidney to check diarrhea

运用温补肾阳的方药，达到止泻目的的治疗方法，适用于肾阳虚所致的泄泻。

同义词：温肾止泻

8.1.8.54 温补心肺 warming and reinforcing the heart and lung

运用具有温补心肺作用的方药，适用于心肺阳虚证的治疗方法。

8.1.8.55 温补脾胃 warming and reinforcing the spleen and stomach

运用具有温补脾胃作用的方药，适用于脾胃阳虚证的治疗方法。

8.1.8.56 温补脾肾 warming and reinforcing the spleen and kidney

运用具有温补脾肾作用的方药，适用于脾肾阳虚证的治疗方法。

8.1.8.57 温补心肾 warming and reinforcing the heart and kidney

运用具有温补心肾作用的方药，适用于心肾阳虚证的治疗方法。

8.1.8.58 温中涩肠 warming the middle and astringing the intestines

运用具有温阳补脾作用的方药,适用于脾阳虚所致泄泻的治疗方法。

同义词:温脾止泻

8.1.8.59 温中止血 warming the middle and stopping bleeding

运用具有温中补脾作用的方药,适用于脾阳虚失于统血所致出血的治疗方法。

同义词:温脾止血

8.1.8.60 温阳行气 warming yang and moving qi

温阳药与理气行滞药并用,适用于阳虚气滞证的治疗方法。

同义词:温阳理气

8.1.8.61 补益气血 tonifying qi and blood

运用具有补益气血作用的方药,适用于气血两虚证的治疗方法。

同义词:补气养血

8.1.8.62 滋阴补阳 enriching yin and reinforcing yang

运用具有滋阴温阳作用的方药,适用于阴阳两虚证的治疗方法

同义词:滋阴温阳

8.1.8.63 益气滋阴 replenishing qi and enriching yin

运用具有滋阴益气作用的方药,适用于气阴两虚证的治疗方法。

同义词:养阴益气

8.1.8.64 补益精髓 reinforcing and replenishing essence and marrow

运用具有滋补强壮、填精益血作用的方药,补肾益精髓,适用于肾虚髓亏证、精气亏虚证的治疗方法。

同义词:补精填髓

8.1.8.65 补益心肾 reinforcing and replenishing the heart and kidney

运用具有补益心肾作用的方药,适用于心肾阴阳两虚证、心肾气阴两虚证、心肾气虚证等的治疗方法。

8.1.8.66 滋阴补血 enriching yin and reinforcing blood

运用具有滋阴补血作用的方药,适用于阴血亏虚证的治疗方法。

8.1.9 八法并用

8.1.9.1 表里双解 simultaneous treatment of both the exterior and the interior

用具有解除表邪和里邪作用的方药,适用于表里同病的治疗方法。

8.1.9.2 解表攻里 releasing the external and purging the internal

解表药与攻下药并用,适用于表里俱实证的治疗方法。

同义词:发表攻下

8.1.9.3 解表清里 releasing the external and clearing the internal

解表药与清里药并用,适用于表里俱热证或表寒里热证的治疗方法。

8.1.9.4 解表温里 releasing the external and warming the internal

发汗解表药与温里祛寒药并用,适用于表里俱寒证或表热里寒证的治疗方法。

同义词:发汗温里

8.1.10 其他内治法

8.1.10.1 理气行滞 regulating qi and moving stagnation

运用具有调理气机、疏通阻滞作用的方药,适用于气滞证的治疗方法。

同义词:理气导滞

8.1.10.2 理气解郁 regulating qi and releasing depression

运用具有理气行滞解郁作用的方药,适用于气机郁滞所致病症的治疗方法。

同义词:顺气开郁

8.1.10.3 疏肝理气 dispersing the stagnated liver qi

运用具有疏肝理气、行滞解郁作用的方药,适用于肝郁气滞证的治疗方法。

同义词:疏肝解郁

8.1.10.4 疏肝利胆 soothing the liver and disinhibiting the gallbladder

运用具有疏肝理气利胆作用的方药,适用于肝胆瘀滞所致病症的治疗方法。

8.1.10.5 宣肺通气 diffusing the lung and unblocking qi

运用具有理气行滞宣肺作用的方药,适用于邪气阻滞,肺气不宣所致病症的治疗方法。

8.1.10.6 宣肺降气 diffusing the lung and downbearing qi

运用具有理气行滞、宣降肺气作用的方药,适用于邪气阻滞,肺失宣降所致病症的治疗方法。

8.1.10.7 宣肺平喘 diffusing the lung and calming panting

用理气宣肺而达平喘目的,适用于邪气阻滞,肺气不宣所致气喘的治疗方法。

8.1.10.8 泻肺平喘 purging the lung and calming panting

用宣泻肺气而达平喘目的,适用于邪气壅滞,肺气不宣所致气喘的治疗方法。

8.1.10.9 降逆平喘 downbearing counterflow and relieving panting

用宣肺降逆而达平喘目的，适用于邪气阻滞，肺气上逆所致气喘的治疗方法。

同义词：下气平喘

8.1.10.10 行气和胃 moving qi and harmonizing the stomach

用理气行滞而达和胃宽中目的，适用于胃失和降，胃气滞证的治疗方法。

同义词：理气健胃

8.1.10.11 理气健脾 regulating qi and fortifying the spleen

理气行滞药与补气健脾药并用，适用于脾失健运所见脾气郁结证、脾虚气滞证、脾气虚证等的治疗方法。

8.1.10.12 行气降逆 moving qi and downbearing counterflow

运用具有行气降逆作用的方药，适用于气逆证的治疗方法。

同义词：顺气降逆

8.1.10.13 和胃降逆 harmonizing the stomach and downbearing counterflow

运用调和胃腑，下降逆气的方药，适用于胃气上逆所致呕吐、呃逆等的治疗方法。

8.1.10.14 降逆止呕 downbearing counterflow and checking vomiting

行气和胃以降逆，适用于胃气上逆所致呕吐的治疗方法。

同义词：和胃止呕

8.1.10.15 理气消痞 regulating qi and dissolving distension

理气行滞以消除痞满结聚，适用于肝胃气滞所致痞满的治疗方法。

同义词：行气散结

8.1.10.16 理气止痛 regulating qi and relieving pain

理气行滞以止痛，适用于气机阻滞所致疼痛的治疗方法。

同义词：行气止痛

8.1.10.17 理气化瘀 regulating qi and resolving stasis

理气行滞药与活血化瘀药并用，适用于气滞血瘀证的治疗方法。

同义词：行气活血

8.1.10.18 行气破血 moving qi and breaking the blood

行气导滞药与破血化瘀药并用，适用于气滞血瘀重证的治疗方法。

8.1.10.19 理气安胎 regulating qi and preventing abortion

调理气机以安胎，适用于气滞所致胎动不安的治疗方法。

同义词：调气安胎

8.1.10.20 活血化瘀 activating blood flow and removing stagnated blood

运用具有活血化瘀作用的方药，适用于血瘀证的治疗方法。

同义词：活血祛瘀

8.1.10.21 破血行瘀 breaking the blood and moving stasis

运用药性峻猛具有活血化瘀作用的方药以破血逐瘀，适用于血瘀重证的治疗方法。

同义词：破血逐瘀

8.1.10.22 化瘀清热 resolving stasis and clearing heat

清热药与化瘀药并用，适用于血瘀化热证、血热瘀滞证的治疗方法。

同义词：清热化瘀

8.1.10.23 凉血化瘀 cooling the blood and resolving stasis

运用具有清热凉血、活血化瘀作用的方药，适用于血热瘀滞证的治疗方法。

同义词：凉血散瘀

8.1.10.24 活血行滞 activating the blood and moving stagnation

活血化瘀药与理气行滞药并用，适用于血瘀气滞证、气滞血瘀证的治疗方法。

同义词：活血行气

8.1.10.25 活血消积 activating the blood and resolving accumulation

活血化瘀以消除癥块，适用于血瘀所致癥积肿块的治疗方法。

同义词：活血消癥

8.1.10.26 祛瘀生新 dispelling stasis and promoting regeneration

活血化瘀而促进新血化生，适用于血瘀兼血虚证的治疗方法。

同义词：化瘀生新

8.1.10.27 活血养血 activating the blood and nourishing the blood

活血祛瘀药与补血药并用，适用于血虚挟瘀证的治疗方法。

同义词：化瘀养血

8.1.10.28 祛瘀通络 dispelling stasis and unblocking the collaterals

运用具有活血化瘀、疏通经络作用的方药，适用于血瘀阻络证的治疗方法。

同义词：化瘀通络

8.1.10.29 化瘀利水 resolving stasis and disinhibiting water

活血化瘀药与利水渗湿药并用，适用于血瘀水停证的治疗方法。

8.1.10.30 化瘀消肿 resolving stasis and dispersing swelling

活血化瘀以散结消肿，适用于血肿之类病症的治疗方法。

同义词：破瘀消肿

8.1.10.31 活血舒筋 activating the blood and relaxing the sinews

运用活血化瘀药与舒筋通络药，治疗血瘀筋脉不利所致病症的方法。

同义词：祛瘀舒筋

8.1.10.32 活血祛风 activating the blood and dispelling wind

活血化瘀药与祛风通络药并用，治疗血瘀风燥证或血瘀风痹所致病症的方法。

同义词：活血搜风

8.1.10.33 活血止痛 activating the blood and relieving pain

活血通络以达止痛目的，适用于血瘀所致疼痛的治疗方法。

同义词：通络止痛

8.1.10.34 化瘀止血 resolving stasis and stopping bleeding

活血化瘀以达止血目的，适用于血瘀出血证的治疗方法。

同义词：祛瘀止血

8.1.10.35 通络下乳 unblocking the collaterals and promoting lactation

活血通络以达催乳目的，适用于血瘀所致乳汁不行的治疗方法。

8.1.10.36 和络 harmonizing the collaterals

用推拿、外治或方药等，达到调和经络的目的，适用于络脉不调的治疗方法。

同义词：通络

8.1.10.37 通经止痒 unblocking the meridians and stopping itching

运用具有通经活络和血作用的方药或其他疗法，达到止痒目的，适用于经气阻痹所致病症的治疗方法。

8.1.10.38 和血安胎 harmonizing the blood and preventing abortion

活血理气以达安胎目的，适用于气血不和所致胎动不安的治疗方法。

8.1.10.39 祛瘀下胎 dispelling stasis and expediting childbirth

活血化瘀以催产下胎，适用于血瘀气滞所致胎儿不下的治疗方法。

同义词：化瘀催产

8.1.10.40 化瘀通脑 resolving stasis and unblocking the brain

活血化瘀以疏通脑络，适用于瘀阻脑络证的治疗方法。

8.1.10.41 化瘀宣肺 resolving stasis and diffusing the lung

活血化瘀以宣通肺气，适用于瘀阻肺络证的治疗方法。

8.1.10.42 化瘀宽胸 resolving stasis and loosening the chest

活血化瘀以宽胸利膈,适用于瘀阻胸膈证、瘀阻胸胁证的治疗方法。

8.1.10.43 化瘀宽心 resolving stasis and loosening the heart

活血化瘀以疏通心脉,适用于心血瘀阻证的治疗方法。

8.1.10.44 化瘀和胃 resolving stasis and harmonizing the stomach

活血化瘀以和胃宽中,适用于瘀阻胃络证的治疗方法。

8.1.10.45 化瘀养胃 resolving stasis and nourishing the stomach

活血化瘀药与养胃和中药并用,适用于胃虚血瘀证的治疗方法。

8.1.10.46 化瘀疏肝 resolving stasis and soothing the liver

活血化瘀以疏肝和络,适用于肝血瘀阻证的治疗方法。

8.1.10.47 化瘀养肝 resolving stasis and nourishing the liver

活血化瘀药与补血养肝药并用,适用于肝虚血瘀证的治疗方法。

8.1.10.48 化瘀健脾 resolving stasis and regulating the spleen

活血化瘀以健脾和中,适用于脾经血瘀证的治疗方法。

8.1.10.49 轻宣润燥 diffusing and moistening dryness

运用具有轻宣散邪、增液润燥作用的方药,适用于外燥袭表的治疗方法。

同义词:轻宣外燥

8.1.10.50 清热润燥 clearing heat and moistening dryness

运用具有清热增液润燥作用的方药,适用于温燥证的治疗方法。

8.1.10.51 清肺润燥 clearing lung and moistening dryness

运用具有清热宣肺、增液润燥作用的方药,适用于燥邪犯肺证、肺热阴虚证、肺燥郁热证等的治疗方法。

同义词:清燥润肺

8.1.10.52 滋阴润燥 nourishing yin and moistening dryness

运用具有滋阴清热、增液润燥作用的方药,适用于阴虚内燥证的治疗方法。

同义词:养阴润燥

8.1.10.53 生津润燥 producing fluid and moistening dryness

生津增液以润燥,适用于津液亏虚所致干燥病症的治疗方法。

8.1.10.54 生津止渴 producing fluid and quenching thirst

运用具有生津作用的方药,增加阴液以止渴,适用于津液亏虚所致口渴的治疗方法。

8.1.10.55 增液润肺 increasing fluid and moistening the lung
生津增液以润肺燥，适用于肺燥阴亏所致病症的治疗方法。

8.1.10.56 补血润燥 nourishing blood and moistening dryness
运用具有补血、增液润燥作用的方药，适用于阴虚血燥证、血虚风燥证的治疗方法。
同义词：养血润燥

8.1.10.57 润燥止渴 moistening dryness and quenching thirst
滋阴增液润燥以止渴，适用于阴虚津亏证的治疗方法。

8.1.10.58 润燥止咳 moistening dryness and suppressing cough
滋阴润燥以止咳，适用于阴虚肺燥证所致咳嗽的治疗方法。
同义词：润肺止咳

8.1.10.59 润燥解毒 moistening dryness and removing toxin
运用具有滋阴、润燥、解毒作用的方药，适用于燥毒证的治疗方法。

8.1.10.60 甘寒润燥 moistening dryness with sweet and cold drugs
用味甘、性寒方药补津亏或阴液不足，适用于燥热所致的阴虚津亏证的方法。

8.1.10.61 疏散风邪 expelling wind
运用具有疏风解表作用的方药，适用于风邪犯表证、风袭表疏证的治疗方法。
同义词：疏散外风

8.1.10.62 祛风解肌 dispelling pathogenic wind from muscles
运用具有疏风散邪作用的方药，适用于风邪侵袭肌表所见证候的治疗方法。
同义词：疏风解肌

8.1.10.63 平肝熄风 subduing liver and inhibiting wind
用重镇潜阳、平肝熄风方药，适用于肝阳化风证、肝阳暴亢证的治疗方法。
同义词：镇肝熄风

8.1.10.64 清热熄风 clearing heat and inhibiting wind
清热泻火而止痉熄风，适用于热极动风证的治疗方法。
同义词：泻火熄风

8.1.10.65 凉血熄风 cooling the blood and inhibiting wind
清热凉血而止痉熄风，适用于血热动风证的治疗方法。

8.1.10.66 清肝熄风 clearing liver-heat and inhibiting wind
清肝泻火而平熄风阳，适用于肝热动风证、肝阳化风证的治疗方法。
同义词：凉肝熄风

8.1.10.67 解毒熄风 removing toxin and inhibiting wind

清热解毒而止痉熄风，适用于风毒证、风毒所致抽搐的治疗方法。

8.1.10.68 潜阳熄风 suppressing yang and inhibiting wind

运用具有平肝潜阳熄风作用的方药，适用于肝阳化风证的治疗方法。

8.1.10.69 柔肝熄风 nourishing liver and inhibiting wind

滋阴养血柔肝而止痉熄风，适用于肝阴血虚动风证的治疗方法。

8.1.10.70 滋阴熄风 nourishing yin and inhibiting wind

滋阴增液而止痉熄风，适用于阴虚动风证的治疗方法。
同义词：养阴熄风

8.1.10.71 和血熄风 regulating blood and inhibiting wind

补养肝血而止痉熄风，适用于血虚动风证的治疗方法。
同义词：养血熄风

8.1.10.72 镇肝潜阳 soothing the liver and relieving yang

用重镇之品平肝潜阳，适用于肝阳上亢证、肝阳暴亢证的治疗方法。
同义词：平肝潜阳

8.1.10.73 熄风解痉 extinguishing wind and relieving spasm

运用具有熄风止痉作用的方药，适用于肝风内动证所见抽搐等的治疗方法。
同义词：镇痉熄风

8.1.10.74 宣肺化痰 ventilating lung and dissipating phlegm

祛除痰浊以宣肺，适用于痰浊阻肺证的治疗方法。
同义词：豁痰宣肺

8.1.10.75 燥湿化痰 drying dampness and dissipating phlegm

运用具有燥湿化痰作用的方药，适用于湿痰证的治疗方法。
同义词：豁痰燥湿

8.1.10.76 清热化痰 clearing heat and dissipating phlegm

运用清化热痰药，或清热药与祛痰药并用，适用于热痰证的治疗方法。
同义词：清热祛痰

8.1.10.77 润肺化痰 moistening lung and dissipating phlegm

运用具有增液润燥化痰作用的方药，适用于燥痰证、燥痰结肺证的治疗方法。
同义词：润燥化痰

8.1.10.78 温化寒痰 warming for resolving cold-phlegm
运用具有温阳、祛寒、化痰作用的方药，适用于寒痰证的治疗方法。
同义词：温阳化痰

8.1.10.79 祛风化痰 dissipating wind-phlegm
疏风药与祛痰药并用，适用于风痰证的治疗方法。
同义词：治风化痰

8.1.10.80 理气化痰 regulating qi-flowing for eliminating phlegm
理气行滞药与祛痰药并用，适用于气滞痰凝证、痰气互结证的治疗方法。
同义词：行气祛痰

8.1.10.81 散寒化饮 dispelling cold and resolving fluid retention
用辛温之药，散寒化饮，适用于寒饮内停的治疗方法。

8.1.10.82 泻肺逐饮 eliminating pathogens from lung for expelling fluid retention
化饮药与泻下药并用，适用于饮邪客肺证的治疗方法。

8.1.10.83 温阳化饮 warming yang to dissipate fluid retention
运用具有辛温助阳化饮作用的方药，适用于阳虚饮停证的治疗方法。
同义词：温散寒饮

8.1.10.84 温化痰饮 warming phlegm and fluiding retention
运用辛温之品，祛痰化饮，适用于阳虚痰饮内停证的治疗方法。

8.1.10.85 化痰消食 relieving phlegm and food
祛痰药与消食导滞药并用，适用于痰食互结证的治疗方法。

8.1.10.86 祛痰化瘀 removing phlegm and removing blood stasis
祛痰药与化瘀药并用，适用于痰瘀互结证的治疗方法。

8.1.10.87 化痰散结 reducing phlegm and resolving masses
祛痰药与消积软坚药并用，适用于痰核留结证的治疗方法。
同义词：祛痰软坚

8.1.10.88 化痰消瘿 dissipating phlegm for eliminating goiter
祛痰药与消散软坚药并用，适用于痰核瘤结证的治疗方法。
同义词：化痰消瘤

8.1.10.89 祛痰宣痹 inducing expectoration and dredging channel blockage
用祛除痰浊的方药，以宣通痹阻，适用于痰浊阻痹所致病症的治疗方法。

8.1.10.90 清热祛痰化瘀 clearing heat, inducing expectoration and removing stagnated blood

清热祛痰与化瘀药并用，适用于痰瘀化热证的治疗方法。

8.1.10.91 祛痰杀虫 inducing expectoration and expelling parasite

祛痰药与杀虫药并用，适用于痰虫互结证的治疗方法。

8.1.10.92 祛痰理气解毒 inducing expectoration, regulating the stagnated qi and removing toxin

祛痰、行气、解毒药并用，适用于痰结毒滞证的治疗方法。

8.1.10.93 开窍通闭 inducing resuscitation and opening closed resistance

运用具有开窍醒神、宣通闭阻作用的方药，适用于各种闭病的治疗方法。
同义词：通窍开闭

8.1.10.94 清热开窍 clearing heat for resuscitation

清热泻火以开窍醒神，适用于热闭心神证的治疗方法。
同义词：泻热开窍

8.1.10.95 清心开窍 clearing away heart-heat for resuscitation

清热泻火以开窍醒神，适用于热闭心包证的治疗方法。

8.1.10.96 解毒开窍 removing toxin for resuscitation

解毒化浊以开窍醒神，适用于浊毒闭神证的治疗方法。

8.1.10.97 凉营开窍 dispelling heat from the ying fen system for resuscitation

清热凉营以开窍醒神，适用于热入心营等的治疗方法。

8.1.10.98 祛暑开窍 dispelling summer for resuscitation

清热祛暑以开窍醒神，适用于暑热闭神证的治疗方法。

8.1.10.99 宁心开窍 tranquilizing mind for resuscitation

运用具有养心宁神定志作用的方药，适用于心厥、血厥及惊恐伤神证所致神昏的治疗方法。
同义词：宁神开窍

8.1.10.100 芳香开窍 inducing resuscitation with aromatics

运用芳香之品，以宣闭开窍，适用于秽浊之邪阻闭心包所致神昏的治疗方法。
同义词：辟秽通窍

8.1.10.101 散寒开窍 dispelling cold for resuscitation

用辛温香窜行气药，以温阳祛寒，宣闭通窍，适用于寒厥及寒凝气滞、窍闭所见肢厥、神昏、疼痛等的治疗方法。

同义词：辛温通窍

8.1.10.102 化湿开窍 dispersing dampness for resuscitation

运用具有化湿通窍作用的方药，适用于湿浊蒙闭心神所见证候的治疗方法。

同义词：化湿通窍；利湿通窍

8.1.10.103 化痰开窍 dissipating phlegm for resuscitation

祛除痰浊以宣闭开窍，适用于痰厥证的治疗方法。

同义词：祛痰开窍

8.1.10.104 祛痰化湿开窍 inducing expectoration and dispersing dampness for resuscitation

祛痰化湿以宣闭开窍，适用于痰湿闭神证的治疗方法。

8.1.10.105 搜风祛痰开窍 removing the wind and inducing expectoration for resuscitation

搜风祛痰以宣闭开窍，适用于风痰闭神证的治疗方法。

8.1.10.106 祛暑开闭 dispelling summer and inducing resuscitation

清热祛暑药与行气开闭药并用，适用于暑闭气机证的治疗方法。

8.1.10.107 解毒开闭 removing toxin and inducing resuscitation

运用具有解毒作用的方药，适用于疫毒内闭证的治疗方法。

8.1.10.108 解毒宣肺开闭 removing toxin to ventilate lung and inducing resuscitation

解毒祛邪以宣肺开闭，适用于热毒闭肺证的治疗方法。

8.1.10.109 开噤通关 opening the locked jaw for resuscitation

将开通关窍药物敷于牙龈上或吹入鼻内，以达苏醒的目的，适用于突然昏倒伴有牙关紧闭的治疗方法。

8.1.10.110 驱蛔杀虫 expelling and killing ascarid

运用具有驱杀蛔虫作用的方药，适用于虫积肠道证的治疗方法。

8.1.10.111 健脾驱虫 strengthening spleen and expelling intestinal parasites

补脾药与驱虫药并用，补脾益气以消除蛔虫，适用于脾虚虫积证的治疗方法。

同义词：补脾驱虫

8.1.10.112 杀虫消积 destroying parasites and dispersing accumulation

运用具有杀虫消积作用的方药，适用于虫积所致疳积的治疗方法。

同义词：杀虫消疳

8.1.10.113 祛风杀虫 dispelling wind and expelling parasites

运用具有祛风杀虫作用的方药，适用于虫毒、风毒蕴结肌肤所致病症的治疗方法。

8.1.10.114 杀虫宁神 expelling parasites and tranquilizing mind

运用具有杀虫作用的方药，以宁神止痛定痫，适用于囊虫侵脑证的治疗方法。

8.1.10.115 杀虫解毒宣肺 expelling parasites, removing toxin and ventilating lung

运用具有杀虫解毒、宣肺理气作用的方药，适用于虫毒犯肺证的治疗方法。

8.1.10.116 养心安神 tranquilizing mind by nourishing the heart

运用具有补心养血作用的方药，适用于心虚神怯所见惊悸、不安等症的治疗方法。
同义词：补心安神

8.1.10.117 养血安神 tranquilizing mind by enriching blood

运用具有补血养心作用的方药，适用于心血虚证所见惊悸、不安、胆怯等症的治疗方法。

8.1.10.118 益肾宁神 tranquilizing mind by nourishing kidney

运用具有补肾益精作用的方药，适用于恐伤神气证或肾虚心肾不交所致惊悸、不安、胆怯等症的治疗方法。

8.1.10.119 益气安神 tranquilizing mind by replenishing qi

补气养心以安神，适用于心气虚所致心神不安的治疗方法。
同义词：益气定神

8.1.10.120 滋阴安神 tranquilizing mind by nourishing yin

滋阴养心以安神，适用于心阴虚所致心神不安的治疗方法。
同义词：养阴安神

8.1.10.121 镇心安神 tranquilizing mind by suppressing heart

用重镇之品安神定志，适用于惊恐伤神证的治疗方法。
同义词：镇惊安神

8.1.10.122 敛汗固表 checking profuse sweating and securing the exterior

收敛固表以止汗，适用于表虚卫气不固而以自汗为主症的治疗方法。
同义词：固表敛汗

8.1.10.123 敛肺止咳 constraining lung to stop cough

补益收敛肺气而达到止咳目的，适用于肺气亏虚而以咳嗽为主症的治疗方法。

8.1.10.124 敛肺平喘 constraining lung to relieve asthma

补益收敛肺气而达到平喘目的，适用于肺气亏虚而以气喘为主症的治疗方法。

8.1.10.125 涩肠止泻 treating diarrhea with astringents
收敛固涩肠道，达到止泻目的的治疗方法，适用于久泻不止的病症。
同义词：固肠止泻

8.1.10.126 涩精止遗 arresting seminal emission
收敛固涩，改善公畜性功能，适用于遗精滑精的治疗方法。

8.1.10.127 固摄止血 inducing astringency to stop bleeding
收敛固涩，达到止血目的的治疗方法，适用于气不摄血所致出血的病症。
同义词：摄血止血

8.1.10.128 固肾安胎 nourishing kidney and soothing the fetus
补益肾气以安胎，适用于肾虚所致胎动不安的治疗方法。

8.1.10.129 止血安胎 stopping bleeding and soothing the fetus
止血药与安胎药并用，以达保胎目的的治疗方法。

8.1.10.130 固涩敛乳 astringing for arresting lactation
运用具有补气收敛作用的方药，适用于气虚不能敛乳而乳汁自出的治疗方法。
同义词：补气摄乳

8.1.10.131 固摄敛泪 inducing astringency to restrain teardrop
运用具有补气收敛作用的方药，适用于气虚不能收敛泪液所致流泪症的治疗方法。
同义词：收敛止泪

8.1.10.132 固脬止遗 reducing urination for preventing enuresis
补益肾气，以固护膀胱，适用于肾气不固所致遗尿、尿失禁等病症的治疗方法。
同义词：固脬止淋

8.1.10.133 补气固摄 replenishing qi to induce astringency
补气增强固摄功能，适用于气虚不摄所致病症的治疗方法。
同义词：益气固摄

8.1.10.134 敛阴固脱 constraining yin fluids and inducing astringency
用收敛固涩和滋阴增液的方药，适用于阴液欲脱所见证候的治疗方法。
同义词：固护阴液

8.1.10.135 温阳固脱 warming yang and inducing astringency
温阳益气药与固涩药并用，适用于阳气欲脱病症的治疗方法。
同义词：温涩固脱

8.1.10.136 升提固涩 elevating and inducing astringency

益气升提药与收敛固涩药并用，适用于气虚不能固摄、精微下泄所见证候的治疗方法。

同义词：升清固涩

8.1.10.137 补肾固涩 nourishing kidney to induce astringency

运用具有补益肾气、增强收摄作用的方药，适用于肾气不固证的治疗方法。

同义词：固摄下元

8.1.10.138 壮水制阳 invigorating water to control yang

采用滋补阴液，适用于阴液充足而能抑制阳气偏亢的治疗方法。

同义词：滋阴抑阳

8.1.10.139 益火消阴 replenishing fire to disperse yin

采用温补阳气为主，使阳气旺以消散阴寒的治疗方法，适用于阳虚而阴寒偏胜的证候。

同义词：壮阳消阴

8.1.10.140 引火归原 directing fire back to its origin

温补阳气，适当加入引经药，使浮越的阳气得以敛藏，亦即使肾阳寓于肾阴之中的治疗方法，适用于命门火衰，虚阳浮越的证候。

同义词：导龙入海

8.1.10.141 和营生新 harmonizing the nutrient to promote regeneration

调和营血、疏通血行，以促进新血化生，或促进新肉生长、创面愈合的治疗方法，适用于因血行不畅而影响新血化生或创面愈合的病症。

8.1.10.142 宣通三焦 diffusing and unblocking the triple energizer

理气祛邪，疏通气机，使三焦得以通利的治疗方法。适用于邪阻气滞而三焦不利的证候。

8.1.10.143 交通心肾 coordinating the heart and kidney

滋肾阴、敛肾阳，降心火、安心神，达到滋阴潜阳、沟通心肾目的的治疗方法，适用于心肾不交、阴阳失调的证候。

8.1.10.144 清利头目 refreshing the head and eyes

清热泻火、潜阳降逆，而使头目清利的治疗方法，适用于火热上扰、阳亢气逆而使头目疼痛之类病症。

8.1.10.145 滑利窍道 lubricating the orifices and canals

运用具有滑利作用的方药等，使窍道通利的治疗方法，适用于因阴液不足所致的窍道干涩病症。

8.1.10.146 宽胸利膈 loosening the chest and disinhibiting the diaphragm

祛邪理气，或升阳降浊，使胸膈得以舒畅的治疗方法，适用于胸膈痞闷、胀满、梗塞的病症。

8.2 外治法

8.2.1 解毒散痈 removing toxin for eliminating carbuncles

运用祛除毒邪的方药及其他疗法，以排出痈疡内蕴之毒邪，达到消散痈疡目的的治疗方法。

8.2.2 活血解毒 promoting blood circulation and detoxication

用具有活血化瘀、和营通络、祛除毒邪作用的方药及其他疗法，治疗痈疡的方法。

8.2.3 解毒护阴 removing toxin for protecting yin

用具有祛除毒邪、滋阴养液作用的方药及其他疗法，治疗热毒伤阴之疮疡病证的方法。

8.2.4 解毒消肿 removing toxin and reducing swelling

运用祛除毒邪的方药及其他疗法，达到消散疮疡目的的治疗方法。
同义词：解毒散结

8.2.5 清热排脓 clearing heat and evacuating pus

运用具有清热解毒作用的方药，达到清解疮疡内蕴之热毒、排除脓液目的的治疗方法。

8.2.6 去腐生肌 eliminating slough and promoting granulation

运用具有祛腐作用的方药及其他疗法，使腐肉脱落，达到促进新肉生长，以加速疮口愈合目的的治疗方法。
同义词：祛腐生肌

8.2.7 消痈散疖 resolving carbuncle and expulsing boil

运用具有消散作用的方药及其他疗法，达到消肿解毒、促进疮疡消散目的的治疗方法。

8.2.8 燥湿敛疮 eliminating dampness and astringing sores

运用具有燥湿和敛疮作用的方药，达到促进新肉生长、加速疮口愈合目的的治疗方法。

8.2.9 清解余毒 expelling retained toxin

运用具有清热解毒作用的方药，为治疗疮疡后期余毒未尽的方法。
同义词：清透伏邪

8.2.10 敛疮止痛 healing sore and relieving pain

运用具有敛疮和止痛作用的方药，达到减轻疼痛、促进疮口愈合的治疗方法。

8.2.11 明目 improving vision

祛风、清热、凉血、化瘀、养血、益气、滋阴、健脾、补肝肾等方法，以祛邪扶正，达到提高视力目的的治疗方法。主要适用于视物不清的内外障眼病。

8.2.12 退翳明目 removing nebula and improving vision

用退翳药与辛散、滋阴、活血等药物配合，治疗黑睛宿翳，使之缩小或变薄，从而起到提高视力作用的治疗方法。

8.2.13 祛风明目 dispelling wind and improving vision

用药物祛散风邪，提高视力，主要适用于风邪犯眼所致的眼病。
同义词：疏搜风明目

8.2.14 清热明目 clearing heat and improving vision

用具有清肝泻火、解毒作用的方药，起到祛邪明目作用的治疗方法。主要适用于热性眼病。
同义词：清肝明目

8.2.15 补肾明目 tonifying the kidney and improving vision

滋肾明目和温肾明目。用具有滋补肾阴作用，或附子、肉桂、巴戟天、肉苁蓉等具有温补肾阳作用的方药，达到补肾明目的治疗方法，主要适用于肾虚所致眼病。

8.2.16 滋肝明目 enriching the liver and improving vision

用具有滋补肝阴作用的方药，达到滋肝明目的治疗方法，主要适用于肝阴亏虚所致的眼病。

8.2.17 养血明目 nourishing the blood and improving vision

用具有补血养肝作用的方药，达到补益肝血而明目的治疗方法。
同义词：养肝明目

8.2.18 通耳 unblocking the ears

祛邪扶正，以通利耳窍的治疗方法。

同义词：利耳

8.2.19 疏风宣耳 dispersing wind and diffusing the ears

运用具有疏风解表、宣通清窍作用的方药，适用于外邪侵犯耳窍所致耳病的治疗方法。

同义词：祛风通耳；疏风通耳

8.2.20 解毒利耳 removing toxin and disinhibiting the ears

运用具有清热解毒作用的方药，适用于热毒上壅耳窍所致耳病的治疗方法。

8.2.21 补血养耳 tonifying the blood and nourishing the ears

运用具有补血养肝作用的方药，适用于血虚耳窍失养所致耳聋等耳部病的治疗方法。

8.2.22 益气通耳 replenishing qi and unblocking the ears

运用具有健脾益气、温中升阳作用的方药，适用于脾虚弱、中气下陷，耳窍失煦所致耳病的治疗方法。

8.2.23 滋阴濡耳 enriching yin to moisten the ears

运用具有滋养阴液作用的方药，适用于阴液亏虚、虚火上炎所致耳聋等耳部病的治疗方法。

同义词：养阴濡耳

8.2.24 滋肝肾濡耳 enriching the liver and kidney to moisten the ears

运用具有滋补肝肾作用的方药，适用于肝肾阴精亏损所致耳聋等耳部病的治疗方法。

8.2.25 通鼻 unblocking the nose

祛邪扶正，以通利鼻窍的治疗方法。

8.2.26 疏风通鼻 dispelling wind and unblock the nose

运用具有疏风宣肺、解表通窍作用的方药，适用于风邪犯肺，肺窍窒塞所致鼻病的治疗方法。

8.2.27 散寒通鼻 dissipating cold and unblocking the nose

运用具有温散寒邪、宣通肺窍作用的方药，适用于寒邪凝滞，鼻窍不利所致鼻病的治疗方法。

8.2.28 清燥润鼻 clearing dryness and moistening the nose

运有清宣燥热、润肺生津作用的方药，适用于燥伤肺津，肺燥阴虚，鼻窍失濡所致鼻病的治疗方法。

8.2.29 祛瘀通鼻 dispelling stasis and unblocking the nose

运用具有活血化瘀、行气通窍作用的方药，适用于瘀血阻滞鼻窍所致鼻病的治疗方法。

8.2.30 芳香通鼻 unblocking the nose with aroma

运用具有辛散走窜、芳香宣散作用的方药，适用下外邪壅塞肺系，肺气失宣，鼻窍不利所致鼻病的治疗方法。

同义词：芳香通窍

8.2.31 利咽 disinhibiting the pharynx

祛邪扶正，以通利咽喉的治疗方法。

8.2.32 疏风利咽 dispersing wind and disinhibiting the pharynx

运有疏风解表作用的方药，适用于风邪侵袭咽喉，咽喉不利所致咽喉病变的治疗方法。

8.2.33 散寒利咽 dissipating cold and disinhibiting the pharynx

运用有温散寒邪作用的方药，适用于寒邪侵袭咽喉，咽喉不利所致咽喉病变的治疗方法。

8.2.34 祛痰利咽 dispelling phlegm and disinhibiting the pharynx

运用具有祛痰化浊作用的方药，适用于痰浊凝滞咽喉，咽喉不利所致咽喉病变的治疗方法。

8.2.35 理气利咽 regulating qi and disinhibiting the pharynx

运用具有疏肝理气、行滞利咽作用的方药，适用于肝郁气滞，咽喉不利所致咽喉病变的治疗方法。

8.2.36 生津利咽 producing fluid and disinhibiting the pharynx

运用具有生津增液作用的方药，适用于燥热伤阴，津液亏虚，咽喉失润所致咽喉病变的治疗方法。

同义词：润燥利咽

8.2.37 固齿 strengthening the teeth

祛邪扶正，以达牙齿健固、牙龈充润目的的治疗方法。

8.2.38 补肾固齿 tonifying the kidney to strengthen the teeth

运用具有补肾培本作用的方药，适用于肾气亏虚，齿失充养所致口齿病症的治疗方法。

8.2.39 益气固齿 replenishing qi to strengthen the teeth

运用具有补脾益气作用的方药，适用于气虚齿牙疏豁等口齿病症的治疗方法。

8.2.40 补血健齿 tonifying the blood and fortifying the teeth

运用具有补血养血作用的方药，适用于血虚齿豁等病症的治疗方法。

同义词：养血固齿

8.2.41 滋阴润齿 enriching yin to moisten the teeth

运用具有滋补阴液作用的方药，适用于阴液亏虚，齿失濡润所致口齿病症的治疗方法。

8.2.42 温阳健齿 warming yang to fortify the teeth

运用具有温补阳气作用的方药，适用于阳气亏虚，齿失温煦所致口齿病症的治疗方法。

8.2.43 降火固齿 downbearing fire to strengthen the teeth

运用具有清热泻火、生津润燥作用的方药，适用于火热炽盛，耗伤津液，齿龈失润所致病症的治疗方法。

同义词：清热固齿

8.2.44 解毒利龈 removing toxin and disinhibiting the gums

运用具有清热泻火解毒作用的方药，适用于火毒结聚齿龈所致口齿病症的治疗方法。

8.2.45 药熨疗法 medicated ironing therapy

将药物（如药袋、药饼、药膏及药酒）加热后置于患病动物体表特定部位，作热熨或往复移动，促使腠理温润、气血调和、经脉舒畅的一种外治疗法。多用于风寒湿痹、气血瘀滞、虚寒证候等。

同义词：热熨疗法

8.2.46 热敷疗法 hot compress therapy

将发热的物体置于身体的患病部位或特定部位（如穴位），以防治疾病的一种方法。

可分为水热敷法、醋热敷法、姜热敷法、葱热敷法、盐热敷法、砂热敷法、蒸饼热敷法、铁末热敷法、砖热敷法等。常用于胃肠疾患、腰腿痛、冻疮、乳痈等。

8.2.47 敷贴疗法 paste application therapy

将药物调成糊状，敷于体表的特定部位，以治疗疾病的一种方法。用于呕泻、自汗盗汗、脱肛、面瘫、风湿痹痛、疮痈癣疹、扭挫伤、口腔糜烂、烫伤等。

8.2.48 膏药疗法 plaster application therapy

以膏药敷贴治疗疾病的一种外治疗法。由于膏药方剂的组成不同，功效主治各异，故应辨证施治。常用于风湿痹痛、跌打损伤、疮疖痈肿、流痰、溃疡等。

同义词：薄贴疗法

8.2.49 药膏疗法 ointment application therapy

将外用药膏敷贴于肌肤，药膏通过皮肤、黏膜的吸收，起到行气活血、疏通经络、祛邪外出的治疗作用。主要用于损伤、骨折、局部感染等。

8.2.50 湿敷疗法 wet compress therapy

用纱布蘸取药液敷于患处以治疗疾病的一种方法。主要适用于疮疡痈肿、皮肤疾患、烧伤、烫伤等。

8.2.51 熏洗疗法 fumigating and washing therapy

利用药物煎汤的热蒸气熏蒸患处，并用温热药液淋洗局部的一种外治疗法。用于风寒感冒、风湿痹痛、湿疹、癣疥、肛门病、阴痒、眼疾、跌打损伤等病症。如眼科的熏洗法、洗眼法（包括结膜囊冲洗法、泪道进冲洗法）等。

8.2.52 冲洗疗法 flushing and washing therapy

用药液反复冲洗患病部位的一种外治疗法。用于五官科疾病、子宫疾病、肛肠疾病等。

8.2.53 浸洗疗法 steeping and washing therapy

用药物煎汤，浸洗患部的一种外治疗法。用于各种癣病、跌损肿痛、脱肛等。

8.2.54 腐蚀疗法 corrosion therapy

选用具有腐蚀作用的药物，敷涂患处，以蚀去恶肉，促使新肉长出的一种外治疗法。适用于体表疮疡、癌瘤、流痰等。

8.2.55 切开疗法 incision therapy

用手术刀切开病灶或脓肿的一种治疗方法，适用于痈疡已形成脓肿等。

8.2.56 引流疗法 drainage therapy

以导管或纱布条、药捻纸条等插入脓疡创孔,使脓液排出畅通的一种治疗方法。适用于外科疮疡、腹腔手术后引流等。

8.2.57 放血疗法 blood-letting therapy

用针具或刀具刺破或划破动物血针穴位或持定部位,放出少量血液。适用于高热、神昏、中暑、感冒、各种疼痛、风眩、急惊风、中毒、毒蛇咬伤等的一种外治疗法。

同义词:针刺放血疗法

8.2.58 火烙疗法 fire cauterizing therapy

用针具烙铁等烙烫病变局部组织,以治疗疾病的一种方法。主要适用于肢蹄病和其他外科及皮肤疾病。

8.2.59 刮痧疗法 scraping therapy

用边缘光滑的刮痧板或自制工具,蘸食油或清水在体表部位进行反复刮动,用以治疗疾病的一种方法。主要用于"痧症"及中暑、感冒、喉痛、腹痛、呕泻等。

8.2.60 点眼疗法 eye drop therapy

将点眼药水、点眼药粉、涂眼药膏、放眼药膜等具有消红肿、去眵泪、止痛痒、除翳膜等作用的药物直接点入眼内,以治疗眼病的一种外治方法。主要适用于外眼疾病。

8.2.61 吹耳疗法 ear-insufflation therapy

将药物吹布于外耳道内或鼓膜上以治疗疾病的一种方法,主要适用于中耳炎、耳螨症等疾病。

同义词:耳内吹粉疗法

8.2.62 滴耳疗法 ear drop therapy

将具有清热解毒、消肿收敛等作用的药液滴入耳道,并适当在耳屏上按压,使药液进入耳道深部,从而治疗疾病的一种外治方法。适用于耳道流脓、耳道糜烂、耳部疮痈疔毒、耳道赘生物及异物入耳等病症。

8.2.63 洗耳疗法 ear washing therapy

用药液清洗耳部患处,或以棉签蘸药液清洁外耳道、耳廓等处的治疗方法。适用于耳道流脓、耳部痈疽疖疮等病症。

8.2.64 塞鼻疗法 nose insertion therapy

将药物塞入鼻内,通过局部作用和鼻黏膜吸收,以治疗疾病的一种方法。多用于鼻部（如嗅觉障碍、鼻甲肥大、鼻塞不通等）、头面部及口腔病症。

8.2.65 吹鼻疗法 nose insufflation therapy

将药物吹喷入鼻腔内，经局部作用或黏膜吸收而治疗疾病的一种方法。适用于冷痛、鼻衄、鼻塞、感冒等疾病。

同义词：吸药疗法

8.2.66 滴鼻疗法 medicated liquid nose drops therapy

将药液滴入鼻腔内，以治疗各种鼻病的方法。常用于伤风鼻塞、鼻窒、鼻鼽、鼻渊、鼻衄等。

8.2.67 取嚏疗法 catching therapy

将芳香辛窜之药末吹入患病动物鼻腔，通过药物对鼻黏膜的刺激，使之引起喷嚏反射，从而达到祛除病邪、治疗疾病的一种方法。多用于冷痛、昏厥及感冒等。

同义词：催嚏疗法；开窍疗法

8.2.68 喷雾疗法 aerosol therapy

将药物的溶液或极细粉末经喷雾器或雾化器等形成药物蒸气、雾粒或气溶胶，供呼吸道吸入或体表喷洒，以治疗疾病的一种方法。常用于治疗呼吸道病或虫症等。

8.2.69 灌肠疗法 coloclysis therapy

以药液灌入肠道内以治疗疾病的一种治疗方法。适用于便秘、肠黄或不适合经口投药的某些疾病。

8.2.70 包扎固定疗法 bandage-fixing therapy

用绷带、夹板、胶布或某些特制器械，将患病部位按要求包扎固定，以利于损伤组织恢复的一种治疗方法。主要用于骨折、脱位及伤筋。

8.2.71 夹板固定疗法 splint-fixing therapy

用扎带或绷带把木板、竹板或钢料制成的夹板固定在骨折已经复位的肢体上，以促进骨折愈合的一种治疗方法。

8.2.72 火烧战船 burning warships

灸熨疗法之一。俗称火烧战船。施术时先用醋直接刷湿术部或搭上醋浸湿的粗布，防止术部皮肤受损，然后喷洒酒精或白酒，点燃后再交替喷洒和浇醋，反复多次直至患畜耳根或腋下出汗为止。能起到发汗解肌、祛风散寒除湿的作用。主治全身风湿、腰胯风湿等。术后要注意保暖。对瘦弱病畜及孕畜应慎用。

同义词：醋酒灸

8.3 针灸疗法

8.3.1 火针疗法 fire needling therapy

将针具烧红后迅速刺入穴位，以治疗疾病的一种方法。临床主要用于各种风寒湿痹、慢性跛行及阳虚泄泻等。

8.3.2 七星针疗法 seven-star needling therapy

运用多支短针（丛针）进行浅刺、速刺而不留针的一种针刺疗法。适用于头痛、脊背痛、胁痛等。

8.3.3 三棱针疗法 three-edged needling therapy

用三棱形不锈钢针，刺破穴位或浅表血络，放出少量血液，以治疗疾病。适用于各种实证、热证、血肿、疼痛等。

8.3.4 点刺疗法 pricking therapy

用锋利的针，在动物体皮肤表面或关节周围、脊椎两旁进行轻刺、点刺、快刺，以治疗疾病。多用于实热证、疼痛或疮疡肿毒等。

同义词：快速浅刺疗法

8.3.5 隔姜灸疗法 moxibustion on ginger therapy

在艾炷与皮肤之间隔一姜片进行施灸，以防病治病。可防治感冒、寒性呕泻、腹痛、咳喘等病症。

8.3.6 隔蒜灸疗法 moxibustion on garlic therapy

在艾炷与皮肤之间衬隔蒜片或蒜饼进行施灸，以防治疾病。多用于治疗疮、疖、乳痈、肢体痹病等。

8.3.7 艾灸疗法 moxibustion therapy

运用艾条或其他药物在动物体表病变部位或穴位上烧灼、温烫，通过温和热力和药力的作用，以达到温通气血，扶正祛邪的治疗方法。

8.3.8 拔罐疗法 cupping therapy

以杯罐为工具，采取一定的措施，排去罐中部分空气，产生负压，使之吸着于皮肤，造成被拔部位的皮肤轻度瘀血现象，以治疗疾病。适用于痹病、腰腿疼痛、跌打损伤等。

同义词：吸筒疗法

8.3.9 电针疗法 electro-acupuncture therapy

应用各种脉冲电针仪在刺入动物体有关穴位的毫针上导入脉冲电流，以加强得气感觉，从而提高针刺疗效的一种方法。其适用范围与毫针疗法相似。

8.3.10 激光针疗法 laser-acupuncture therapy

应用医用激光仪输出的激光束替代毫针来刺激穴位，以治疗疾病的方法。其适应证与毫针疗法类同，清热作用更为明显。

8.3.11 水针疗法 hydro-acupuncture therapy

选用药物注入有关穴位，集针刺与药物治疗的共同作用以治疗疾病的方法。适用于风湿痹痛、脏腑实热或积滞等病变。

8.3.12 穴位埋线疗法 catgut embedding therapy

将羊肠线等埋入穴位，利用埋植物对穴位持续刺激作用以治疗疾病的方法。多用于风湿痹痛、面瘫、癫痫及跌打损伤等病症。

8.3.13 针刺麻醉法 acupuncture anesthesia

运用针灸针刺入穴位而产生镇痛或麻醉的方法。适用于多种外科手术。

8.3.14 耳针疗法 auriculo-acupuncture therapy

以毫针、皮内针、激光照射等器具，通过对耳廓穴位的刺激以防治疾病。可应用于便秘、冷痛、皮肤瘙痒等多种病症。

8.4 推拿疗法

8.4.1 推拿疗法 massage therapy

在中兽医学理论指导下，通过在动物体体表一定的部位或穴位施以各种手法，或配合某些特定肢体活动，以防治疾病的方法。适用于多种疾病，尤其对跌打损伤及各种疼痛性疾患更为适宜。

同义词：按摩疗法

8.5 饮食疗法

8.5.1 食疗 dietotherapy

应用具有药理作用的食物防治疾病的一种方法。适用于气血虚弱、营养缺乏、生长发育迟缓、生产性能低下等。

同义词：食物疗法；食治

8.5.2 药酒疗法 medicated liquor therapy

用药酒内服或外用以防治疾病的一种方法。适用于草慢不食、宿草不转、跌打损伤、风湿痹痛与疥癣顽疮等。

8.6 杂疗法

8.6.1 热蜡疗法 hot wax therapy

将液态或半固态的黄蜡、石蜡或地蜡涂布或热敷患处，以治疗疾病的一种方法。用于腰腿疼痛、冻伤、某些皮肤病变等。

8.6.2 药浴疗法 medicinal bath therapy

将动物浸泡在药液中，以此来治疗疾病的一种方法。多用于皮肤疮癣、皮肤瘙痒等。

8.6.3 烟熏疗法 fumigation therapy

利用药物燃烧后的烟气熏蒸来防治疾病的一种方法。可用于急救晕厥、肛肠与口唇耳鼻疾病，或作为杀虫避秽、预防疾病之用。

8.6.4 夹板固定疗法 splint-fixing therapy

用扎带或绷带把木板、竹板或钥料制成的夹板固定在骨折已经复位的肢体上，以促进骨折愈合的一种治疗方法。

第九章 针 灸

9.1 基础术语

9.1.1 针灸 acupuncture and moxibustion

针法和灸法的合称。

9.1.2 针灸学 science of acupuncture and moxibustion

以中医理论为指导，研究经络、腧穴及刺灸方法，探讨运用针灸防治疾病规律的一门学科。

9.1.3 兽医针灸学 science of veterinary acupuncture and moxibustion

以中兽医理论为指导，研究动物经络、腧穴及刺灸方法，探讨运用针术和灸术防治动物病证及其作用原理的一门学科。

9.1.4 经络学 subject of meridian and collateral

以经络知识和经络现象为依据，阐述机体各部联系，以及相关生理功能、病理变化、疾病诊断和防治的一门学科。

9.1.5 腧穴学 subject of acupuncture points

研究腧穴的定位、特点、主治、应用及其机理的一门学科。

9.1.6 刺法灸法学 subject of acupuncture and moxibustion technique

研究针灸治疗疾病的各种方法、操作技术、临床应用及其基本原理的一门学科。

9.1.7 针灸治疗学 subject of acupuncture and moxibustion therapy

运用中（兽）医基础理论和经络、腧穴、刺灸方法等基础知识来研究针灸防治疾病的一门学科。

9.1.8 针刺疗法 acupuncture therapy

应用针具刺激腧穴或机体的某些部位以防治疾病的方法。

9.1.9 针刺手法 needling technique

指针刺过程中的操作方法。有时总指针刺时（包括进针、运针及出针）所使用的各种操作方法。有时专指运针时所使用的各种促使针刺得气或保持与加强针感以及各种针刺补泻的方法。

9.1.10 灸法 moxibustion

将艾绒或其他药物放置在体表的腧穴或特定部位上烧灼、温熨等，借灸火的温和热力以及药物的作用，通过经络的传导，起到温通气血、扶正祛邪、协调脏腑、平衡阴阳作用，以治疗疾病和预防保健的方法。

9.1.11 针灸刺激量 stimulating quantity of acupuncture and moxibustion

针灸时对动物机体的刺激程度。针刺刺激量由针具、手法、强度、频率、幅度、方向、时间、深度等多种因素决定。艾灸刺激量由艾炷大小、壮数和时间等多种因素决定。

9.1.12 针灸刺激强度 stimulus intensity of acupuncture and moxibustion

针灸刺激的强弱程度。一般分强刺激、中刺激和弱刺激三种。

9.1.12.1 强刺激 strong stimulation

刺激强度较大的针灸刺激。针刺以粗长针具、高频率、大幅度及长时间的捻转提插，患畜反应较为强烈为强刺激；灸治则以大艾炷、多壮数或较长时间熏灸为强刺激。

9.1.12.2 中刺激 moderate stimulation

介于强、弱刺激之间的针灸刺激强度。针刺以适中长短粗细针具，均匀中等强度捻转提插，患畜反应明显但不强烈；灸治则以中等量的艾炷灸或艾条熏灸。

9.1.12.3 弱刺激 mild stimulation

刺激强度较小的针灸刺激。针刺以细短针具，低频率、小幅度及短时间捻转提插，使患畜反应较为微弱为弱刺激；灸治则以小艾炷、少壮数或短时间熏灸为弱刺激。

9.1.13 得气 obtaining qi

针刺后获得的特殊经气效应。兽医常可根据进针和运针时动物产生轻微的肢体反应以及术者手下的沉紧感来判定。

同义词：气至；针感

9.1.14 调气 regulating qi

在取得针刺感应的基础上适当调节其感应，以起到调整机体功能，实现防病治病的作用。

9.1.15 候气 awaiting qi

针刺入腧穴后，等待气至。

9.1.15.1 留针候气 retaining needle to await qi

针刺入腧穴一定深度后,将针留于穴内,等待气至。

9.1.15.2 行针候气 manipulating needle to await qi

进针后刺手施以各种行针手法,以促使气至和等待气至。

9.1.16 疏通经络 unblocking the meridian and collateral

通过各种针灸手法的作用,使经络通畅、气血运行正常,以治疗经络阻塞的针灸治疗方法。

9.1.17 针灸治则

9.1.17.1 盛则泻之 treating excess with expelling

邪气壅盛、正气未衰,属实证者,针刺可用泻的方法治疗。

9.1.17.2 虚则补之 treating deficiency with reinforcement

正气虚弱、体质较差者,针刺可用补的方法治疗。

9.1.17.3 不盛不虚以经取之 treating inapparent asthenia and sthenia with point on the meridian

非他经所犯而本经有病,且虚实不显著者,按本经循经取穴的方法治疗。

9.1.17.4 热则疾之 quick needling for heat

热证应当用浅刺、疾刺、不留针或以三棱针点刺出血的方法治疗。

9.1.17.5 寒则留之 retained needling for cold

寒证应当用深刺而久留针的方法治疗。

9.1.17.6 陷下则灸之 moxibustion for sunken pulse condition

阳气不足而脉陷下者,用灸法治疗。

9.1.17.7 菀陈则除之 bloodletting for chronic blood stasis

凡气血郁久不散,积于脉络所引起的疾病,以祛除瘀滞的方法治疗。

9.1.17.8 随变而调气 regulating qi according to changes

根据病变部位的深浅和病情的轻重等情况,分别采用适当的针灸治法以调气。

9.1.17.9 上病下取 treating upper disorder with lower point

上部(前部)的病证选取下部(后部)腧穴。

同义词:上病取下

9.1.17.10 下病上取 treating lower disorder with upper point

下部(后部)的病证选取其高位(前部)的腧穴。

同义词：下病取上

9.1.17.11 以左治右 treating right disease with left point
右侧病证针刺左侧的腧穴。

9.1.17.12 以右治左 treating left disease with right point
左侧病证针刺右侧的腧穴。

9.1.17.13 实则泻其子 reducing child-element in excess condition
实证，泻其所属的子经或子穴的方法。

9.1.17.14 虚则补其母 reinforcing mother-element in deficiency condition
虚证，补其所属的母经或母穴的方法。

9.1.17.15 脏病取原 selection of source point for zang-viscus disease
五脏有病取其原穴进行治疗。

9.1.18 选穴法 point selection
根据病情选取有效穴位进行治疗的方法。
同义词：取穴法；腧穴取穴

9.1.18.1 局部选穴法 local point selection
按患病所在部位选取本经或邻经穴位治疗的选穴方法。
同义词：局部取穴

9.1.18.2 邻近选穴法 adjacent point selection
在靠近病变部位的周围选取有关穴位治疗的选穴方法。
同义词：近取法

9.1.18.3 远道选穴法 distant point selection
在远离患病部位的经络上选取穴位治疗的选穴方法。
同义词：远取法

9.1.18.4 循经选穴法 point selection along the meridian
脏腑的病证，选本经循行路径上穴位治疗的选穴方法。
同义词：本经取穴法

9.1.18.5 同经相应选穴法 point selection according to corresponding portion of same name meridian
选取前后肢经脉名称相同，部位相应，功能相近的腧穴治疗的选穴方法。

9.1.18.6 异经选穴法 other meridian point selection
本经患病而取用他经穴位治疗的选穴方法。

同义词：他经选穴法

9.1.18.7 对症选穴法 point selection according to symptom

根据症状表现，直接取用对症状有特定疗效的穴位以治疗疾病的选穴方法。

同义词：随症选穴法

9.1.18.8 表里选穴法 exterior-interior meridian points selection

在表里相合的经脉上选穴用以治疗本脏、本腑有关疾病的选穴方法。

9.1.18.9 对应取穴法 point selection according to corresponding portion

在与病痛部相对应的远部选取穴位治疗的选穴方法。包括前后对应、左右对应等。

9.1.18.10 辨证选穴法 point selection according to syndrome differentiation

根据辨证施治的原则，分析病证与脏腑、经络之间的关系，选取有关穴位治疗的选穴方法。

同义词：对证选穴

9.1.18.11 交叉选穴法 contralateral meridian point selection

根据经脉相互交贯的理论取用健侧的穴位以治疗疾病的选穴方法。

同义词：交叉取穴法

9.1.18.12 交会选穴法 crossing point selection

选用交会穴治疗的选穴方法。

同义词：交会取穴法

9.1.18.13 以痛为腧 tender point selection

以疼痛局部或压痛点作为腧穴以治疗疾病的选穴方法。

同义词：以痛为输

9.1.19 配穴法 points combination

按照辨证论治法则，根据经络腧穴的主治特性，在临床中应用腧穴配伍治疗的方法。

同义词：配穴

9.1.19.1 本经配穴法 same meridian points combination

选取同一条经脉穴位的配穴方法。

9.1.19.2 表里经配穴法 exterior-interior meridian points combination

除选取本经腧穴外，再选取与本经表里相合经脉的腧穴组成配穴，治疗本经、本脏病及与其有关疾病的配穴方法。

同义词：表里配穴法

9.1.19.3 原络配穴法 source-connecting points combination

以本经原穴与其表里经的络穴相配合，用以治疗本脏、本腑有关疾病的方法。

同义词：主客配穴法

9.1.19.4 同名经配穴法 same name meridian points combination

四肢同名经所属腧穴配合同用的配穴方法。

9.1.19.5 子母配穴法 mother-son points combination

脏腑、经络发生病变时，视病情之虚实，予以补母或泻子的配穴方法。

同义词：子母补泻法

9.1.19.6 交会经配穴法 crossing meridian points combination

选取与本经交会的经脉上的穴位治疗疾病的配穴方法。

9.1.19.7 阴阳配穴法 yin-yang points combination

阴经穴与阳经穴配伍使用的配穴方法。

9.1.19.8 上下配穴法 upper-lower points combination

将前肢、后肢或头面、四肢的穴位上下（前后）相配，用以治疗同一病证的配穴法。

9.1.19.9 前后配穴法 anterior-posterior points combination

胸腹、背腰前后穴位相配，治疗疾病的方法。

同义词：腹背配穴法

9.1.19.10 左右配穴法 left-right points combination

以经络循行交叉的特点为取穴依据而进行左右相配的配穴方法。

9.1.19.11 内外配穴法 medial-lateral points combination

选取内侧穴位与外侧穴位进行配伍的方法。内为阴，外为阳，此法是以调整内外阴阳为主的方法。

9.1.19.12 远近配穴法 distant-local points combination

病变局部穴位与远隔穴位相配，用以提高治疗效果的配穴方法。

9.1.19.13 主应配穴法 corresponding master points combination

以远离病痛部位的穴为"主"，邻近病痛部位的穴为"应"的配穴方法。

9.2 针术

9.2.1 九针 nine kinds of classical needles

古代九种针具，包括镵针、圆针、鍉针、锋针、铍针、圆利针、毫针、长针、大针。

9.2.1.1 镵针 arrowhead-like needle

九针之一。针头大，针尖锐利如箭头。针体长 9.2 cm，直径为 0.3 cm，针体的末端延伸为 0.5 cm 长的箭头状锋利针头。

同义词：箭头针

9.2.1.2 圆针 round-pointed needle

九针之一。针尖卵圆形。

9.2.1.3 鍉针 spoon-like needle

九针之一。针体长 10~13 cm，前端有黍粟样大小圆钝针头，直径 0.2~0.3 cm。

9.2.1.4 锋针 lance needle

九针之一。针体为三棱形，针锋三面有刃，十分锐利。

9.2.1.5 铍针 stiletto needle

九针之一。长而宽，形如剑。

同义词：剑针

9.2.1.6 圆利针 round-sharp needle

九针之一。圆而且锐，针尾微大，针体较毫针粗，长 8~9.2 cm。

9.2.1.7 毫针 filiform needle

九针之一。针尖锋利，针柄长 3~9.2 cm，针体较细。

9.2.1.8 长针 long needle

九针之一。针体较长，约 17.5 cm，针锋锐利。

9.2.1.9 大针 large needle

九针之一。针体粗而长，其锋微圆。

9.2.2 兽用传统针具

9.2.2.1 毫针 filiform needle

针尖锐利，针体细长，针体直径一般在 0.45~1.5 mm，针体长有 3 cm、4 cm、5 cm、6 cm、9 cm、12 cm、15 cm、18 cm、20 cm、25 cm、30 cm 等多种类型，而以直径为 1~1.25 mm、长 6~12 cm 为最常用。适用于深刺或透刺。

9.2.2.2 圆利针 round-sharp needle

形状结构与毫针相似，但针体较粗，针体直径约 2 mm，针体长有 9.2~10 cm 数种，针柄有平头式、盘龙式、八角式、圆珠式等。针尖呈三角形，较为锋利，进针容易，起针迅速，适合留针、运针。

9.2.2.3 宽针 wide needle

头端呈矛尖状，针刃锋利，针长 7.5~10 cm，头宽 9.2~8 mm。依据针头的宽度分为大、中、小三种。

9.2.2.4 箭针 arrow needle

形状与宽针类似，针尖锐利，针头宽约 3 mm，针体直径约 2 mm。

9.2.2.5 三棱针 three-edged needle

针体呈圆柱状（笔杆状），针尖为三棱形，尖端三面有利刃。依据针体大小，分为大、小两种。

9.2.2.6 火针 fire needle

针尖圆而略钝，针体光滑，直径约 3 mm，针柄呈盘龙柄式、木柄式或胶木式的针具。

9.2.2.7 穿黄针 chuan huang needle

形状与大宽针相似，针尾部有一小孔，可穿系马尾（约 20 根）或棕绳的针具。

9.2.2.8 夹气针 jia qi needle

针端呈圆弧状，针体扁平，长 25~30 cm，用竹片或合金制成的长针。

9.2.2.9 眉刀针 mei dao needle

形似眉毛，刀刃薄而锋利的针具。

9.2.2.10 宿水管 su shui guan

原用鹅翎管制作，现多采用铜、铝或薄铁皮制成，长约 5 cm，形似毛笔帽，尖端密封，扁圆而钝，粗端管口有一圈凸出的唇形缘，管身周围钻满小孔的针具。

9.2.2.11 针锤 zhen chui

用硬质木料旋成，长约 35 cm，含锤头、锤柄、持针孔、锯缝、活动箍等用于固定针宽的持针器。

9.2.2.12 针棒（杖）zhen bang (zhang)

用硬木制成，棒长 29.2~30 cm，杖长约 100 cm，直径约 9.2 cm，前端约 7 cm 锯去一半，中间挖槽，用于固定针宽的持针器。

9.2.2.13 走索子 zou suo zi

绳长约 1.5 m，直径 0.3~0.5 cm，绳质光滑而柔韧的丝绳或尼龙绳。

同义词：大血绳

9.2.3 兽用现代针具

9.2.3.1 电针仪 electroacupuncture apparatus

可以输出脉冲电流，并满足电针疗法要求的电子仪器，包括主机、电极线、电源适配器等附件。

9.2.3.2 电热针灸仪 electrothermal needle apparatus

利用直流电通过特制的针具产生热量以治疗疾病的针灸仪器。

9.2.3.3 激光针灸仪 laser acupuncture apparatus

利用激光照射、灸熨、烧烙穴位或患部治疗疾病的仪器。可分为氦氖激光仪、二氧化碳激光仪等。

9.2.3.4 微波针灸仪 microwave acupuncture apparatus

由主机和辐射器组成，通过特制的毫针向穴位注入微波以治疗疾病的针灸仪器。

9.2.4 押手法

9.2.4.1 押手 pressing hand

针刺操作时，兽医用来按压进针部位以固定穴位或辅助进针的手。

9.2.4.2 指切押手法 fingernail-pressing method

以押手拇指尖切押穴位及近旁皮肤的押手法。

9.2.4.3 舒张押手法 skin spreading pressing method

用押手的拇指、食指或食指、中指贴按在穴位的皮肤上，并向两侧撑开，使穴位皮肤绷紧，以利进针的押手法。

9.2.4.4 提捏押手法 pinching pressing method

用押手的拇指、食指把穴位的皮肤捏起的押手法。

9.2.4.5 骈指押手法 side-by-side finger pressing method

用押手拇指、食指夹捏消毒干棉球，裹住针尖部，按在穴位处的押手法。

9.2.5 持针法

9.2.5.1 刺手 needle-holding hand

针刺操作时，兽医持针的手。

9.2.5.2 执笔式持针法 holding needle like using pen

刺手拇、食、中三指挟持针柄，以无名指抵住针体，似执毛笔状的持针方法。

9.2.5.3 二指持针法 holding needle with two fingers

刺手拇、食指持针柄,中指抵住针体,针体与拇指呈 90° 的持针方法。

9.2.5.4 双手持针法 holding needle with both hands

刺手拇、食、中三指挟持针柄,另一只手拇、食指捏一消毒干棉球挟持针体下端,针尖露出 3~6mm。

9.2.5.5 全握式持针法 holding needle with full grip

刺手拇指和食指捏住针尖部,根据刺入的深度留出针刃,用其余三指握住针身,针尾握在手心的持针方法。

9.2.5.6 弹琴式持针法 holding needle with

以拇、食指夹持针锋,留出适当长度,其余三指护住针身,针尾露出手心的持针方法。

9.2.5.7 针锤持针法 needle hammer holding method

先将针具(宽针)夹在锤头针缝内,针尖露出适当的长度,推上活动箍以固定针体的持针法。

9.2.5.8 手代针锤持针法 hand replace needle hammer holding method

以持针手的食、中、无名指握紧针体,针尖放置在小指中节的外侧,留出适当长度,拇指抵压针尾上端,以代针锤持针的方法。

9.2.6 进针法

9.2.6.1 针刺方向 needling direction

进针时和进针后针尖所朝向的方向。

9.2.6.2 针刺角度 angle of needle insertion

进针时,针体与穴位或病灶所在部位皮肤之间的角度。

9.2.6.2.1 直刺 perpendicular insertion

针体与穴位所在皮肤约呈 90° 刺入的方法。

9.2.6.2.2 斜刺 oblique insertion

针体与穴位所在皮肤约呈 45° 刺入的方法。

9.2.6.2.3 平刺 transverse insertion

针体与穴位所在皮肤约呈 15° 刺入的方法。

同义词:沿皮刺;横刺

9.2.6.3 透刺 penetration needling

用长毫针同时透过两个或两个以上穴位的针刺方法。

9.2.6.3.1 直透 direct penetrating

在四肢内外侧或前后侧相对穴位之间透刺的方法。

9.2.6.3.2 横透 transverse penetration

在一定部位上下方或前后方邻近穴位之间透刺的方法。

9.2.6.4 针刺深度 needling depth

针刺时，针体进入体内的程度。

9.2.6.4.1 浅刺 shallow needling

针刺时，针体进入机体皮下较浅部位，或仅在皮下。

9.2.6.4.2 深刺 deep needling

针刺时，针体进入深部组织。

9.2.6.5 急刺进针法 thrusting insertion

针尖抵于腧穴皮肤上，运用指力不加捻转或其他术式，直接推送至适宜深度的方法。

9.2.6.6 捻转进针法 twirling insertion

先将针尖快速刺入皮下，然后捻转进针至穴位一定深度的进针法。

9.2.6.7 飞针法 needle-flying insertion

以刺手点穴并快速进针至穴位适宜深度，可减轻进针时的刺皮痛，常用于不太配合的患畜。

9.2.6.8 管针进针法 insertion of needle with tube

押手持针管放在穴位上，将毫针放于管内，刺手拍击或弹击针尾，将针刺入皮下的进针法。

9.2.7 运针法

9.2.7.1 运针 needle manipulation

毫针或圆利针刺入腧穴一定深度后，为了求得针感（得气）或增强针感，以提高疗效而运动针体的方法。

同义词：行针

9.2.7.2 提法 lifting method

将针刺入一定深度后，向上抽提的行针手法。

9.2.7.3 插法 thrusting method

将针刺入一定深度后，向下按插的行针手法。

9.2.7.4 提插法 method of lifting-thrusting needle

将针刺入腧穴一定深度后，施以上提下插动作的行针手法。

9.2.7.5 捣法 continuous lifting-thrusting method

将针刺入一定深度后，刺手捏持针柄，作较大幅度的快速提插动作的行针手法。

9.2.7.6 捻转法 twirling method

将针刺入腧穴一定深度后，施以向前向后捻转动作的行针手法。

9.2.7.7 弹法 handle-flicking method

将针刺入一定深度后，以手指轻弹针尾或针柄，使针体微微振动的行针手法。

同义词：弹柄法

9.2.7.8 刮法 handle-scraping method

将针刺入一定深度后，用指甲上下刮动针柄以加强针感的行针手法。

同义词：刮柄法

9.2.7.9 摇法 handle-waggling method

将针刺入一定深度后，手持针柄，将针轻轻摇动的运针手法。

同义词：摇柄法

9.2.7.10 搓法 needle-twisting method

将针刺入一定深度后，用刺手拇食指持针柄，单向地捻动针体为搓法。大幅度搓针，一搓一放，使针体自动向回退旋，称为飞法。

9.2.8 留针法

9.2.8.1 留针 needle retention

针刺得气后，将针留置于体内，停留一段时间的方法。

9.2.8.2 静留针法 action less needle retention

针下气至后，将针自然地留置于体内，留针时不再运针。

9.2.8.3 动留针法 active needle retention

将针刺入腧穴先行针，待气至后，留置一定时间，留针时反复运针。

9.2.9 出针法

9.2.9.1 出针 needle withdrawal

针刺完毕，将针拔出体外。

同义词：起针

9.2.9.2 捻转出针法 twirling needle withdrawal method

用刺手拇、食二指或拇、食、中三指挟持针柄，轻捻慢提，将针拔出体外的出针法。

同义词：捻转起针法

9.2.9.3 抽拔出针法 withdraw needle quickly

一手拇食指捏住针柄，快速将针拔出。

同义词：快速起针法

9.2.9.4 退法 needle-withdrawing method

针刺出针时，将针先提至浅部稍作停留，当针下不觉沉紧时再拔出体外的方法。

9.2.10 针刺异常情况

9.2.10.1 晕针 fainting during acupuncture treatment

由于针刺而产生的晕厥现象。

9.2.10.2 滞针 stuck needle

针刺后发生的针下滞涩而捻转提插不便等运针困难的现象。

9.2.10.3 弯针 bending of needle

针刺时针体在体内发生弯曲的现象。

9.2.10.4 断针 needle breakage

针刺时针体折断于体内的现象。

同义词：折针

9.2.11 针刺补泻法

9.2.11.1 针刺补泻 reinforcing and reducing manipulations of needle

针刺得气后，根据动物病证的不同情况采用相应的针刺操作以补虚泻实。

9.2.11.2 针刺补法 reinforcing manipulation

使机体虚弱的功能状态恢复到正常生理状态的针刺方法。

9.2.11.3 三进一退 reinforcing needle by three inserting and one lifting

分三层进针，逐步由浅、中进到深层，一次退针至浅层，可反复施行的针刺补法。

9.2.11.4 推而纳之 pushing and thrusting needle

得气后将针推进按纳的针刺补法。

9.2.11.5 针刺泻法 reducing needle manipulation

使机体亢盛的功能状态恢复到正常生理状态的针刺方法。

9.2.11.6 一进三退 reducing needle by one inserting and three lifting

一次进针至深层，分三层退针，可反复进行的针刺泻法。

9.2.11.7 三出三入 three times needle of one-inserting and three-lifting

一进三退手法操作，反复操作三次，即一度。

9.2.11.8 大泻 heavy reduction
手法较重,刺激量较大的针刺泻法。

9.2.11.9 单式补泻手法 simple reinforcing and reducing method
用单一手法进行补泻的操作方法。

9.2.11.9.1 迎随补泻法 directional reinforcing and reducing method
以针尖的顺逆经脉循行方向,进行针刺补泻的操作方法。

9.2.11.9.2 捻转补泻法 twirling reinforcing and reducing method
以捻转用力的方向、角度的大小、频率的快慢和次数的多少来分别补泻的方法。

9.2.11.9.3 提插补泻法 lifting-thrusting reinforcing and reducing method
以针上下进退的快慢和用力的轻重来分别补泻的方法。

9.2.11.9.4 重提轻插 swift lifting and slow insertion
实施提插补泻法时,慢而轻地插针和快而重地提针,为泻法。

9.2.11.9.5 重插轻提 swift insertion and slow lifting
实施提插补泻法时,快而重地插针和轻而缓地提针,为补法。

9.2.11.9.6 徐疾补泻法 quick-slow reinforcing and reducing method
通过掌握毫针进针、出针以及行针的速度快慢进行针刺补泻的操作方法。

9.2.11.9.7 呼吸补泻法 respiratory reinforcing and reducing method
配合动物的呼吸进行针刺补泻的操作方法。

9.2.11.9.8 开阖补泻法 open-closed reinforcing and reducing method
以出针时是否按压针孔或摇大针孔来进行针刺补泻的操作方法。

9.2.11.9.9 平补平泻法 neutral reinforcing and reducing method
得气后均匀地提插、捻转的针刺补泻操作方法。

9.2.11.9.10 子母补泻 mother-son reinforcing and reducing method
脏腑经络发生病变时,视病情之虚实予以补母或泻子的针刺补泻方法。

9.2.11.9.11 九六补泻 ninereinforcing and six-reducing
以提插、捻转手法结合九六数的补泻方法,补法用九数,泻法用六数。

9.2.11.10 复式补泻法 compound reinforcing and reducing method
多种手法配合应用,操作较为复杂的针刺补泻方法。

9.2.12 练针法

9.2.12.1 纸垫练针法 puncturing drills on a paper pad
在纸垫上反复练习毫针刺入、提插、捻转等操作,以锻炼指力和熟练针刺的练针法。

9.2.12.2 棉团练针法 puncturing drills with a cotton ball
在棉团上反复练习毫针刺入、提插、捻转等操作,以锻炼指力和熟练针刺手法的练针法。

9.2.12.3 水中漂果练习法 puncturing drills with floating fruit

用漂浮在水面上的水果练习针刺腕力和准确性的方法。

9.2.12.4 速刺金钱法 puncturing drills with copper coin

用悬吊在空中的古代带有方孔的圆形铜钱练习进针准确性的方法。

9.2.13 常用针术

9.2.13.1 白针疗法 ordinary needling therapy

使用毫针、圆利针或箭针等，在血针穴位以外的穴位上施针，借以调整机体功能活动，以治疗畜禽各种病证的一种方法。

9.2.13.1.1 毫针疗法 filiform needle therapy

应用毫针的白针疗法。

9.2.13.1.2 圆利针疗法 round-sharp needle therapy

应用圆利针的白针疗法。

9.2.13.1.3 箭针疗法 arrow needle therapy

应用箭针的白针疗法。

9.2.13.2 血针疗法 blood injection therapy

使用三棱针或宽针等，在血针穴位上施针，刺破浅表静脉（丛）使之出血，达到泻热排毒、活血消肿，防治疾病目的的方法。

9.2.13.2.1 宽针刺血法 wide needle blood therapy

以宽针针刺体表静脉治疗疾病的方法。

9.2.13.2.2 三棱针刺血法 three-edged needle therapy

用三棱针刺破皮肤浅表部小静脉或静脉丛，使之少量出血以治疗疾病的方法。

9.2.13.3 火针疗法 fire needling therapy

将火针加热至针体通红，按一定刺法迅速刺入穴位，以治疗疾病方法。

9.2.13.4 电针疗法 electroacupuncture therapy

在毫针针刺得气的基础上，应用电针仪输出脉冲电流，通过毫针作用于机体以达到防治疾病目的的一种针刺疗法。

同义词：脉冲电针疗法

9.2.13.5 水针疗法 liquid injection therapy

选用药物注入穴位，利用针刺与药物的双重作用以治疗疾病的方法，包括穴位液压疗法、穴位药物疗法、穴位免疫疗法和穴位封闭疗法。

同义词：穴位注射疗法

9.2.13.5.1 穴位液压疗法 acupoint fluid pressure therapy

将生理盐水，葡萄糖注射液和注射用水等注入穴位，取其对穴位的压迫刺激作用以治疗疾病的方法。

9.2.13.5.2 穴位药物疗法 acupoint drug therapy

将抗生素、维生素等药物注入穴位以治疗疾病的方法。

9.2.13.5.3 穴位封闭疗法 acupoint blocking therapy

将局部麻醉剂或镇静止痛剂注入穴位以治疗疾病的方法。

9.2.13.5.4 穴位免疫疗法 acupoint immunity therapy

将抗原物质注入穴位以增强机体免疫力，治疗和预防疾病的方法。

9.2.13.6 自家血疗法 self-blood injection therapy

将动物静脉抽出的血液重新注入其腧穴中，以防治疾病的方法。

9.2.13.7 气针疗法 air acupuncture therapy

向穴位内送入适当的气体，以治疗疾病的方法，包括夹气针疗法、提皮进气法和穴位注气法。

9.2.13.7.1 夹气针刺法 jia qi needle acupuncture

用夹气针针刺马、牛等大牲畜夹气穴的方法。

9.2.13.7.2 提皮进气法 lift skin air acupuncture therapy

用大宽针刺破皮肤，用手力提起穴位周围皮肤，一提一松反复数次，将空气引到皮下并挤压到病变部位的方法。

9.2.13.7.3 穴位注气法 air injection method

用注射器向穴位深部注入空气或氧气等气体以治疗疾病的方法。

9.3 灸术

9.3.1 灸具

9.3.1.1 艾绒 moxa floss

艾叶经加工制成的淡黄色细软绒状物。

9.3.1.2 艾炷 moxa cone

用艾绒制成的小圆柱或圆锥体。

9.3.1.3 艾条 moxa stick

用艾绒或艾绒与其他药物混合制成的圆柱形长条。

9.3.1.4 温灸器 moxa burner

由手柄和布满小孔的金属筒体组成，专门用于施灸的器具。

9.3.2 常用灸法

9.3.2.1 艾灸疗法 moxa floss moxibustion therapy
以艾绒为主要材料制成艾炷或艾条，点燃后熏熨或温灼体表腧穴的灸法。

9.3.2.1.1 艾炷灸 moxa cone moxibustion
将艾炷直接或间接置于施灸部位上的灸法。

9.3.2.1.1.1 壮 cone
艾炷灸时计数的单位。每燃一个艾炷，称灸一壮。

9.3.2.1.1.2 麦粒灸 wheat-grain size cone moxibustion
用麦粒大小的艾炷施灸。

9.3.2.1.1.3 直接灸 direct moxibustion
将艾炷直接置放在皮肤上施灸的方法。根据对皮肤刺激程度不同，分为化脓灸法（瘢痕灸法）和非化脓灸法（无瘢痕灸法）。

9.3.2.1.1.4 瘢痕灸 scarring moxibustion
将艾炷直接置于施灸部位上点燃施灸，以使局部皮肤起泡，化脓，形成永久性瘢痕的直接灸法。

9.3.2.1.1.5 无瘢痕灸 non-scarring moxibustion
将艾炷直接置于施灸部位上点燃施灸，但不灼伤皮肤，不使局部皮肤起泡化脓，不留瘢痕的直接灸法。

9.3.2.1.1.6 间接灸 indirect moxibustion
艾炷与施灸部位皮肤之间衬隔物品的灸法。
同义词：隔物灸

9.3.2.1.1.7 隔姜灸 ginger interposed moxibustion
在艾炷与皮肤之间隔以生姜片实施的隔物灸法。

9.3.2.1.1.8 隔蒜灸 garlic interposed moxibustion
在艾炷与皮肤之间隔以蒜片或蒜饼实施的隔物灸法。

9.3.2.1.1.9 隔附子饼灸 monkshood cake interposed moxibustion
在艾炷与皮肤之间隔以附子饼实施的隔物灸法。
同义词：附子灸；隔附子灸

9.3.2.1.1.10 隔盐灸 indirect moxibustion on navel
将粗盐粒敷在脐部或穴位表面，再于其上实施放置艾炷的隔物灸法。

9.3.2.1.2 艾条灸 moxa stick moxibustion therapy
用艾条在施灸部位或病变部位实施的灸法。
同义词：艾卷灸

9.3.2.1.2.1 悬起灸 suspended moxibustion
将艾条悬于腧穴上方实施的灸法。

同义词：悬灸

9.3.2.1.2.2 温和灸 gentle moxibustion
将艾条点燃的一端与施灸部位的皮肤保持一定距离，使动物有温热感而无灼痛的悬起灸。

9.3.2.1.2.3 回旋灸 circling moxibustion
施灸时，艾条点燃的一端与施灸部位的皮肤保持一定距离，左右移动、反复旋转的悬起灸。

9.3.2.1.2.4 雀啄灸 sparrow pecking-like moxibustion
将艾条点燃一端接近施灸部位，待其有灼痛感后迅速提起，如此一上一下如同雀啄的悬起灸。

9.3.2.1.2.5 实按灸 pressing moxibustion
施灸时，先在施灸部位垫上布或纸数层，然后将药物艾条的一端点燃，趁热按到施灸部位上，使热力透达深部的艾条灸。

同义词：药物艾条实按灸

9.3.2.1.2.6 雷火针 thunder-fire moxibustion
用含有沉香、木香、乳香、茵陈、羌活、干姜、穿山甲、麝香等药末的艾条施实按灸。

同义词：雷火神针

9.3.2.1.2.7 太乙针 tai yi moxa stick moxibustion
用含有檀香、山柰、羌活、桂枝、木香、雄黄、白芷、细辛等药末的艾条施实按灸。

同义词：太乙神针

9.3.2.1.3 温灸器灸 moxa burner moxibustion
在温灸器中放入艾绒点燃后实施的艾灸法。

9.3.2.1.3.1 温筒灸 moxibustion with moxa tube
将艾绒点燃，放在由金属制成的筒状灸治器械内，在施灸部位上来回滚动温熨的灸法。

9.3.2.1.4 灸法补泻 reinforcing and reducing with moxibustion
艾灸时以火力的大小区分补泻的方法。

9.3.2.1.4.1 以火补之 reinforcing with ignited moxa
用艾火补虚时，勿吹其火，使其缓慢燃烧，待其自灭。

9.3.2.1.4.2 以火泻之 reducing with ignited moxa
用艾火泻实时，急吹其火，使其急燃，待其自灭。

9.3.2.2 熨灸疗法 hot compress therapy
应用温热物体对家畜患部或穴位以温热刺激借以疏通经络，驱散寒湿治疗疾病的一种灸法。

9.3.2.2.1 醋麸灸 oat-vinegar moxibustion

应用醋拌麸皮炒热搭于患部的一种疗法。

9.3.2.2.2 醋酒灸 vinegar-liquor moxibustion

用醋浸湿被毛，盖上醋浸的粗布或多层纱布，在布上连续浇洒白酒，点燃白酒进行灸熨的一种疗法。

同义词：火烧战船；被火鞍

9.3.2.2.3 酒灸 liquor moxibustion

点燃穴位上的面碗内的酒或酒精施灸的一种方法。

9.3.2.2.4 软烧术 soft burning method

通过点燃白酒或酒精的火焰熏烤患部的一种疗法。

9.3.2.3 烧烙疗法 cauterization therapy

用烧红的不同形状烙铁在家畜患部或穴位上进行画烙或熨烙的一种疗法，包括画烙法和熨烙法。

9.3.2.3.1 画烙法 pyrography

用烧红的烙铁直接在穴位或患部皮肤上直接画烙的一种疗法。

9.3.2.3.2 熨烙法 ironing and cauterization technique

用烧红的大方形烙铁熨烙覆盖在穴位或患部皮肤上的浸醋纱布棉垫的一种疗法。

9.3.3 其他针灸疗法

9.3.3.1 温针灸 warming needling

毫针留针时在针柄上置以艾绒（艾团或艾条段）施灸，是针刺与艾灸结合应用的一种疗法。

9.3.3.2 电热针疗法 electrothermal needle therapy

在针刺穴位产生针感后在针上连接电热针仪，综合针刺、火针、灸疗作用以治疗疾病的一种方法。

9.3.3.3 皮肤针 dermal needle

多枚不锈钢短针集束固定，用以浅刺皮肤的针具。分梅花针和七星针两种。

9.3.3.3.1 皮肤针疗法 cutaneous needle therapy

用皮肤针叩打浅表皮肤以治疗疾病的一种方法。

9.3.3.3.1.1 叩刺 knock needling

运用腕力，使针尖刺到皮肤后迅速弹起的一种刺法。

9.3.3.3.1.2 局部叩刺 local knock needling

由患处周围向中心叩刺的一种刺法。

9.3.3.3.1.3 循经叩刺 knock needling along the meridian
按经脉辨证，沿经脉循行路线叩刺的一种刺法。

9.3.3.4 臭氧针刺疗法 oxone point-injection therapy
将臭氧注入相应穴位以治疗疾病的一种方法。

9.3.3.5 头针疗法 scalp acupuncture therapy
针刺头皮上特定刺激区，以治疗疾病的一种方法。

9.3.3.6 耳针疗法 ear acupuncture therapy
以毫针、皮内针、艾灸、激光照射等器具，通过对耳廓穴位的刺激以防治疾病的一种方法。

9.3.3.7 拔罐疗法 cupping therapy
利用罐内的负压，使罐吸附于施术部位，引起局部皮肤充血或瘀血，以治疗疾病的一种方法。

9.3.3.7.1 火罐法 fire cupping method
利用点火燃烧排除罐内空气造成负压的一种拔罐疗法。

9.3.3.7.1.1 火罐投火法 fire throwing cupping method
用小纸片点燃后投入罐内，纸片烧尽时，迅速将火罐扣下，使其吸拔于施术部位的一种拔罐方法。

9.3.3.7.1.2 火罐闪火法 flash-fire cupping method
用镊子夹住酒精棉球，点燃后迅速伸入罐内片刻，随即退出，迅速将火罐扣下，使其吸拔于施术部位的一种拔罐方法。

9.3.3.7.1.3 火罐贴棉法 cotton-adhibiting method
将大小适宜的酒精棉贴在罐内壁下 1/3 处，点燃酒精棉后，迅速扣在施术部位的一种拔罐方法。

同义词：贴棉法

9.3.3.7.1.4 火罐滴酒法 alcohol cupping method
将 95% 酒精滴入罐内 1～3 滴，沿罐内壁摇匀，用火点燃后，迅速扣在施术部位的一种拔罐方法。

9.3.3.7.1.5 火罐架火法 alcohol fire-separated method
将酒精棉球置于不易燃烧、传热的物体内，如瓶盖、小酒盅等，然后置于施术部位上，点燃酒精棉球后，将罐迅速扣下的一种拔罐方法。

9.3.3.7.1.6 闪罐法 flash cupping method
当火罐吸着体表后，立刻除去，随拔随除，反复吸拔多次，直至皮肤潮红的一种拔罐方法。

9.3.3.7.1.7 留罐法 retained cupping method

拔罐后根据病情需要将罐留置一定时间的方法。

9.3.3.7.2 刺血拔罐法 pricking-cupping bloodletting method

用皮肤针叩刺患处，再在局部拔罐，以治疗疾病的一种疗法。

同义词：刺络拔罐法

9.3.3.7.3 抽气罐法 suction cupping method

用特制的罐，利用罐底的橡皮活塞接通吸引器，抽去罐内空气，形成负压，使罐吸附于皮肤上的一种拔罐疗法。

9.3.3.7.4 针罐法 needling associated with cupping

针刺留针时，在针刺部位配合拔罐的一种疗法。

9.3.3.7.5 贮药罐法 cupping with medicinal liquid inside

拔罐前在罐内贮放一定的药液，然后使罐吸拔于所选部位的一种拔罐疗法。

9.3.3.7.6 针药罐法 needle-medicated cupping method

针罐法和药罐法结合运用以治疗疾病的一种拔罐疗法。

9.3.3.7.7 起罐法 cup removing method

拔罐后，将罐从吸拔部位取下的方法。一手握住罐体腰底部稍倾斜，另一手拇指或食指按压罐口边缘的皮肤，使罐口与皮肤之间产生空隙，空气进入罐内后即可将罐取下。

9.3.3.8 刮痧疗法 scraping therapy

用刮痧板，蘸刮痧油等介质在体表部位进行反复刮拭，以治疗疾病的一种方法。

9.3.3.9 穴位埋线疗法 point catgut-embedding therapy

将羊肠线埋入穴位，利用其对穴位的持续刺激作用以治疗疾病的一种方法。

同义词：埋线法

9.3.3.9.1 穿刺针埋线法 catgut-embedding with lumbar puncture needle

用穿刺针将羊肠线埋入穴位皮下组织或肌层以治疗疾病的一种方法。

9.3.3.9.2 三角针埋线法 catgut-embedding with suture needle

用三角缝针将羊肠线埋植穴位皮下组织或肌层以治疗疾病的一种方法。

9.3.3.10 穴位激光照射疗法 laser irradiation therapy of point

应用医用激光仪输出的低功率激光束直接照射穴位以治疗疾病的一种方法。

同义词：激光穴位照射法；激光针疗法

9.3.3.11 穴位磁疗法 acupoint magnetic therapy

以磁场作用于腧穴或患病部位来治疗疾病的一种方法。

同义词：磁穴疗法

9.3.3.12 紫外线穴位照射疗法 ultraviolet irradiation therapy of point

利用人工紫外线照射腧穴以治疗疾病的一种方法。

9.3.3.13 红外线灸　infrared moxibustion

利用红外线辐射器在机体上照射，使经穴产生温热效应，以疏通经络，宣导气血，扶正祛邪的一种灸法。

9.3.3.14 微波针疗法　microwave acupuncture therapy

应用微波针治疗仪输出的微波，通过特制的辐射器与刺入穴位的毫针连接，将针刺和微波的热效应结合起来以治疗疾病的一种方法。

9.3.3.15 药物贴敷疗法　plaster therapy

在穴位上贴敷某些药物以治疗疾病的一种方法。

同义词：穴位敷药疗法；腧穴贴敷疗法

9.3.3.16 敷脐疗法　therapy of medicinal application on navel

将药物做成适当剂型（糊、散、丸、膏等）敷于脐部，或在脐部给予某些物理刺激（艾灸、针刺、热熨、拔罐等）以治疗疾病的一种方法。

参考文献

国家中医药管理局, 1997. 中医临床诊疗术语 疾病部分 非书资料: GBT 16751.1-1997［S］. 北京: 国家技术监督局.

国家中医药管理局, 1997. 中医临床诊疗术语 治法部分 非书资料: GBT 16751.3-1997［S］. 北京: 国家技术监督局.

国家中医药管理局, 2006. 中医基础理论术语 非书资料: GB/T 20348-2006［S］. 北京: 国家质量监督检验检疫总局 中国国家标准化管理委员会.

国家中医药管理局, 2013. 针灸学通用术语 非书资料: GB/T 30232-2013［S］. 北京: 国家质量监督检验检疫总局 中国国家标准化管理委员会.

国家中医药管理局, 2021. 中医病证分类与代码 非书资料: GB/T 15657-2021［S］. 北京: 国家市场监督管理总局 国家标准化管理委员会.

国家中医药管理局, 2021. 中医临床诊疗术语 第2部分: 证候 非书资料: GBT 16751.2-2021［S］. 北京: 国家市场监督管理总局 国家标准化管理委员会.

胡元亮, 2013. 中兽医学［M］. 北京: 科学出版社.

蒋次昇, 2003. 汉英中兽医辞典［M］. 北京: 中国农业出版社.

刘钟杰, 许剑琴, 2010. 中兽医学［M］. 北京: 中国农业出版社.

农业农村部, 2023. 中兽医基本术语 非书资料: GB/T 42953-2023［S］. 北京: 国家市场监督管理总局 国家标准化管理委员会.

上海中医药大学编写委员会, 2022. WHO中医药术语国际标准［M］. 日内瓦: 世界卫生组织.

石学敏, 2002. 针灸学［M］. 北京: 中国中医药出版社.

世界中医药学会联合会, 2007. 中医基本名词术语中英对照国际标准［M］. 北京: 人民卫生出版社.

王雪苔, 1989. 中国医学百科全书·针灸学［M］. 上海: 上海科学技术出版社.

谢观, 1998. 中国医学大辞典［M］. 天津: 天津科学技术出版社.

杨英, 芒来, 2016. 马病针灸学［M］. 北京: 中国农业出版社.

于船, 2002. 中兽医学大辞典［M］. 成都: 四川科学技术出版社.

郑洪新, 2016. 中医基础理论［M］. 北京: 中国中医药出版社.

郑继方, 2005. 中兽医诊疗手册［M］. 北京: 金盾出版社.

郑继方, 罗永江, 辛蕊华, 2016. 传统中兽医诊病技巧［M］. 北京: 中国农业出版社.

钟秀会, 陈玉库, 2007. 中兽医学［M］. 北京: 中国农业科学技术出版社.

钟秀会, 马爱团, 2011. 中兽医基础理论［M］. 北京: 中国农业出版社.

汉语拼音索引

A

艾灸疗法…………………………… 8.3.7
艾灸疗法…………………………… 9.3.2.1
艾绒………………………………… 9.3.1.1
艾条………………………………… 9.3.1.3
艾条灸……………………………… 9.3.2.1.2
艾炷………………………………… 9.3.1.2
艾炷灸……………………………… 9.3.2.1.1

B

八法并用…………………………… 8.1.9
八纲辨证…………………………… 1.1.1.1.5
八纲病证病机……………………… 4.2.2
拔罐疗法…………………………… 8.3.8
拔罐疗法…………………………… 9.3.3.7
白针疗法…………………………… 9.2.13.1
百骸………………………………… 2.1.10.3.1
百叶干……………………………… 6.3.6
败血凝蹄…………………………… 6.7.26
瘢痕灸……………………………… 9.3.2.1.1.4
半表半里病机……………………… 4.2.5.3.1
包扎固定疗法……………………… 8.2.70
胞宫………………………………… 2.1.10.5
胞宫湿热证………………………… 7.9.22
胞宫虚寒证………………………… 7.9.21
胞宫血热证………………………… 7.9.23
胞黄………………………………… 6.5.23
胞虚………………………………… 6.5.24
胞转………………………………… 6.5.9
暴喜伤心…………………………… 4.1.5.3
悲忧伤肺…………………………… 4.1.5.7
悲则气消…………………………… 4.1.5.8
背黄………………………………… 6.6.17
本经配穴法………………………… 9.1.19.1

鼻黄………………………………… 6.6.13
鼻窍证……………………………… 7.12.9
鼻血………………………………… 6.2.9
痹证………………………………… 6.11.13
便秘………………………………… 6.3.20
便血………………………………… 6.3.21
遍身黄……………………………… 6.8.1
辨证论治…………………………… 1.1.1.1.2
辨证求因…………………………… 4.1.1.1
辨证选穴法………………………… 9.1.18.10
标本缓急…………………………… 5.2.4
标本同治…………………………… 5.2.4.3
表闭水停证………………………… 7.11.21
表寒………………………………… 4.2.2.5.1.1
表寒肺热证………………………… 7.6.25
表寒里热…………………………… 4.2.2.5.3.1
表寒里热证………………………… 7.15.10
表里病机…………………………… 4.2.2.5
表里分消…………………………… 8.1.4.15
表里寒热…………………………… 4.2.2.5.1
表里经配穴法……………………… 9.1.19.2
表里俱寒…………………………… 4.2.2.5.3.3
表里俱寒证………………………… 7.15.12
表里俱热…………………………… 4.2.2.5.3.4
表里俱热证………………………… 7.15.13
表里俱实…………………………… 4.2.2.5.3.7
表里俱虚…………………………… 4.2.2.5.3.8
表里双解…………………………… 8.1.9.1
表里同病…………………………… 4.2.2.5.3
表里虚实…………………………… 4.2.2.5.2
表里选穴法………………………… 9.1.18.8
表里转化…………………………… 4.2.2.5.3.9
表热………………………………… 4.2.2.5.1.2

· 245 ·

表热里寒	4.2.2.5.3.2
表热里寒证	7.15.11
表实	4.2.2.5.2.1
表实里虚	4.2.2.5.3.5
表虚	4.2.2.5.2.2
表虚里实	4.2.2.5.3.6
表证入里	4.2.2.5.3.10
禀赋不足	4.1.14
并病	4.2.1.4.10
病机	4.2.1.1
病机学说	4.2.1
病因	4.1.1.2
病因学说	4.1.1
膊尖痛	6.7.2
补法	8.1.8
补气	8.1.8.1
补气固摄	8.1.10.133
补气固脱	8.1.8.10
补肾固齿	8.2.38
补肾固涩	8.1.10.137
补肾明目	8.2.15
补肾止泻	8.1.8.53
补血	8.1.8.11
补血固脱	8.1.8.14
补血健齿	8.2.40
补血润燥	8.1.10.56
补血养耳	8.2.21
补血养肝	8.1.8.13
补血养神	8.1.8.19
补血养心	8.1.8.12
补阳	8.1.8.43
补阳益气	8.1.8.42
补益肺气	8.1.8.3
补益肝气	8.1.8.5
补益精髓	8.1.8.64
补益脾肺	8.1.8.8
补益脾肾	8.1.8.9
补益气血	8.1.8.61
补益肾气	8.1.8.6

补益心肺	8.1.8.7
补益心气	8.1.8.2
补益心肾	8.1.8.65
补益中气	8.1.8.4
不盛不虚以经取之	9.1.17.3

C

仓廪之本	2.1.5.10
藏精气而不泻	2.1.4.7
藏象	2.1.1
槽结	6.10.5.2
草噎	6.3.2
插法	9.2.7.3
缠腕痛	6.7.12
镵针	9.2.1.1
产后恶露不尽	6.9.3
产后发热	6.9.9
产后腹痛	6.9.5
肠道寒湿证	7.7.49
肠道津亏证	7.7.44
肠道气滞证	7.7.50
肠道湿热证	7.7.40
肠道实热证	7.7.41
肠道瘀滞证	7.7.75
肠风下血	6.3.16
肠黄	6.3.14
肠积沙	6.3.12
肠绞痛	6.3.29
肠嵌闭	6.3.32
肠热气滞证	7.7.41.3
肠热阴虚证	7.7.41.2
肠入阴	6.3.28
肠痈	6.3.15
常用灸法	9.3.2
常用针术	9.2.13
乘重痛	6.7.9
持针法	9.2.5
齿为骨之余	2.1.10.3.2
齿龈证	7.12.11
冲脉	3.1.8.3

冲任不固证	7.9.25
冲任失调证	7.9.24
冲任瘀阻证	7.9.26
冲洗疗法	8.2.52
虫毒犯肺证	7.6.35
虫毒侵袭肌肤证	7.11.19
虫毒蕴肤证	7.11.14
虫毒证	7.3.14.10
虫积肠道证	7.7.51
虫积化疳证	7.12.2.4
虫积证	7.3.16
虫扰胆膈证	7.8.36
虫扰魄门证	7.7.52
虫湿壅络证	7.13.1.11
虫兽伤	4.1.12
抽拔出针法	9.2.9.3
抽气罐法	9.3.3.7.3
臭氧针刺疗法	9.3.3.4
出针	9.2.9.1
出针法	9.2.9
除湿通络	8.1.7.31
穿刺针埋线法	9.3.3.9.1
穿黄针	9.2.2.7
传化物而不藏	2.1.5.3.1
传化之腑	2.1.5.3
传经痛	6.7.3
喘证	6.2.3
疮	6.6.21
疮黄疔毒	6.6.10
创伤	6.6.1
吹鼻疗法	8.2.65
吹耳疗法	8.2.61
垂缕不收	6.5.22
刺法灸法学	9.1.6
刺手	9.2.5.1
刺血拔罐法	9.3.3.7.2
醋麸灸	9.3.2.2.1
醋酒灸	9.3.2.2.2
搓法	9.2.7.10

D

大肠	2.1.7
大肠病机	4.2.4.4.2
大肠寒结	4.2.4.4.2.2
大肠热结	4.2.4.4.2.1
大肠热结证	7.7.42
大肠湿热	4.2.4.4.2.3
大肠虚寒	4.2.4.4.2.5
大肠液亏	4.2.4.4.2.6
大肠主传导	2.1.7.1
大肠主津	2.1.7.2
大肚结	6.3.26
大怒伤肝	4.1.5.1
大泻	9.2.11.8
大针	9.2.1.9
带脉	3.1.8.4
带下	6.9.2.1
丹毒	6.6.29
单式补泻手法	9.2.11.9
胆	2.1.5.4
胆病病机	4.2.4.6.2
胆经郁热	4.2.4.6.2.1
胆经郁热证	7.8.39
胆气	2.1.5.4.2
胆气虚证	7.8.33
胆热痰扰证	7.8.35
胆郁痰扰	4.2.4.6.2.2
胆郁痰扰证	7.8.34
胆胀	6.4.2
胆者精之腑	2.1.5.4.1
胆汁	2.1.5.4.3
胆主决断	2.1.5.4.4
淡渗利湿	8.1.7.8
捣法	9.2.7.5
得气	9.1.13
滴鼻疗法	8.2.66
滴耳疗法	8.2.62
鍉针	9.2.1.3
点刺疗法	8.3.4

点眼疗法	8.2.60
电热针灸仪	9.2.3.2
电热针疗法	9.3.3.2
电针疗法	8.3.9
电针疗法	9.2.13.4
电针仪	9.2.3.1
调和肝脾	8.1.4.2
调和气血	5.2.10
调和气血	8.1.4.10
调和营卫	8.1.1.8
调理肠胃	8.1.4.7
调理冲任	8.1.4.16
调理脾胃	8.1.4.8
调理阴阳	5.2.3.5
调气	9.1.14
调气和营	8.1.4.11
调整阴阳	5.2.9
钉伤	6.7.19.2
疔疮	6.6.23
动留针法	9.2.8.3
窦道	6.6.7
督脉	3.1.8.1
毒火犯耳证	7.12.8.2
毒火攻唇证	7.12.12.2
毒入营血证	7.3.14.9
毒陷心肝证	7.10.64
毒邪流窜证	7.3.14.1
独阴不长	1.2.1.6.2
肚底黄	6.6.19
肚胀	6.3.33
断针	9.2.10.4
对应取穴法	9.1.18.9
对症选穴法	9.1.18.7

E

恶虫叮咬伤	6.8.19
耳黄	6.6.15
耳窍证	7.12.8
耳针疗法	8.3.14
耳针疗法	9.3.3.6

| 二指持针法 | 9.2.5.3 |

F

发病	4.2.1.4
翻胃吐草	6.3.31
反治	5.2.3
芳香化湿	8.1.7.1
芳香开窍	8.1.10.100
芳香通鼻	8.2.30
放血疗法	8.2.57
飞针法	9.2.6.7
肺	2.1.4.9
肺把胸膊痛	6.2.11
肺败	6.2.8
肺病病机	4.2.4.4.1
肺朝百脉	2.1.4.9.9
肺恶寒	2.1.4.9.18
肺风毛燥	6.2.7
肺风毛燥	6.8.2
肺寒吐沫	6.2.10
肺合大肠	2.1.4.9.19
肺华在毛	2.1.4.9.14
肺黄	6.2.12
肺火流鼻	6.2.4.1
肺经风热证	7.6.12
肺经郁火证	7.6.13
肺开窍于鼻	2.1.4.9.15
肺脾两虚	4.2.4.8.12
肺气	2.1.4.9.1
肺气不宣	4.2.4.4.1.4
肺气上逆	4.2.4.4.1.6
肺气虚	4.2.4.4.1.1
肺气虚咳嗽	6.2.2.6
肺气虚证	7.6.1
肺气阴两虚证	7.6.2
肺热鼻血	6.2.9.1
肺热炽盛证	7.6.7
肺热咳嗽	6.2.2.4
肺热血瘀证	7.6.14
肺热熏鼻证	7.12.9.9

肺热移肠证	7.6.9
肺热阴虚证	7.6.8
肺热饮停证	7.6.23
肺肾气虚	4.2.4.8.14
肺肾气虚证	7.10.59
肺肾相生	2.1.4.9.20
肺肾阳虚证	7.10.60
肺肾阴虚	4.2.4.8.13
肺肾阴虚证	7.10.58
肺失清肃	4.2.4.4.1.5
肺司呼吸	2.1.4.9.5.1
肺卫气虚证	7.6.5
肺为华盖	2.1.4.9.11
肺为娇脏	2.1.4.9.12
肺为水之上源	2.1.4.9.8.1
肺为贮痰之器	2.1.4.9.8.2
肺胃风热证	7.10.61
肺胃火热证	7.10.62
肺胃阴虚证	7.10.63
肺系	2.1.4.9.4
肺虚肠脱证	7.10.65
肺阳	2.1.4.9.3
肺阳虚	4.2.4.4.1.2
肺阳虚证	7.6.3
肺阴	2.1.4.9.2
肺阴虚	4.2.4.4.1.3
肺阴虚咳嗽	6.2.2.7
肺阴虚证	7.6.4
肺痈	6.2.5
肺与大肠病证病机	4.2.4.4
肺郁水停证	7.6.22
肺在液为涕	2.1.4.9.17
肺在志为悲	2.1.4.9.16
肺燥肠闭证	7.6.32
肺燥肠热证	7.6.31
肺燥郁热证	7.6.30
肺主皮毛	2.1.4.9.13
肺主气	2.1.4.9.5
肺主声	2.1.4.9.5.2
肺主肃降	2.1.4.9.7
肺主通调水道	2.1.4.9.8
肺主宣发	2.1.4.9.6
肺主治节	2.1.4.9.10
分消上下	8.1.4.14
分消走泄	8.1.4.13
风毒犯表证	7.11.8
风毒入络证	7.13.1.4
风毒蕴肤证	7.11.11
风毒证	7.3.14.2
风寒犯头证	7.12.1.1
风寒感冒	6.2.1.1
风寒咳嗽	6.2.2.2
风寒化热证	7.3.1.8
风寒湿凝滞筋骨证	7.13.2.6
风寒束表证	7.11.3
风寒袭鼻证	7.12.9.1
风寒袭肺	4.2.4.4.1.8
风寒袭肺证	7.6.20
风寒袭咽证	7.12.10.1
风火犯齿证	7.12.11.1
风火热毒证	7.3.14.8
风轮风热证	7.12.6.1
风轮热毒证	7.12.6.3
风轮湿热证	7.12.6.2
风轮阴虚证	7.12.6.4
风轮证	7.12.6
风热闭肺证	7.6.11
风热犯鼻证	7.12.9.2
风热犯表证	7.11.4
风热犯耳证	7.12.8.3
风热犯肺	4.2.4.4.1.7
风热犯肺证	7.6.10
风热犯目证	7.12.2.1
风热犯头证	7.12.1.2
风热感冒	6.2.1.2
风热咳嗽	6.2.2.3
风热侵咽证	7.12.10.2
风热痰毒证	7.3.1.4

风热外袭证	7.3.1.3
风热挟湿证	7.3.1.6
风热血燥证	7.4.7
风热郁滞肌肤证	7.11.18
风热中络证	7.13.1.2
风伤肠络证	7.7.53
风胜行痹证	7.13.2.1
风湿犯表证	7.3.1.2
风湿犯头证	7.12.1.3
风湿化热证	7.3.1.7
风湿凌目证	7.12.2.2
风湿袭表证	7.11.5
风湿挟毒证	7.3.1.5
风湿蕴肤证	7.11.10
风水	6.5.14
风水证	7.11.20
风瘫	6.11.12
风痰	4.1.8.1.1.5
风痰闭神证	7.5.27
风痰入络证	7.13.1.1
风痰上扰证	7.12.1.4
风痰证	7.3.7.1
风蹄	6.7.21
风土疮	6.8.14
风为百病之长	4.1.2.1.5
风袭表疏证	7.11.2
风邪	4.1.2.1
风邪犯表证	7.3.1.1
风邪犯唇证	7.12.12.1
风性开泄	4.1.2.1.1
风性善行数变	4.1.2.1.4
风性主动	4.1.2.1.3
风易伤阳位	4.1.2.1.2
风中经络证	7.13.1
锋针	9.2.1.4
敷脐疗法	9.3.3.16
敷贴疗法	8.2.47
伏邪	4.2.1.4.2
扶正固本	5.2.5.1
扶正解表	8.1.1.11
扶正祛邪	5.2.5
浮络	3.1.11
腑病治脏	5.2.10.6
腐蚀疗法	8.2.54
腐蹄病	6.7.27
复发	4.2.1.4.6
复式补泻法	9.2.11.10

G

甘寒润燥	8.1.10.60
肝	2.1.4.11
肝病病机	4.2.4.6.1
肝藏血	2.1.4.11.9
肝肠气滞证	7.10.68
肝胆风	6.4.7
肝胆寒湿	4.2.4.8.29
肝胆寒湿证	7.8.40
肝胆火旺证	7.8.38
肝胆气滞证	7.8.42
肝胆湿热	4.2.4.8.27
肝胆湿热证	7.8.37
肝胆实热	4.2.4.8.28
肝胆瘀滞证	7.8.41
肝恶风	2.1.4.11.17
肝肺风热证	7.10.50
肝肺热盛证	7.10.51
肝风内动	4.2.4.6.1.12
肝风内动证	7.8.32
肝合胆	2.1.4.11.18
肝华在爪	2.1.4.11.14
肝黄	6.4.8
肝火炽盛证	7.8.21
肝火燔耳证	7.12.8.1
肝火犯肺	4.2.4.8.22
肝火犯肺证	7.10.49
肝火犯头证	7.12.1.10
肝火犯胃证	7.10.41
肝火上炎	4.2.4.6.1.7
肝火上炎证	7.8.22

肝经风热	4.2.4.6.1.8	肝胃气滞阴虚证	7.10.45
肝经风热	6.4.5	肝胃气滞证	7.10.42
肝经风热证	7.8.24	肝胃热盛证	7.10.40
肝经火旺证	7.8.23	肝胃虚寒证	7.10.48
肝经湿热	4.2.4.6.1.10	肝胃阴虚血瘀证	7.10.47
肝经湿热证	7.8.25	肝胃阴虚证	7.10.46
肝经郁热	4.2.4.6.1.9	肝系	2.1.4.11.5
肝开窍于目	2.1.4.11.15	肝虚血瘀证	7.8.14
肝脾不和	4.2.4.8.26	肝血	2.1.4.11.2
肝脾两虚证	7.10.30	肝血虚	4.2.4.6.1.2
肝脾气血两虚证	7.10.31	肝血虚动风证	7.8.32.4
肝脾气阴两虚证	7.10.32	肝血虚证	7.8.2
肝脾气滞证	7.10.36	肝阳	2.1.4.11.4
肝脾湿热证	7.10.35	肝阳暴亢证	7.8.6
肝脾血瘀证	7.10.37	肝阳化风	4.1.4.1.1
肝气	2.1.4.11.1	肝阳化风证	7.8.32.1
肝气犯脾	4.2.4.8.23	肝阳上亢	4.2.4.6.1.11
肝气犯胃	4.2.4.8.24	肝阳上亢证	7.8.5
肝气横逆	4.2.4.6.1.6	肝阳上扰证	7.12.1.6
肝气虚	4.2.4.6.1.1	肝阳虚	4.2.4.6.1.4
肝气虚血瘀证	7.8.16	肝阳虚证	7.8.4
肝气虚证	7.8.3	肝阴	2.1.4.11.3
肝气郁结	4.2.4.6.1.5	肝阴虚	4.2.4.6.1.3
肝热鼻血	6.2.9.2	肝阴虚动风证	7.8.32.3
肝热传眼	6.4.3	肝阴虚血瘀证	7.8.15
肝热动风证	7.8.32.2	肝阴虚阳亢证	7.8.7
肝热脾虚证	7.10.38	肝阴虚证	7.8.1
肝热气滞证	7.8.28	肝瘀化热证	7.8.12
肝热血瘀证	7.8.29	肝瘀痰结证	7.8.20
肝热阴虚证	7.8.30	肝瘀证	7.8.13
肝肾亏虚证	7.10.27	肝与胆病证病机	4.2.4.6
肝肾同源	2.1.4.11.19	肝郁化火证	7.8.17
肝肾阴虚阳亢证	7.10.29	肝郁脾虚	4.2.4.8.25
肝肾阴虚证	7.10.28	肝郁脾虚证	7.10.33
肝体阴用阳	2.1.4.11.12	肝郁肾虚证	7.10.52
肝为刚脏	2.1.4.11.11	肝郁湿热证	7.8.26
肝胃不和证	7.10.39	肝郁痰火证	7.8.19
肝胃气虚血瘀证	7.10.44	肝郁血热证	7.8.18
肝胃气滞血瘀证	7.10.43	肝郁血虚证	7.8.9

肝郁血瘀证	7.8.10
肝郁阴虚证	7.8.11
肝郁证	7.8.8
肝在液为泪	2.1.4.11.16
肝在志为怒	2.1.4.11.8
肝胀	6.4.6
肝主筋	2.1.4.11.13
肝主谋虑	2.1.4.11.7
肝主升发	2.1.4.11.10
肝主疏泄	2.1.4.11.6
感冒	6.2.1
肛门热毒证	7.7.54
肛门湿热证	7.7.55
膏淋	6.5.7.5
膏药疗法	8.2.48
隔附子饼灸	9.3.2.1.1.9
隔姜灸	9.3.2.1.1.7
隔姜灸疗法	8.3.5
隔蒜灸	9.3.2.1.1.8
隔蒜灸疗法	8.3.6
隔盐灸	9.3.2.1.1.10
攻补兼施	5.2.5.5
孤阳不生	1.2.1.6.1
骨	2.1.10.3
骨眼	6.11.15
骨折	6.6.8
固齿	8.2.37
固脬止遗	8.1.10.132
固涩敛乳	8.1.10.130
固摄敛泪	8.1.10.131
固摄止血	8.1.10.127
固肾安胎	8.1.10.128
刮法	9.2.7.8
刮痧疗法	8.2.59
刮痧疗法	9.3.3.8
管针进针法	9.2.6.8
灌肠疗法	8.2.69
滚蹄	6.7.22
过饱	4.1.6.1.1
过饥	4.1.6.1.2
过劳	4.1.7.1
过逸	4.1.7.2

H

寒结	6.11.1
寒秘	6.3.20.4
寒凝胞宫证	7.9.18
寒凝气滞证	7.3.2.2
寒凝血涩肌肤证	7.11.23
寒凝血虚证	7.4.7.2
寒凝血瘀证	7.3.2.3
寒凝阳虚证	7.4.7.1
寒凝证	7.3.2
寒热错杂	4.2.2.4.7
寒热错杂证	7.3.2.7
寒热失调	4.2.2.4
寒热真假	4.2.2.4.8
寒热转化	4.2.2.4.9
寒伤肩膊痛	6.7.1.1
寒伤脾胃	4.2.4.8.16
寒伤腰胯	6.7.14.1
寒胜痛痹证	7.13.2.2
寒湿犯腰证	7.13.2.7
寒湿化热证	7.3.2.4
寒湿困脾	4.2.4.5.1.10
寒湿困脾证	7.7.20
寒湿泄泻	6.3.19.4
寒湿瘀滞证	7.3.2.5
寒湿蕴肤证	7.11.15
寒湿阻络证	7.13.1.7
寒湿阻滞证	7.3.2.1
寒实结胸证	7.15.1.3
寒痰	4.1.8.1.1.1
寒痰证	7.3.7.2
寒痰阻肺证	7.6.24
寒邪	4.1.2.2
寒邪犯胃证	7.7.36
寒性凝滞	4.1.2.2.2
寒性收引	4.1.2.2.3

寒易伤阳	4.1.2.2.1
寒因寒用	5.2.3.1
寒饮停肺证	7.6.21
寒饮停胃证	7.7.37
寒则留之	9.1.17.5
寒者热之	5.2.2.1
寒滞肠道证	7.7.39
寒滞肝经	4.2.4.6.1.13
寒滞肝脉证	7.8.31
寒滞经脉证	7.13.1.3
寒滞胃肠证	7.7.73
寒滞心脉证	7.5.14
汗法	8.1.1
汗证	6.1.3
毫针	9.2.1.7 或 9.2.2.1
毫针疗法	9.2.13.1.1
合病	4.2.1.4.9
合邪	4.1.2.7
合子骨肿痛	6.7.18
和法	8.1.4
和解少阳	8.1.4.1
和络	8.1.10.36
和胃降逆	8.1.10.13
和胃消痞	8.1.7.40
和血安胎	8.1.10.38
和血熄风	8.1.10.71
和营生新	8.1.10.141
和中缓急	8.1.4.9
黑疔	6.6.24
黑汗风	6.1.8.2
横透	9.2.6.3.2
红外线灸	9.3.3.13
喉骨胀	6.10.5.1
后天之精	2.2.4.2
后天之气	2.2.1.5
后肢厥阴肝经	3.1.4.4.2
后肢三阳经	3.1.4.3
后肢三阴经	3.1.4.4
后肢少阳胆经	3.1.4.3.2
后肢少阴肾经	3.1.4.4.3
后肢太阳膀胱经	3.1.4.3.3
后肢太阴脾经	3.1.4.4.1
后肢阳明胃经	3.1.4.3.1
候气	9.1.15
呼吸补泻法	9.2.11.9.7
滑精	6.5.22.1
滑利窍道	8.1.10.145
化气利水	8.1.7.30
化湿和营	8.1.7.5
化湿和中	8.1.7.6
化湿开窍	8.1.10.102
化痰开窍	8.1.10.103
化痰散结	8.1.10.87
化痰消食	8.1.10.85
化痰消瘿	8.1.10.88
化瘀和胃	8.1.10.44
化瘀健脾	8.1.10.48
化瘀宽心	8.1.10.43
化瘀宽胸	8.1.10.42
化瘀利水	8.1.10.29
化瘀清热	8.1.10.22
化瘀疏肝	8.1.10.46
化瘀通脑	8.1.10.40
化瘀消肿	8.1.10.30
化瘀宣肺	8.1.10.41
化瘀养肝	8.1.10.47
化瘀养胃	8.1.10.45
化瘀止血	8.1.10.34
画烙法	9.3.2.3.1
缓则治本	5.2.4.2
黄	6.6.11
黄疸	6.4.1
黄水疮	6.8.9
回旋灸	9.3.2.1.2.3
回阳救逆	8.1.5.1
豁鼻	6.6.2
活血化瘀	8.1.10.20
活血解毒	8.2.2

活血祛风	8.1.10.32
活血舒筋	8.1.10.31
活血消积	8.1.10.25
活血行滞	8.1.10.24
活血养血	8.1.10.27
活血止痛	8.1.10.33
火乘金	1.3.13.4
火赤疮	6.8.22
火毒窜络证	7.13.1.5
火毒犯鼻证	7.12.9.8
火毒犯齿证	7.12.11.5
火毒犯龈证	7.12.11.6
火毒流窜证	7.3.14.4
火毒证	7.3.14.3
火罐滴酒法	9.3.3.7.1.4
火罐法	9.3.3.7.1
火罐架火法	9.3.3.7.1.5
火罐闪火法	9.3.3.7.1.2
火罐贴棉法	9.3.3.7.1.3
火罐投火法	9.3.3.7.1.1
火克金	1.3.12.6
火烙疗法	8.2.58
火热炽盛证	7.3.6
火热伤阴证	7.4.9.2
火烧战船	8.2.72
火生土	1.3.11.4
火侮水	1.3.14.3
火邪	4.1.2.6
火邪易致疮痈	4.1.2.6.8
火性炎上	4.1.2.6.3
火易动血	4.1.2.6.5
火易耗气伤津	4.1.2.6.6
火易扰心	4.1.2.6.7
火易生风	4.1.2.6.4
火针	9.2.2.6
火针疗法	8.3.1
火针疗法	9.2.13.3

J

肌肤失养证	7.11.24

激光针灸仪	9.2.3.3
激光针疗法	8.3.10
急肠黄	6.3.14.1
急刺进针法	9.2.6.5
急惊风	6.11.7.1
急则治标	5.2.4.1
既病防变	5.1.1.2
继发	4.2.1.4.5
寄生虫	4.1.10
夹板固定疗法	8.2.71
夹板固定疗法	8.6.4
提捏押手法	9.2.4.4
夹气痛	6.7.8
夹气针	9.2.2.8
夹气针刺法	9.2.13.7.1
颊黄	6.6.14
间接灸	9.3.2.1.1.6
肩膊痛	6.7.1
肩痈	6.6.35
茧唇	6.8.20
健脾化湿	8.1.7.7
健脾利水	8.1.7.24
健脾驱虫	8.1.10.111
箭针	9.2.2.4
箭针疗法	9.2.13.1.3
降火固齿	8.2.43
降逆平喘	8.1.10.9
降逆止呕	8.1.10.14
交叉选穴法	9.1.18.11
交会经配穴法	9.1.19.6
交会选穴法	9.1.18.12
交通心肾	8.1.10.143
角折	6.6.9
疖	6.6.20
结石	4.1.8.3
结症	6.3.11
解表攻里	8.1.9.2
解表利水	8.1.7.28
解表清里	8.1.9.3

解表温里	8.1.9.4	惊恐伤神证	7.5.32
解毒护阴	8.2.3	惊恐伤肾	4.1.5.9
解毒开闭	8.1.10.107	惊恐伤肾证	7.9.31
解毒开窍	8.1.10.96	惊则气乱	4.1.5.11
解毒利耳	8.2.20	精	2.2.4
解毒利龈	8.2.44	精气	2.2.4.3
解毒散痈	8.2.1	精气夺则虚	4.2.2.2.2
解毒熄风	8.1.10.67	精气亏虚证	7.2.16
解毒消肿	8.2.4	精脱	4.2.4.7.1.8
解毒宣肺开闭	8.1.10.108	精血亏虚证	7.2.17
疥疮	6.8.12	精血同源	2.2.12
金乘木	1.3.13.5	静留针法	9.2.8.2
金克木	1.3.12.7	九六补泻	9.2.11.9.11
金生水	1.3.11.6	九针	9.2.1
金侮火	1.3.14.2	灸法	9.1.10
津	2.2.3.1	灸法补泻	9.3.2.1.4
津枯血燥	4.2.3.2.9	灸具	9.3.1
津亏热结证	7.4.5	酒灸	9.3.2.2.3
津亏血瘀	4.2.3.2.10	疽	6.6.31
津亏证	7.2.14.1	局部叩刺	9.3.3.3.1.2
津能载气	2.2.10	局部选穴法	9.1.18.1
津气亏虚证	7.2.15	决渎之官	2.1.9.4
津脱	4.2.3.2.3	厥热胜复	4.2.5.6.1
津血同源	2.2.11	厥阴病机	4.2.5.6
津液	2.2.3	厥阴病证	7.14.9
津液不足	4.2.3.2.1	厥阴寒化证	7.14.9.1
津液亏虚证	7.2.14	厥阴热化证	7.14.9.2
津液失常	4.2.3.2	峻下逐水	8.1.3.8
筋疖	6.6.25	**K**	
筋断	6.7.20	开关涌吐	8.1.2.5
进针法	9.2.6	开阖补泻法	9.2.11.9.8
浸洗疗法	8.2.53	开噤通关	8.1.10.109
经络	3.1.1.1	开窍通闭	8.1.10.93
经络学	9.1.4	亢害承制	1.3.19
经络学说	3.1.1	咳嗽	6.2.2
经脉	3.1.1.1.1	可致病的病理产物	4.1.8
经气	3.1.2	恐则气下	4.1.5.10
经气不利证	7.13.2	口唇证	7.12.12
经隧	3.1.3	口僻	6.11.18

口舌生疮	6.1.1	痢疾	6.3.13
叩刺	9.3.3.3.1.1	敛疮止痛	8.2.10
胯瓦痛	6.7.15	敛肺平喘	8.1.10.124
宽胸利膈	8.1.10.146	敛肺止咳	8.1.10.123
宽针	9.2.2.3	敛汗固表	8.1.10.122
宽针刺血法	9.2.13.2.1	敛阴固脱	8.1.10.134
宽中利湿	8.1.7.9	练针法	9.2.12
困水膈痰	6.3.35	凉血化瘀	8.1.10.23

L

		凉血熄风	8.1.10.65
劳复	4.2.1.4.6.2	凉营开窍	8.1.10.97
劳力过度	4.1.7.1.1	凉燥袭肺证	7.6.27
劳淋	6.5.7.2	凉燥证	7.3.5.2
劳伤咳嗽	6.2.2.8	两感	4.2.1.4.7
劳伤流鼻	6.2.4.2	料伤	6.3.3
劳神过度	4.1.7.1.2	邻近选穴法	9.1.18.2
劳逸失度	4.1.7	淋证	6.5.7
劳则气耗	4.2.3.1.1.1.2	留罐法	9.3.3.7.1.7
雷火针	9.3.2.1.2.6	留针	9.2.8.1
冷痛	6.3.27	留针法	9.2.8
冷拖竿	6.7.17	留针候气	9.1.15.1
里寒	4.2.2.5.1.3	流鼻	6.2.4
里热	4.2.2.5.1.4	流产	6.9.2
里实	4.2.2.5.2.3	流产与死胎	6.9.13
里虚	4.2.2.5.2.4	流皮漏	6.8.24
里证出表	4.2.2.5.3.11	流痰	6.6.39
理法方药	1.1.1.1.4	流注	6.6.38
理气安胎	8.1.10.19	六腑	2.1.5
理气化痰	8.1.10.80	六腑病机	4.2.4.2
理气化瘀	8.1.10.17	六腑以降为顺	2.1.5.2
理气健脾	8.1.10.11	六腑以通为用	2.1.5.1
理气解表	8.1.1.10	六经辨证	1.1.1.1.8
理气解郁	8.1.10.2	六经病证病机	4.2.5
理气利咽	8.2.35	六淫	4.1.2
理气消痞	8.1.10.15	癃闭	6.5.10
理气行滞	8.1.10.1	瘘管	6.6.6
理气止痛	8.1.10.16	漏瘘	6.6.41
利湿化浊	8.1.7.2	瘰疬	6.6.40
利水消肿	8.1.7.22	络脉	3.1.1.1.2
利咽	8.2.31	络伤出血证	7.13.2.13

掠草痛	6.7.16	内湿	4.1.4.3

M

马肠臌气	6.3.17	内外配穴法	9.1.19.11
麦粒灸	9.3.2.1.1.2	内燥	4.1.4.4
脉	2.1.10.4	内障眼	6.11.16
脉舍神	2.1.10.4.2	逆传心包	4.2.6.3.2
脉为血府	2.1.10.4.1	捻转补泻法	9.2.11.9.2
慢肠黄	6.3.14.2	捻转出针法	9.2.9.2
慢惊风	6.11.7.2	捻转法	9.2.7.6
毛边漏	6.7.23	捻转进针法	9.2.6.6
眉刀针	9.2.2.9	尿崩	6.5.12
泌别清浊	2.1.6.4	尿闭	6.5.25
棉团练针法	9.2.12.2	尿不禁	6.5.11
面游风	6.8.16	尿血	6.5.26
明目	8.2.11	尿浊	6.5.8
命门	2.1.4.12.18	宁心开窍	8.1.10.99
母行	1.3.11.1	牛红眼病	6.10.6
木乘土	1.3.13.1	牛羊气胀	6.3.18
木克土	1.3.12.3	脓毒蕴肾证	7.9.12
木舌	6.1.4	脓毒证	7.3.14.13
木生火	1.3.11.3	怒则气上	4.1.5.2
木侮金	1.3.14.1	暖肝散寒	8.1.5.8

N

O

		呕吐	6.3.1
难产	6.9.6		

P

囊虫侵脑证	7.5.31	盘肠结	6.3.30
脑	2.1.10.1	膀胱	2.1.8
脑黄	6.1.12	膀胱病机	4.2.4.7.2
脑颡黄	6.6.36	膀胱气闭	4.2.4.7.2.4
脑颡流鼻	6.2.4.3	膀胱气化	2.1.8.2
脑为髓海	2.1.10.1.1	膀胱湿热	4.2.4.7.2.1
内闭外脱证	7.4.14	膀胱湿热气滞证	7.9.13.1
内风	4.1.4.1	膀胱湿热血瘀证	7.9.13.2
内寒	4.1.4.2	膀胱湿热证	7.9.13
内火	4.1.4.5	膀胱虚寒	4.2.4.7.2.3
内伤病因	4.1.1.2.2	膀胱虚寒证	7.9.17
内伤咳嗽	6.2.2.5	膀胱蓄水证	7.9.15
内伤七情	4.1.5	膀胱蓄血证	7.9.16
内肾黄	6.5.20	膀胱蕴热证	7.9.14
内生五邪	4.1.4	膀胱主藏津液	2.1.8.1

配穴法	9.1.19
配种过度	4.1.7.1.3
喷雾疗法	8.2.68
皮肤瘙痒症	6.8.5
皮肤针	9.3.3.3
皮肤针疗法	9.3.3.3.1
皮水	6.5.15
铍针	9.2.1.5
脾	2.1.4.10
脾（胃）为气血生化之源	2.1.4.10.5.2
脾病病机	4.2.4.5.1
脾不统血	4.2.4.5.1.7
脾不统血证	7.7.4
脾藏营	2.1.4.10.5.4
脾恶湿	2.1.4.10.15
脾肺两虚证	7.10.53
脾肺气阴两虚证	7.10.54
脾华在唇	2.1.4.10.11
脾经热毒证	7.7.19
脾开窍于口	2.1.4.10.12
脾气	2.1.4.10.1
脾气不升	4.2.4.5.1.2
脾气下陷证	7.7.2
脾气虚	4.2.4.5.1.1
脾气虚证	7.7.1
脾气郁结证	7.7.3
脾肾两虚证	7.10.57
脾肾气虚水停证	7.10.57.1
脾肾气虚证	7.10.56
脾肾阳虚	4.2.4.8.20
脾肾阳虚水停证	7.10.57.2
脾肾阳虚证	7.10.55
脾失健运	4.2.4.5.1.6
脾旺不受邪	2.1.4.10.8
脾为后天之本	2.1.4.10.5.1
脾为生痰之源	2.1.4.10.5.3
脾胃不和证	7.7.66
脾胃气虚证	7.7.57
脾胃气阴两虚证	7.7.61

脾胃湿热	4.2.4.8.18
脾胃湿热证	7.7.63
脾胃实热证	7.7.62
脾胃虚寒	4.2.4.8.17
脾胃虚弱	4.2.4.8.15
脾胃阳虚气滞证	7.7.60
脾胃阳虚证	7.7.59
脾胃阴虚	4.2.4.8.19
脾胃阴虚证	7.7.58
脾系	2.1.4.10.4
脾虚不磨	6.3.8
脾虚肠脱证	7.10.66
脾虚虫积证	7.7.17
脾虚带下	6.3.34
脾虚浮肿	6.3.9
脾虚慢草	6.3.5
脾虚气滞证	7.7.10
脾虚生风	4.2.4.5.1.11
脾虚生痰	4.2.4.5.1.12
脾虚湿困	4.2.4.5.1.8
脾虚湿困证	7.7.13
脾虚湿热证	7.7.14
脾虚食积证	7.7.16
脾虚水泛证	7.7.11
脾虚痰湿证	7.7.15
脾虚泄泻	6.3.19.2
脾虚泻	6.11.8.3
脾虚血亏证	7.7.8
脾虚血燥证	7.7.9
脾虚营亏证	7.7.7
脾阳	2.1.4.10.3
脾阳虚	4.2.4.5.1.5
脾阳虚水泛证	7.7.12
脾阳虚证	7.7.5
脾阴	2.1.4.10.2
脾阴虚	4.2.4.5.1.4
脾阴虚证	7.7.6
脾与胃病证病机	4.2.4.5
脾与胃相表里	2.1.4.10.16

脾在液为涎	2.1.4.10.14
脾在志为思	2.1.4.10.13
脾之大络	3.1.9.1
脾主肌肉	2.1.4.10.10
脾主升清	2.1.4.10.7
脾主四肢	2.1.4.10.9
脾主统血	2.1.4.10.6
脾主运化	2.1.4.10.5
骈指押手法	9.2.4.5
平补平泻法	9.2.11.9.9
平刺	9.2.6.2.3
平调寒热	8.1.4.12
平肝熄风	8.1.10.63
破伤风	6.10.4
破血行瘀	8.1.10.21

Q

七星针疗法	8.3.2
其他内治法	8.1.10
其他针灸疗法	9.3.3
奇恒之腑	2.1.10
奇经八脉	3.1.8
脐疝	6.11.5.2
起罐法	9.3.3.7.7
起卧症	6.3.10
气	2.2.1
气闭	4.2.3.1.1.2.1
气闭神厥证	7.5.29
气闭证	7.3.12
气病治血	5.2.10.1
气不摄血	4.2.3.1.3.4
气不摄血证	7.2.10.2
气疗	6.6.26
气分病机	4.2.6.2
气分湿热证	7.3.4.3
气分证	7.3.6.1
气化	2.2.1.2
气化无权	4.2.3.2.6
气机	2.2.1.3
气机不畅	4.2.3.1.1.5

气机失调	4.2.3.1.1.2
气轮风热证	7.12.3.1
气轮热毒证	7.12.3.4
气轮湿热证	7.12.3.2
气轮血瘀证	7.12.3.3
气轮阴虚证	7.12.3.5
气轮证	7.12.3
气能摄津	2.2.9
气能摄血	2.2.5.3
气能生津	2.2.7
气能生血	2.2.5.1
气能行津	2.2.8
气能行血	2.2.5.2
气逆	4.2.3.1.1.2.2
气逆证	7.3.11
气失调	4.2.3.1.1
气随血脱	4.2.3.1.3.5
气随血脱证	7.2.10.1
气随液脱	4.2.3.2.5
气脱	4.2.3.1.1.4
气脱证	7.2.3
气为血帅	2.2.5
气陷	4.2.3.1.1.3
气陷证	7.2.2
气虚	4.2.3.1.1.1
气虚鼻窍失充证	7.12.9.10
气虚便血	6.3.21.2
气虚齿动证	7.12.11.12
气虚毒滞证	7.4.6.2
气虚耳窍失充证	7.12.8.11
气虚发热证	7.4.1.1
气虚寒凝证	7.4.1.5
气虚气滞证	7.4.1.6
气虚湿困证	7.4.1.3
气虚水停证	7.4.1.4
气虚痰结证	7.4.1.2
气虚邪恋证	7.4.6.1
气虚挟实证	7.4.1
气虚血瘀	4.2.3.1.3.3

气虚血瘀证	7.4.1.7	前肢三阴经	3.1.4.1
气虚咽喉失充证	7.12.10.11	前肢少阳三焦经	3.1.4.2.2
气虚余热证	7.4.6.3	前肢少阴心经	3.1.4.1.3
气虚证	7.2.1	前肢太阳小肠经	3.1.4.2.3
气虚中满	4.2.3.1.1.1.1	前肢太阴肺经	3.1.4.1.1
气血关系失调	4.2.3.1.3	前肢阳明大肠经	3.1.4.2.1
气血津液辨证	1.1.1.1.7	潜阳熄风	8.1.10.68
气血津液病证病机	4.2.3	浅刺	9.2.6.4.1
气血两燔	4.2.6.7	强刺激	9.1.12.1
气血两燔证	7.3.6.7	抢风痛	6.7.6
气血两虚	4.2.3.1.3.1	鞘管疝	6.11.5.3
气血两虚动风证	7.2.10.3	切开疗法	8.2.55
气血两虚证	7.2.10	轻宣润燥	8.1.10.49
气血失调	4.2.3.1	清胆利湿	8.1.7.11
气血瘀滞肛门证	7.7.56	清法	8.1.6
气一元论	2.2.1.1	清肺润燥	8.1.10.51
气阴两虚	4.2.3.2.11	清肝熄风	8.1.10.66
气阴两虚证	7.2.11	清肝泻火	8.1.6.18
气营两燔	4.2.6.6	清解余毒	8.2.9
气营两燔证	7.3.6.6	清利三焦	8.1.7.20
气郁	4.2.3.1.1.6	清利头目	8.1.10.144
气郁化火	4.2.3.1.1.7	清里热	8.1.6.7
气郁化火证	7.3.10.4	清脾泻热	8.1.6.17
气郁伤肺证	7.6.36	清热安胎	8.1.6.32
气针疗法	9.2.13.7	清热除蒸	8.1.6.34
气滞	4.2.3.1.1.8	清热导滞	8.1.7.37
气滞耳窍证	7.12.8.7	清热攻下	8.1.3.1
气滞热壅证	7.3.10.3	清热化湿	8.1.7.18
气滞声带证	7.12.10.8	清热化痰	8.1.10.76
气滞湿阻证	7.3.10.2	清热解毒	8.1.6.2
气滞水停证	7.3.10.5	清热解暑	8.1.6.35
气滞痰凝咽喉证	7.12.10.4	清热开窍	8.1.10.94
气滞血瘀	4.2.3.1.3.2	清热利湿	8.1.7.17
气滞血瘀证	7.3.10.1	清热凉血	8.1.6.3
气滞证	7.3.10	清热明目	8.2.14
气主煦之	2.2.1.12	清热排脓	8.2.5
前后配穴法	9.1.19.9	清热祛湿	8.1.7.16
前肢厥阴心包经	3.1.4.1.2	清热祛痰化瘀	8.1.10.90
前肢三阳经	3.1.4.2	清热润燥	8.1.10.50

清热生津	8.1.6.33	祛风明目	8.2.13
清热通淋	8.1.6.31	祛风杀虫	8.1.10.113
清热熄风	8.1.10.64	祛风燥湿	8.1.7.14
清热泻肺	8.1.6.15	祛湿解表	8.1.1.9
清热泻火	8.1.6.1	祛暑解表	8.1.6.36
清热燥湿解毒	8.1.7.21	祛暑开闭	8.1.10.106
清暑解毒	8.1.6.37	祛暑开窍	8.1.10.98
清暑利湿	8.1.6.38	祛痰化湿开窍	8.1.10.104
清暑宣肺	8.1.6.41	祛痰化瘀	8.1.10.86
清暑益气	8.1.6.39	祛痰理气解毒	8.1.10.92
清胃泻热	8.1.6.16	祛痰利咽	8.2.34
清泄胆热	8.1.6.19	祛痰杀虫	8.1.10.91
清泄膈热	8.1.6.28	祛痰宣痹	8.1.10.89
清泻湿热	8.1.7.19	祛邪扶正	5.2.5.2
清泻相火	8.1.6.30	祛瘀生新	8.1.10.26
清泻肠热	8.1.6.29	祛瘀通鼻	8.2.29
清泻肺胃	8.1.6.27	祛瘀通络	8.1.10.28
清泻肝胆	8.1.6.20	祛瘀下胎	8.1.10.39
清心安神	8.1.6.14	取嚏疗法	8.2.67
清心导赤	8.1.6.9	去腐生肌	8.2.6
清心涤暑	8.1.6.40	全握式持针法	9.2.5.5
清心解毒	8.1.6.10	缺乳	6.9.10
清心开窍	8.1.10.95	雀啄灸	9.3.2.1.2.4
清心利湿	8.1.7.10	**R**	
清心凉血	8.1.6.12	热闭心包证	7.5.20
清心凉营	8.1.6.11	热痹	6.11.13.4
清心泻肺	8.1.6.23	热炽阴伤	4.2.6.2.2
清心泻肝	8.1.6.22	热毒闭肺证	7.6.34
清心泻火	8.1.6.8	热毒攻喉证	7.12.10.7
清心泻脾	8.1.6.21	热毒攻舌证	7.12.12.7
清心泻肾	8.1.6.24	热毒内陷证	7.3.14.6
清心养阴	8.1.6.13	热毒入营证	7.3.14.5
清宣郁热	8.1.6.5	热毒壅聚头面证	7.12.1.9
清虚热	8.1.6.6	热毒淤肝证	7.8.27
清营凉血	8.1.6.4	热毒蕴肠证	7.7.43
清燥润鼻	8.2.28	热毒蕴结肌肤证	7.11.13
驱蛔杀虫	8.1.10.110	热敷疗法	8.2.46
祛风化痰	8.1.10.79	热极生风	4.1.4.1.5
祛风解肌	8.1.10.62	热结膀胱	4.2.4.7.2.2

热厥证	7.3.6.11
热蜡疗法	8.6.1
热淋	6.5.7.1
热秘	6.3.20.3
热迫大肠	4.2.4.4.2.4
热气疮	6.8.4
热扰神明	4.2.4.3.1.8
热扰心神证	7.5.21
热扰胸膈证	7.15.1.1
热入心营证	7.5.22
热入血分	4.2.6.4.1
热入营血证	7.3.6.4
热伤营阴证	7.4.9.4
热深厥深	4.2.5.6.2
热盛动风证	7.3.6.8
热盛动血证	7.3.6.9
热盛酿脓证	7.3.6.12
热盛气分	4.2.6.2.1
热盛气滞证	7.3.6.10
热盛伤津证	7.4.9.3
热实结胸证	7.15.1.2
热痰	4.1.8.1.1.2
热痰证	7.3.7.4
热痛	6.1.8.1
热微厥微	4.2.5.6.3
热陷心包	4.2.6.3.1
热邪	4.1.2.6.1
热邪阻痹证	7.13.2.4
热因热用	5.2.3.2
热则疾之	9.1.17.4
热者寒之	5.2.2.2
任脉	3.1.8.2
柔肝熄风	8.1.10.69
肉轮风热证	7.12.5.2
肉轮气虚证	7.12.5.6
肉轮热毒证	7.12.5.3
肉轮湿热证	7.12.5.4
肉轮痰湿证	7.12.5.5
肉轮血虚证	7.12.5.7
肉轮血瘀证	7.12.5.1
肉轮证	7.12.5
乳痈	6.9.8
褥疮	6.6.22
软坚润燥	8.1.3.7
软坚散结	8.1.7.41
软烧术	9.3.2.2.4
润肠泄热	8.1.3.6
润肺化痰	8.1.10.77
润燥解毒	8.1.10.59
润燥通便	8.1.3.4
润燥止咳	8.1.10.58
润燥止渴	8.1.10.57
弱刺激	9.1.12.3

S

腮黄	6.6.16
塞鼻疗法	8.2.64
塞因塞用	5.2.3.3
三出三入	9.2.11.7
三喉症	6.10.5
三焦	2.1.9
三焦辨证	1.1.1.1.10
三焦病证病机	4.2.7
三焦湿热	4.2.7.4
三角针埋线法	9.3.3.9.2
三进一退	9.2.11.3
三棱针	9.2.2.5
三棱针刺血法	9.2.13.2.2
三棱针疗法	8.3.3
三因制宜	5.2.8
散寒除湿	8.1.7.15
散寒化饮	8.1.10.81
散寒开窍	8.1.10.101
散寒利水	8.1.7.29
散寒利咽	8.2.33
散寒通鼻	8.2.27
散寒止痛	8.1.5.4
颡黄	6.10.5.3
臊疣	6.8.8

涩肠止泻	8.1.10.125	少腹瘀滞证	7.15.6.5
涩精止遗	8.1.10.126	少阳病机	4.2.5.3
杀虫解毒宣肺	8.1.10.115	少阳病证	7.14.6
杀虫宁神	8.1.10.114	少阴病机	4.2.5.5
杀虫消积	8.1.10.112	少阴病证	7.14.8
兽用现代针具	9.2.3	少阴寒化	4.2.5.5.2
晒疮	6.8.18	少阴寒化证	7.14.8.1
闪挫	6.6.5	少阴热化	4.2.5.5.1
闪罐法	9.3.3.7.1.6	少阴热化证	7.14.8.2
闪伤	6.6.4	蛇毒内攻证	7.3.14.18
闪伤肩膊痛	6.7.1.2	深刺	9.2.6.4.2
闪伤腰胯	6.7.14.2	肾	2.1.4.12
疝	6.11.5	肾病病机	4.2.4.7.1
伤津	4.2.3.2.2	肾不纳气	4.2.4.7.1.3
伤乳泻	6.11.8.1	肾藏精	2.1.4.12.5
伤食呕吐	6.3.1.1	肾恶燥	2.1.4.12.16
伤食泄泻	6.3.19.1	肾寒	6.5.2
伤暑证	7.3.3.1	肾华在齿	2.1.4.12.11
伤损筋骨证	7.13.2.11	肾火症	6.5.13
伤阳	4.2.2.3.7.4	肾精	2.1.4.12.1
伤阴	4.2.2.3.7.2	肾精不足	4.2.4.7.1.7
上病下取	9.1.17.9	肾精亏虚证	7.9.7
上寒下热	4.2.2.4.7.1	肾厥	6.5.6
上寒下热证	7.15.16	肾开窍于耳	2.1.4.12.12
上焦	2.1.9.1	肾冷拖腰	6.7.13
上焦如雾	2.1.9.1.1	肾气	2.1.4.12.2
上焦湿热	4.2.7.1	肾气不固	4.2.4.7.1.2
上焦湿热证	7.15.3	肾气不固证	7.9.2
上焦主纳	2.1.9.1.2	肾气虚	4.2.4.7.1.1
上热下寒	4.2.2.4.7.2	肾气虚水泛证	7.9.3.1
上热下寒证	7.15.15	肾气虚证	7.9.1
上盛下虚证	7.15.14	肾生髓	2.1.4.12.10
上下配穴法	9.1.19.8	肾水	6.5.17
烧烙疗法	9.3.2.3	肾司二阴	2.1.4.12.13
少腹气滞证	7.15.6.2	肾痛	6.5.1
少腹热滞证	7.15.6.3	肾虚肠脱证	7.10.67
少腹湿热阻滞证	7.15.6.4	肾虚带下	6.5.4
少腹血瘀证	7.15.6.1	肾虚骨痿	6.5.5
少腹瘀热证	7.15.6.6	肾虚寒凝证	7.9.10.1

肾虚寒湿证	7.9.10
肾虚寒痰证	7.9.10.2
肾虚水泛	4.2.4.7.1.5
肾虚水泛证	7.9.3
肾虚髓亏证	7.9.8
肾虚腿肿	6.5.3
肾虚腿肿	6.5.27
肾虚泄泻	6.3.19.5
肾虚血瘀证	7.9.10.3
肾阳	2.1.4.12.4
肾阳虚	4.2.4.7.1.4
肾阳虚水泛证	7.9.3.2
肾阳虚证	7.9.4
肾阴	2.1.4.12.3
肾阴虚	4.2.4.7.1.6
肾阴虚火旺	7.9.6
肾阴虚证	7.9.5
肾阴阳两虚证	7.9.9
肾与膀胱病证病机	4.2.4.7
肾与膀胱相表里	2.1.4.12.17
肾在液为唾	2.1.4.12.15
肾在志为恐	2.1.4.12.14
肾主封藏	2.1.4.12.8
肾主骨	2.1.4.12.9
肾主纳气	2.1.4.12.6
肾主生殖	2.1.4.12.5.2
肾主水	2.1.4.12.7
肾主先天之本	2.1.4.12.5.1
渗湿利水	8.1.7.23
升提固涩	8.1.10.136
生津利咽	8.2.36
生津润燥	8.1.10.53
生津止渴	8.1.10.54
生杀之本	1.2.1.13
盛则泻之	9.1.17.1
湿疮	6.8.13
湿毒	6.8.3
湿毒蕴结肌肤证	7.11.12
湿毒证	7.3.14.7

湿敷疗法	8.2.50
湿困脾胃证	7.7.64
湿热便血	6.3.21.1
湿热毒蕴证	7.3.4.6
湿热犯耳证	7.12.8.4
湿热犯腰证	7.13.2.9
湿热痢	6.3.13.1
湿热弥漫三焦证	7.15.2
湿热呕吐	6.3.1.2
湿热侵淫证	7.3.4.5
湿热下注证	7.15.5.1
湿热泄泻	6.3.19.3
湿热泻	6.11.8.2
湿热壅滞证	7.3.4.4
湿热瘀阻证	7.3.4.7
湿热蕴脾	4.2.4.5.1.9
湿热蕴脾证	7.7.18
湿热蕴肾证	7.9.11
湿热蒸鼻证	7.12.9.6
湿热蒸齿证	7.12.11.4
湿热蒸喉证	7.12.10.6
湿热蒸口证	7.12.12.3
湿热蒸舌证	7.12.12.6
湿热证	7.3.4.2
湿热阻痹证	7.13.2.5
湿热阻络证	7.13.1.6
湿热阻滞精室证	7.9.27
湿热组结肌肤证	7.11.16
湿胜着痹证	7.13.2.3
湿痰	4.1.8.1.1.3
湿痰蕴结肌肤证	7.11.17
湿痰证	7.3.7.3
湿邪	4.1.2.4
湿性黏滞	4.1.2.4.4
湿性趋下	4.1.2.4.5
湿性重浊	4.1.2.4.3
湿易伤阳	4.1.2.4.2
湿壅鼻窍证	7.12.9.3
湿阻肠道证	7.7.48

湿阻气机	4.1.2.4.1	食滞胃肠证	7.7.69
湿阻气滞证	7.3.4.1	食滞胃热证	7.7.70
湿阻证	7.3.4	食滞胃脘	4.2.4.5.2.12
十二经别	3.1.5	手代针锤持针法	9.2.5.8
十二经筋	3.1.6	兽医针灸学	9.1.3
十二经脉	3.1.4	兽用针具	9.2.2
十二经脉流注次序	3.1.14	舒张押手法	9.2.4.3
十二皮部	3.1.7	疏表通经	8.1.1.6
十四经脉	3.1.1.1.3	疏风利水	8.1.7.27
十五络脉	3.1.9	疏风利咽	8.2.32
石疽	6.8.25	疏风通鼻	8.2.26
石淋	6.5.7.3	疏风透疹	8.1.1.5
石水	6.5.16	疏风宣耳	8.2.19
石阻气闭证	7.3.17.2	疏肝和胃	8.1.4.6
石阻气机证	7.3.17.1	疏肝健脾	8.1.4.4
石阻证	7.3.17	疏肝理脾	8.1.4.3
时饥时饱	4.1.6.1.3	疏肝理气	8.1.10.3
时行感冒	6.10.3	疏肝利胆	8.1.10.4
时疫	6.10.2	疏散风邪	8.1.10.61
实按灸	9.3.2.1.2.5	疏通经络	9.1.16
实喘	6.2.3.1	疏邪解表	8.1.1.4
实寒	4.2.2.4.1	暑闭气机证	7.3.3.5
实火	4.2.2.4.4	暑热闭神证	7.5.24
实秘	6.3.20.1	暑热动风证	7.3.3.4
实热	4.2.2.4.3	暑热证	7.3.3
实热尿闭	6.5.25.2	暑伤肺络证	7.6.15
实热跳脓	6.11.11.2	暑伤津气证	7.4.8
实邪犯目证	7.12.2	暑湿袭表证	7.11.6
实邪犯头证	7.12.1	暑湿证	7.3.3.3
实则泻其子	9.1.17.13	暑邪	4.1.2.3
实则泻之	5.2.2.4	暑性升散	4.1.2.3.2
实者泻其子	5.2.10.4	暑性炎热	4.1.2.3.1
实证病机	4.2.2.2.1.1	暑易扰心	4.1.2.3.4
实中夹虚	4.2.2.2.3.2	暑易伤津耗气	4.1.2.3.5
实中挟虚证	7.4.12	暑易挟湿	4.1.2.3.3
食毒证	7.3.14.14	腧穴学	9.1.5
食复	4.2.1.4.6.1	双手持针法	9.2.5.4
食积证	7.3.15	水不涵木	4.2.4.8.30
食疗	8.5.1	水不化气	4.2.3.2.7

水乘火	1.3.13.3	饲喂失宜	4.1.6
水疗	6.6.27	松皮癣	6.8.23
水寒射肺	4.2.4.8.9	搜风祛痰开窍	8.1.10.105
水火烫伤	6.6.3	速刺金钱法	9.2.12.4
水克火	1.3.12.5	宿草不转	6.3.23
水轮火邪伤络证	7.12.7.10	宿水管	9.2.2.10
水轮络痹精亏证	7.12.7.12	宿水停脐	6.3.22
水轮气虚血亏证	7.12.7.5	随变而调气	9.1.17.8
水轮气虚血瘀证	7.12.7.6	髓	2.1.10.2
水轮气虚证	7.12.7.4	孙络	3.1.10
水轮气滞血瘀证	7.12.7.7	损伤尿闭	6.5.25.1
水轮实热证	7.12.7.1	所不胜	1.3.12.2
水轮水湿停聚证	7.12.7.8	所胜	1.3.12.1
水轮痰火证	7.12.7.2	锁喉风	6.2.15
水轮痰瘀互结证	7.12.7.9	锁口黄	6.6.12
水轮血脉痹阻证	7.12.7.11	**T**	
水轮阴亏证	7.12.7.3	踏伤	6.7.19.1
水轮证	7.12.7	胎动不安	6.9.1
水气凌心	4.2.4.8.8	胎毒	4.1.11
水气凌心证	7.10.4	胎毒蕴热证	7.15.8
水伤	6.3.4	胎毒证	7.3.14.16
水生木	1.3.11.7	胎风	6.9.11
水停气阻	4.2.3.2.8	胎气	6.9.12
水停证	7.3.9	胎衣不下	6.9.4
水土不服	4.1.13	太阳病机	4.2.5.1
水侮土	1.3.14.4	太阳病证	7.14.1
水饮内停证	7.3.8	太阳腑证	7.14.2
水针疗法	8.3.11	太阳经证	7.14.1.1
水针疗法	9.2.13.5	太阳伤寒证	7.14.1.1.2
水中漂果练习法	9.2.12.3	太阳蓄水证	7.14.2.1
水肿	6.11.2	太阳蓄血证	7.14.2.2
思虑伤脾	4.1.5.5	太阳中风证	7.14.1.1.1
思伤脾气证	7.7.21	太乙针	9.3.2.1.2.7
思则气结	4.1.5.6	太阴病机	4.2.5.4
四季药	5.1.2	太阴病证	7.14.7
四肢神经麻痹症	6.7.7	太阴寒湿	4.2.5.4.1
饲喂不节	4.1.6.1	太阴虚寒	4.2.5.4.2
饲喂不洁	4.1.6.2	弹法	9.2.7.7
饲喂偏嗜	4.1.6.3	弹琴式持针法	9.2.5.6

痰	4.1.8.1.1
痰虫互结证	7.3.7.15
痰毒壅喉证	7.12.10.5
痰核留结证	7.3.7.17
痰火扰神证	7.5.25
痰火扰心	4.2.4.3.1.10
痰结毒滞证	7.3.7.13
痰聚鼻窍证	7.12.9.4
痰蒙心窍	4.2.4.3.1.11
痰迷心窍证	7.5.26
痰凝胞宫证	7.9.19
痰气互结证	7.3.7.6
痰气阻膈证	7.15.1.5
痰热闭肺证	7.6.17
痰热动风证	7.3.7.12
痰热犯鼻证	7.12.9.5
痰热结胸证	7.15.1.4
痰热内闭证	7.3.7.11
痰热内扰证	7.3.7.10
痰热气滞证	7.3.7.9
痰热阴虚证	7.4.10
痰热壅肺	4.2.4.4.1.11
痰热壅肺证	7.6.16
痰湿犯腰证	7.13.2.8
痰湿泛耳证	7.12.8.5
痰湿流注证	7.13.2.15
痰湿凝阻咽喉证	7.12.10.3
痰湿瘀滞证	7.3.7.16
痰湿中阻证	7.7.74
痰湿阻络证	7.13.1.8
痰湿阻滞精室证	7.9.28
痰食互结证	7.3.7.14
痰饮	4.1.8.1
痰瘀互结证	7.3.7.7
痰瘀化热证	7.3.7.8
痰瘀阻肺证	7.6.19
痰瘀阻膈证	7.15.1.6
痰证	7.3.7
痰浊犯头证	7.12.1.5
痰浊阻肺	4.2.4.4.1.10
痰浊阻肺证	7.6.18
痰阻心脉证	7.5.13
提插补泻法	9.2.11.9.3
提插法	9.2.7.4
提法	9.2.7.2
提皮进气法	9.2.13.7.2
蹄伤	6.7.19
蹄头痛	6.7.24
体虚感冒	6.2.1.3
体虚胎动	6.9.1.1
跳脓	6.11.11
通鼻	8.2.25
通耳	8.2.18
通经止痒	8.1.10.37
通络下乳	8.1.10.35
通因通用	5.2.3.4
同病异治	5.2.6
同经相应选穴法	9.1.18.5
同名经配穴法	9.1.19.4
痛痹	6.11.13.2
头针疗法	9.3.3.5
透刺	9.2.6.3
土乘水	1.3.13.2
土克水	1.3.12.4
土生金	1.3.11.5
土侮木	1.3.14.5
土壅木郁	4.2.4.8.21
土壅侮木证	7.10.34
吐法	8.1.2
兔流涎病	6.10.7
推而纳之	9.2.11.4
推拿疗法	8.4.1
退法	9.2.9.4
退翳明目	8.2.12
脱膊	6.7.5
脱白	6.7.4
脱毛症	6.8.6

W

外风证	7.3.1
外感病因	4.1.1.2.1
外感咳嗽	6.2.2.1
外伤	4.1.9
外伤鼻血	6.2.9.5
外伤目络证	7.12.2.3
外伤胎动	6.9.1.3
外伤瘀滞证	7.13.2.14
外肾黄	6.5.19
外燥袭表证	7.11.7
外燥证	7.3.5
外障眼	6.11.17
弯针	9.2.10.3
顽湿结聚	6.8.15
菀陈则除之	9.1.17.7
亡阳	4.2.2.3.7.3
亡阳证	7.2.8
亡阴	4.2.2.3.7.1
亡阴证	7.2.6
微波针灸仪	9.2.3.4
微波针疗法	9.3.3.14
腲腿风	6.5.22.2
痿证	6.11.14
卫出上焦	2.2.1.9.1
卫出下焦	2.2.1.9.3
卫出中焦	2.2.1.9.2
卫分病机	4.2.6.1
卫气	2.2.1.9
卫气不固	4.2.5.1.2
卫气同病	4.2.6.5
卫气同病证	7.3.6.5
卫气营血辨证	1.1.1.1.9
卫气营血病证病机	4.2.6
卫强营弱	4.2.5.1.1.1
卫弱营强	4.2.5.1.1.2
卫虚证	7.2.18
卫营同病	4.2.6.8
未病先防	5.1.1.1
胃	2.1.5.5
胃病病机	4.2.4.5.2
胃肠气滞证	7.7.71
胃肠湿热证	7.7.67
胃肠实热证	7.7.68
胃寒	4.2.4.5.2.4
胃寒	6.3.24
胃寒气逆证	7.7.25.2
胃火炽盛	4.2.4.5.2.6
胃火燔齿证	7.12.11.3
胃火燔龈证	7.12.11.2
胃火证	7.7.31
胃家实	4.2.5.2.3
胃津	2.1.5.5.5.1
胃纳呆滞	4.2.4.5.2.11
胃气	2.1.5.5.3
胃气不和	4.2.4.5.2.8
胃气不降	4.2.4.5.2.9
胃气上逆	4.2.4.5.2.10
胃气上逆证	7.7.25
胃气虚	4.2.4.5.2.1
胃气虚血瘀证	7.7.24
胃气虚证	7.7.22
胃气阴两虚证	7.7.23
胃气滞血瘀证	7.7.27
胃热	4.2.4.5.2.5
胃热	6.3.25
胃热鼻血	6.2.9.3
胃热津伤证	7.7.33
胃热脾虚证	7.7.65
胃热气逆证	7.7.25.1
胃热气滞证	7.7.32
胃热消谷	4.2.4.5.2.7
胃热阴虚证	7.7.34
胃脘	2.1.5.5.2
胃喜柔润	2.1.5.9
胃阳	2.1.5.5.4
胃阳虚	4.2.4.5.2.2
胃阳虚气滞证	7.7.26.1

胃阳虚血瘀证	7.7.26.2
胃阳虚证	7.7.26
胃阴	2.1.5.5.5
胃阴虚	4.2.4.5.2.3
胃阴虚气滞证	7.7.29
胃阴虚血瘀证	7.7.30
胃阴虚证	7.7.28
胃燥津伤证	7.7.35
胃滞气逆证	7.7.25.3
胃主腐熟	2.1.5.7
胃主降浊	2.1.5.8.1
胃主受纳	2.1.5.6
胃主通降	2.1.5.8
温病	6.10.1
温补肺阳	8.1.8.45
温补肝阳	8.1.8.48
温补命火	8.1.8.50
温补纳气	8.1.8.51
温补脾肾	8.1.8.56
温补脾胃	8.1.8.55
温补脾阳	8.1.8.46
温补肾阳	8.1.8.49
温补胃阳	8.1.8.47
温补下元	8.1.8.52
温补心肺	8.1.8.54
温补心肾	8.1.8.57
温补心阳	8.1.8.44
温毒袭表证	7.11.9
温法	8.1.5
温肺散寒	8.1.5.6
温和灸	9.3.2.1.2.2
温化寒痰	8.1.10.78
温化痰饮	8.1.10.84
温经活血	8.1.5.11
温经散寒	8.1.5.3
温经止血	8.1.5.12
温灸器	9.3.1.4
温灸器灸	9.3.2.1.3
温肾利水	8.1.7.25
温肾散寒	8.1.5.7
温通小肠	8.1.5.10
温筒灸	9.3.2.1.3.1
温胃散寒	8.1.5.9
温下寒积	8.1.3.3
温邪	4.1.2.6.2
温阳固脱	8.1.10.135
温阳化饮	8.1.10.83
温阳健齿	8.2.42
温阳散寒	8.1.5.5
温阳行气	8.1.8.60
温燥袭肺证	7.6.28
温燥证	7.3.5.1
温针灸	9.3.3.1
温中散寒	8.1.5.2
温中涩肠	8.1.8.58
温中止血	8.1.8.59
瘟毒下注证	7.15.5.2
无瘢痕灸	9.3.2.1.1.5
无名肿毒	6.6.37
无头疽	6.6.33
五方（宫）	1.3.10
五官	1.3.9
五华	2.1.4.3
五化	1.3.5
五气	1.3.4
五窍	2.1.4.5
五色	1.3.6
五时	1.3.3
五实	4.2.2.2.1.2
五体	2.1.4.4
五味	1.3.7
五行	1.3.2
五行乘侮	1.3.16
五行归类	1.3.2.1
五行生克	1.3.15
五行胜复	1.3.18
五行相乘	1.3.13
五行相克	1.3.12

五行相生	1.3.11
五行相侮	1.3.14
五行学说	1.3.1
五行制化	1.3.17
五虚	4.2.2.2.2.2
五液	2.1.4.2
五音	1.3.8
五攒痛	6.7.25
五脏	2.1.4
五脏病机	4.2.4.1
五脏六腑之海	2.1.5.5.1
五脏所恶	2.1.4.6
五脏所主	2.1.4.1

X

息脉	2.1.10.4.3
熄风解痉	8.1.10.73
洗耳疗法	8.2.63
喜则气缓	4.1.5.4
下病上取	9.1.17.10
下法	8.1.3
下焦	2.1.9.3
下焦如渎	2.1.9.3.1
下焦湿热	4.2.7.3
下焦湿热证	7.15.5
下焦主出	2.1.9.3.2
先补后攻	5.2.5.3
先攻后补	5.2.5.4
先天之精	2.2.4.1
先天之气	2.2.1.4
痫病	6.1.9
陷下则灸之	9.1.17.6
项脊怯	6.11.4
项痛	6.6.34
消法	8.1.7
消痞化积	8.1.7.38
消食和胃	8.1.7.33
消食化滞	8.1.7.32
消食解毒	8.1.7.36
消食止呕	8.1.7.34

消食止痛	8.1.7.35
消痈散疖	8.2.7
小肠	2.1.6
小肠病机	4.2.4.3.2
小肠实热	4.2.4.3.2.2
小肠实热证	7.7.41.1
小肠虚寒	4.2.4.3.2.1
小肠主化物	2.1.6.2
小肠主受盛	2.1.6.1
小肠主液	2.1.6.3
邪毒炽盛证	7.3.14
邪犯清窍证	7.12.13
邪犯少腹证	7.15.6
邪恋耳窍证	7.12.8.6
邪气	4.2.1.3
邪气盛则实	4.2.2.2.1
邪去正虚	4.2.2.2.7
邪扰胸膈证	7.15.1
邪热伤阴证	7.4.9
邪入少阳证	7.15.7
邪盛正衰	4.2.2.2.8
邪袭卫表证	7.11.1
邪陷正脱证	7.4.13
邪郁肺卫	4.2.6.1.1
邪郁少阳	4.2.5.3.2
邪正盛衰	4.2.2.2
斜刺	9.2.6.2.2
泄泻	6.3.19
泻热消痞	8.1.7.39
泻肺平喘	8.1.10.8
泻肺逐饮	8.1.10.82
泻肝清肺	8.1.6.25
泻肝清胃	8.1.6.26
泻结行滞	8.1.3.2
心	2.1.4.8
心包络	2.1.4.8.14
心病病机	4.2.4.3.1
心藏神	2.1.4.8.6
心胆气虚	4.2.4.8.11

心恶热	2.1.4.8.12
心肺气虚	4.2.4.8.1
心肺气虚证	7.10.8
心肺气阴两虚证	7.10.10
心肺热盛证	7.10.13
心肺阳虚证	7.10.12
心肺阴虚血瘀证	7.10.11
心肺阴虚证	7.10.9
心肝火旺	4.2.4.8.3
心肝火旺证	7.10.20
心肝气虚血瘀证	7.10.26
心肝气血两虚证	7.10.25
心肝血虚	4.2.4.8.4
心肝血虚挟瘀证	7.10.24
心肝血虚证	7.10.22
心肝血瘀证	7.10.21
心肝阴虚证	7.10.23
心合小肠	2.1.4.8.13
心华在面	2.1.4.8.9
心黄	6.1.10
心火炽盛证	7.5.17
心火亢盛	4.2.4.3.1.6
心火上炎	4.2.4.3.1.7
心火上炎证	7.5.18
心开窍于舌	2.1.4.8.10
心冷吐水	6.1.6
心脉气滞证	7.5.15
心脾积热证	7.10.19
心脾两虚	4.2.4.8.2
心脾两虚证	7.10.15
心脾气虚证	7.10.16
心脾气血两虚证	7.10.18
心脾阳虚证	7.10.17
心气	2.1.4.8.1
心气虚	4.2.4.3.1.1
心气虚	6.1.5.1
心气虚血瘀证	7.5.2
心气虚证	7.5.1
心气血两虚证	7.5.3

心气阴两虚证	7.5.4
心气滞血瘀证	7.5.12.1
心热风邪	6.1.7
心热血瘀证	7.5.12.2
心热阴虚证	7.5.19
心神不宁证	7.5.33
心肾不交	4.2.4.8.7
心肾不交证	7.10.2
心肾火热证	7.10.14
心肾气虚证	7.10.5
心肾气阴两虚证	7.10.6
心肾相交	2.1.4.8.15
心肾阳虚	4.2.4.8.6
心肾阳虚证	7.10.3
心肾阴虚	4.2.4.8.5
心肾阴虚证	7.10.1
心肾阴阳两虚证	7.10.7
心痛	6.1.11
心胃火燔	4.2.4.8.10
心系	2.1.4.8.5
心虚	6.1.5
心虚神怯证	7.5.30
心血	2.1.4.8.2
心血虚	4.2.4.3.1.2
心血虚	6.1.5.2
心血虚证	7.5.7
心血瘀阻	4.2.4.3.1.9
心血瘀阻证	7.5.12
心阳	2.1.4.8.4
心阳暴脱	4.2.4.3.1.4
心阳暴脱证	7.5.5.1
心阳虚	4.2.4.3.1.3
心阳虚脱证	7.5.5.2
心阳虚血瘀证	7.5.6
心阳虚证	7.5.5
心阴	2.1.4.8.3
心阴虚	4.2.4.3.1.5
心阴虚火旺证	7.5.9
心阴虚血瘀证	7.5.10

词条	编号	词条	编号
心阴血虚证	7.5.8	虚阳上浮	4.2.2.3.6.2
心阴阳两虚证	7.5.11	虚则补其母	9.1.17.14
心营过耗	4.2.6.3.3	虚则补之	9.1.17.2
心与小肠病证病机	4.2.4.3	虚者补其母	5.2.10.3
心在液为汗	2.1.4.8.11	虚证病机	4.2.2.2.1
心志喜	2.1.4.8.7	虚中夹实	4.2.2.2.3.1
心主脉	2.1.4.8.8.2	徐发	4.2.1.4.4
心主血	2.1.4.8.8.1	徐疾补泻法	9.2.11.9.6
心主血脉	2.1.4.8.8	宣肺化痰	8.1.10.74
辛凉解表	8.1.1.2	宣肺降气	8.1.10.6
辛凉清热	8.1.1.3	宣肺解表	8.1.1.7
辛温解表	8.1.1.1	宣肺利水	8.1.7.26
新感	4.2.1.4.1	宣肺平喘	8.1.10.7
新驹奶泻	6.11.8	宣肺通气	8.1.10.5
行痹	6.11.13.1	宣散湿邪	8.1.7.3
行气和胃	8.1.10.10	宣通三焦	8.1.10.142
行气化湿	8.1.7.4	悬起灸	9.3.2.1.2.1
行气降逆	8.1.10.12	选穴法	9.1.18
行气破血	8.1.10.18	穴位磁疗法	9.3.3.11
行针候气	9.1.15.2	穴位封闭疗法	9.2.13.5.3
胸黄	6.6.18	穴位激光照射疗法	9.3.3.10
胸水	6.2.13	穴位埋线疗法	8.3.12
胸痛	6.2.6	穴位埋线疗法	9.3.3.9
虚喘	6.2.3.2	穴位免疫疗法	9.2.13.5.4
虚寒	4.2.2.4.2	穴位药物疗法	9.2.13.5.2
虚寒痢	6.3.13.3	穴位液压疗法	9.2.13.5.1
虚寒呕吐	6.3.1.3	穴位注气法	9.2.13.7.3
虚寒跳欣	6.11.11.1	血	2.2.2
虚火	4.2.2.4.6	血病治气	5.2.10.2
虚火上炎	4.2.2.3.2.3.4	血不归经	4.2.3.1.2.6
虚火灼口证	7.12.12.4	血不养筋	4.2.3.1.2.7
虚火灼龈证	7.12.11.9	血疗	6.6.28
虚劳病	6.11.3	血分病机	4.2.6.4
虚秘	6.3.20.2	血分热毒	4.2.6.4.4
虚热	4.2.2.4.5	血分瘀热	4.2.6.4.3
虚实错杂	4.2.2.2.3	血分证	7.3.6.3
虚实真假	4.2.2.2.4	血滚毒症	6.11.10.1
虚实转化	4.2.2.2.5	血寒	4.2.3.1.2.3
虚阳浮越证	7.2.9	血寒证	7.3.2.8

血汗同源	2.2.13
血淋	6.5.7.4
血轮实热证	7.12.4.1
血轮虚热证	7.12.4.2
血轮证	7.12.4
血能养气	2.2.6.2
血能载气	2.2.6.1
血热	4.2.3.1.2.4
血热肠燥证	7.7.47
血热扰神证	7.5.23
血热伤阴证	7.4.9.1
血热胎动	6.9.1.2
血热妄行	4.2.6.4.2
血失调	4.2.3.1.2
血随气逆	4.2.3.1.3.6
血脱	4.2.3.1.2.5
血为气母	2.2.6
血虚	4.2.3.1.2.1
血虚鼻窍失养证	7.12.9.11
血虚肠燥证	7.7.45
血虚齿槽失养证	7.12.11.11
血虚唇燥证	7.12.12.5
血虚动风证	7.2.4.1
血虚耳窍失养证	7.12.8.12
血虚耳燥证	7.12.8.13
血虚风燥证	7.2.4.2
血虚寒凝证	7.4.2.3
血虚津亏证	7.2.4.3
血虚内热证	7.4.2.4
血虚生风	4.1.4.1.3
血虚邪恋证	7.4.6.7
血虚挟实证	7.4.2
血虚挟痰证	7.4.2.2
血虚挟瘀证	7.4.2.1
血虚龈肉失养证	7.12.11.10
血虚证	7.2.4
血瘀	4.2.3.1.2.2
血瘀齿龈证	7.12.11.7
血瘀动血证	7.3.13.2
血瘀耳窍证	7.12.8.8
血瘀风燥证	7.4.11
血瘀化热证	7.3.13.3
血瘀气滞证	7.3.13.1
血瘀舌下证	7.12.12.8
血瘀水停证	7.3.13.4
血瘀证	7.3.13
血燥生风	4.1.4.1.4
血针疗法	9.2.13.2
血证	6.1.2
血主濡之	2.2.2.1
熏洗疗法	8.2.51
循经叩刺	9.3.3.3.1.3
循经选穴法	9.1.18.4

Y

押手	9.2.4.1
押手法	9.2.4
烟熏疗法	8.6.3
咽喉证	7.12.10
阳	1.2.1.1.2
阳病治阴	5.2.9.1
阳化气	1.2.1.1.5
阳黄	6.4.1.1
阳脉	3.1.12
阳明病机	4.2.5.2
阳明病证	7.14.3
阳明腑实	4.2.5.2.2
阳明腑证	7.14.5
阳明经证	7.14.4
阳明虚寒	4.2.5.2.4
阳明燥热	4.2.5.2.1
阳气	1.2.1.1.4
阳跷脉	3.1.8.8
阳杀阴藏	1.2.1.12
阳肾黄	6.5.19.2
阳生阴长	1.2.1.11
阳盛	4.2.2.3.1.1
阳盛格阴	4.2.2.3.6.3
阳盛则热	4.2.2.3.1.1.2

阳盛则阴病	4.2.2.3.1.1.1
阳衰	4.2.2.3.2.1
阳水	6.11.2.2
阳损及阴	4.2.2.3.3.1
阳损及阴证	7.2.13.2
阳维脉	3.1.8.6
阳痿	6.5.21
阳虚鼻窍失煦证	7.12.9.13
阳虚耳窍失煦证	7.12.8.10
阳虚寒凝证	7.4.4.7
阳虚气滞证	7.4.4.1
阳虚湿困证	7.4.4.2
阳虚水泛证	7.4.4.4
阳虚痰凝证	7.4.4.6
阳虚外感证	7.4.4.8
阳虚邪恋证	7.4.6.8
阳虚挟实证	7.4.4
阳虚血瘀证	7.4.4.5
阳虚饮停证	7.4.4.3
阳虚则寒	4.2.2.3.2.2
阳虚证	7.2.7
阳中求阴	5.2.9.3
阳中之阳	1.2.1.10
阳中之阴	1.2.1.9
养心安神	8.1.10.116
养血安神	8.1.10.117
养血安胎	8.1.8.17
养血调血	8.1.8.16
养血明目	8.2.17
养血润肠	8.1.8.15
养血止血	8.1.8.18
腰胯痛	6.7.14
摇法	9.2.7.9
药毒	6.8.17
药复	4.2.1.4.6.3
药膏疗法	8.2.49
药酒疗法	8.5.2
药物贴敷疗法	9.3.3.15
药浴疗法	8.6.2

药熨疗法	8.2.45
夜盲	6.4.10
液	2.2.3.2
液亏证	7.2.14.2
液脱	4.2.3.2.4
一进三退	9.2.11.6
以火补之	9.3.2.1.4.1
以火泻之	9.3.2.1.4.2
以痛为输	9.1.18.13
以右治左	9.1.17.12
以左治右	9.1.17.11
异病同治	5.2.7
异经选穴法	9.1.18.6
异物呛肺	6.2.14
异物伤胃	6.3.7
抑肝扶脾	8.1.4.5
疫毒痢	6.3.13.2
疫毒内闭证	7.3.14.12
疫毒证	7.3.14.11
疫疠	4.1.3
益火消阴	8.1.10.139
益气安神	8.1.10.119
益气固齿	8.2.39
益气通耳	8.2.22
益气通下	8.1.3.5
益气滋阴	8.1.8.63
益肾宁神	8.1.10.118
因畜（禽）制宜	5.2.8.3
因地制宜	5.2.8.2
因时制宜	5.2.8.1
因虚致实	4.2.2.2.5.2
阴	1.2.1.1.1
阴病治阳	5.2.9.2
阴成形	1.2.1.1.6
阴道脱出	6.9.7
阴毒证	7.3.14.17
阴黄	6.4.1.2
阴竭阳脱	4.2.2.3.7.5
阴竭阳脱证	7.2.13.3

阴脉	3.1.13
阴囊疝	6.11.5.1
阴平阳秘	1.2.1.3
阴气	1.2.1.1.3
阴跷脉	3.1.8.7
阴肾黄	6.5.19.1
阴盛	4.2.2.3.1.2
阴盛格阳	4.2.2.3.6.1
阴盛则寒	4.2.2.3.1.2.2
阴盛则阳衰	4.2.2.3.1.2.1
阴衰	4.2.2.3.2.3
阴水	6.11.2.1
阴损及阳	4.2.2.3.3.2
阴损及阳证	7.2.13.1
阴维脉	3.1.8.5
阴虚鼻窍失濡证	7.12.9.12
阴虚鼻血	6.2.9.4
阴虚肠燥证	7.7.46
阴虚齿燥证	7.12.11.8
阴虚动风证	7.2.5.4
阴虚动血证	7.2.5.3
阴虚耳窍失濡证	7.12.8.9
阴虚肺燥证	7.6.6
阴虚风动	4.1.4.1.2
阴虚火旺	4.2.2.3.2.3.3
阴虚火旺证	7.4.3.2
阴虚津亏证	7.2.5.5
阴虚内热证	7.4.3.1
阴虚气滞证	7.4.3.7
阴虚热盛证	7.4.3.13
阴虚热郁证	7.4.3.6
阴虚湿热证	7.4.3.10
阴虚水停证	7.4.3.9
阴虚痰热证	7.4.3.11
阴虚痰湿证	7.4.3.12
阴虚外感证	7.4.3.5
阴虚邪恋证	7.4.6.6
阴虚挟实证	7.4.3
阴虚血热证	7.4.3.4
阴虚血瘀证	7.4.3.8
阴虚血燥证	7.2.5.2
阴虚咽喉失濡证	7.12.10.12
阴虚阳亢证	7.2.5.1
阴虚阳亢证	7.4.3.3
阴虚余热证	7.4.6.9
阴虚则热	4.2.2.3.2.3.1
阴虚则阳亢	4.2.2.3.2.3.2
阴虚证	7.2.5
阴癣	6.8.11
阴血亏虚证	7.2.12
阴阳	1.2.1.1
阴阳对立	1.2.1.2.1
阴阳格拒	4.2.2.3.6
阴阳互根	1.2.1.2.2
阴阳互损	4.2.2.3.3
阴阳交感	1.2.1.2
阴阳离决	1.2.1.6
阴阳离决	4.2.2.3.8
阴阳两虚	4.2.2.3.3.3
阴阳两虚证	7.2.13
阴阳配穴法	9.1.19.7
阴阳偏盛	4.2.2.3.1
阴阳偏衰	4.2.2.3.2
阴阳平衡	1.2.1.4
阴阳胜复	4.2.2.3.4
阴阳失调	4.2.2.3
阴阳亡失	4.2.2.3.7
阴阳消长	1.2.1.2.3
阴阳学说	1.2.1
阴阳转化	1.2.1.2.4
阴阳转化	4.2.2.3.5
阴阳自和	1.2.1.5
阴中求阳	5.2.9.4
阴中之阳	1.2.1.8
阴中之阴	1.2.1.7
引火归原	8.1.10.140
引流疗法	8.2.56
饮	4.1.8.1.2

饮停心包证	7.5.16
饮停胸胁证	7.15.1.7
迎随补泻法	9.2.11.9.1
营出中焦	2.2.1.8.1
营分病机	4.2.6.3
营分证	7.3.6.2
营气	2.2.1.8
营气不从	4.2.5.1.3
营卫不和	4.2.5.1.1
营卫不和证	7.15.9
营行脉中	2.2.1.8.2
营虚证	7.2.19
营阴损伤	4.2.6.3.4
痈	6.6.30
涌吐风痰	8.1.2.2
涌吐宿食	8.1.2.4
涌吐痰食	8.1.2.3
涌吐痰涎	8.1.2.1
由寒化热	4.2.2.4.9.1
由热化寒	4.2.2.4.9.2
由实转虚	4.2.2.2.5.1
由阳转阴	4.2.2.3.5.1
由阴转阳	4.2.2.3.5.2
疣	6.8.7
有头疽	6.6.32
幼畜惊风	6.11.7
幼畜胎粪不下	6.11.9
幼驹尿血	6.11.10
幼驹血尿	6.11.10.2
瘀热犯头证	7.12.1.8
瘀热入络证	7.13.1.9
瘀血	4.1.8.2
瘀血败精尿闭	6.5.25.3
瘀血犯头证	7.12.1.7
瘀血犯腰证	7.13.2.10
瘀血阻膈证	7.15.1.9
瘀血阻络证	7.13.1.10
瘀血阻滞精室证	7.9.29
瘀滞肌肤证	7.11.22

瘀滞筋骨证	7.13.2.12
瘀滞胃肠证	7.7.72
瘀浊阻滞精室证	7.9.30
瘀阻胞宫证	7.9.20
瘀阻肺络证	7.6.33
瘀阻清窍证	7.12.13.2
瘀阻声带证	7.12.10.9
瘀阻胃络证	7.7.38
瘀阻下焦证	7.15.5.3
瘀阻胸胁证	7.15.1.8
瘀阻咽喉证	7.12.10.10
郁火	4.2.2.5.1.5
寓补于攻	5.2.5.7
寓攻于补	5.2.5.6
愈后防复	5.1.1.3
元气	2.2.1.6
元神之府	2.1.10.1.2
原络配穴法	9.1.19.3
圆利针	9.2.1.6 或 9.2.2.2
圆利针疗法	9.2.13.1.2
圆癣	6.8.10
圆针	9.2.1.2
远道选穴法	9.1.18.3
远近配穴法	9.1.19.12
月盲	6.4.9
晕针	9.2.10.1
云翳遮睛	6.4.4
运针	9.2.7.1
运针法	9.2.7
熨灸疗法	9.3.2.2
熨烙法	9.3.2.3.2

Z

攒筋痛	6.7.10
脏病取原	9.1.17.15
脏病治腑	5.2.10.5
脏腑	2.1.2
脏腑辨证	1.1.1.1.6
脏腑病证病机	4.2.4
脏腑兼病病机	4.2.4.8

脏腑学说	2.1.3
脏腑之气	2.2.1.10
燥毒证	7.3.14.15
燥干清窍证	7.12.13.1
燥热伤肺	4.2.4.4.1.9
燥伤鼻窍证	7.12.9.7
燥湿和胃	8.1.7.13
燥湿化痰	8.1.10.75
燥湿敛疮	8.2.8
燥湿行气	8.1.7.12
燥痰	4.1.8.1.1.4
燥痰结肺证	7.6.29
燥痰证	7.3.7.5
燥邪	4.1.2.5
燥邪犯肺证	7.6.26
燥性干涩	4.1.2.5.1
燥易伤肺	4.1.2.5.2
增液润肺	8.1.10.55
长夏	1.3.3.1
长针	9.2.1.8
掌骨痛	6.7.11
着痹	6.11.13.3
针棒（杖）	9.2.2.12
针锤	9.2.2.11
针锤持针法	9.2.5.7
针刺补法	9.2.11.2
针刺补泻	9.2.11.1
针刺补泻法	9.2.11
针刺方向	9.2.6.1
针刺角度	9.2.6.2
针刺疗法	9.1.8
针刺六脉血	5.1.3
针刺麻醉法	8.3.13
针刺深度	9.2.6.4
针刺手法	9.1.9
针刺泻法	9.2.11.5
针刺异常情况	9.2.10
针罐法	9.3.3.7.4
针灸	9.1.1
针灸刺激量	9.1.11
针灸刺激强度	9.1.12
针灸学	9.1.2
针灸治疗学	9.1.7
针灸治则	9.1.17
针药罐法	9.3.3.7.6
真寒假热	4.2.2.4.8.2
真寒假热证	7.3.2.6
真热假寒	4.2.2.4.8.1
真实假虚	4.2.2.2.4.2
真实假虚证	7.3.18
真虚假实	4.2.2.2.4.1
镇肝潜阳	8.1.10.72
镇心安神	8.1.10.121
整体观念	1.1.1.1.1
正气	4.2.1.2
正胜邪退	4.2.2.2.6
正水	6.5.18
正邪相争	4.2.2.1
正虚毒恋证	7.4.6.4
正虚毒陷证	7.4.6.5
正虚脓毒证	7.4.6.10
正虚邪恋	4.2.2.2.9
正虚邪恋证	7.4.6
正治	5.2.2
证	7.1.1
证候	1.1.1.1.3
执笔式持针法	9.2.5.2
直肠脱出	6.11.6
直刺	9.2.6.2.1
直接灸	9.3.2.1.1.3
直透	9.2.6.3.1
直中	4.2.1.4.8
止血安胎	8.1.10.129
纸垫练针法	9.2.12.1
指切押手法	9.2.4.2
治病求本	5.2.1.1
治未病	5.1.1
治则	5.2.1

滞针	9.2.10.2
中刺激	9.1.12.2
中风	6.1.13
中寒虫扰证	7.15.4.2
中焦	2.1.9.2
中焦如沤	2.1.9.2.1
中焦湿热	4.2.7.2
中焦湿热证	7.15.4
中焦实热证	7.15.4.1
中焦主化	2.1.9.2.2
中气	2.2.1.11
中气下陷	4.2.4.5.1.3
中兽医基础理论	1.1.1.1
中兽医学	1.1.1
中暑	6.1.8
中暑证	7.3.3.2
重插轻提	9.2.11.9.5
重提轻插	9.2.11.9.4
重阳必阴	1.2.1.2.4.2
重阴必阳	1.2.1.2.4.1
主应配穴法	9.1.19.13
贮药罐法	9.3.3.7.5
壮	9.3.2.1.1.1
壮水制阳	8.1.10.138
浊毒闭神证	7.5.28
滋补肺阴	8.1.8.22
滋补肝阴	8.1.8.25
滋补脾阴	8.1.8.24
滋补肾阴	8.1.8.29
滋补心肺	8.1.8.31
滋补心阴	8.1.8.21
滋肝明目	8.2.16
滋肝肾濡耳	8.2.24

滋肾纳气	8.1.8.30
滋阴	8.1.8.20
滋阴安神	8.1.10.120
滋阴补血	8.1.8.66
滋阴补阳	8.1.8.62
滋阴降火	8.1.8.37
滋阴凉血	8.1.8.34
滋阴凉营	8.1.8.33
滋阴平肝	8.1.8.28
滋阴潜阳	8.1.8.38
滋阴清热	8.1.8.32
滋阴柔肝	8.1.8.26
滋阴濡耳	8.2.23
滋阴润齿	8.2.41
滋阴润燥	8.1.10.52
滋阴生津	8.1.8.35
滋阴疏肝	8.1.8.27
滋阴通便	8.1.8.41
滋阴熄风	8.1.10.70
滋阴益胃	8.1.8.23
滋阴止咳	8.1.8.39
滋阴止汗	8.1.8.40
滋阴止渴	8.1.8.36
子母补泻	9.2.11.9.10
子母配穴法	9.1.19.5
子行	1.3.11.2
紫癜风	6.8.21
紫外线穴位照射疗法	9.3.3.12
自家血疗法	9.2.13.6
宗气	2.2.1.7
走索子	9.2.2.13
卒发	4.2.1.4.3
左右配穴法	9.1.19.10

英文对应词索引

A

abdominal fullness ················· 6.3.33
abdominal swelling ················· 6.6.19
abnormal transmission of warm pathogen
　　to pericardium ··············· 4.2.6.3.2
abortion ······························· 6.9.2
acceleration of digestion by stomach heat
　　··································· 4.2.4.5.2.7
accumulated heat due to fetal toxicity pattern
　　······································· 7.15.8
accumulation of dampness-heat in spleen
　　··································· 4.2.4.5.1.9
acquired essence ····················· 2.2.4.2
acquired qi ···························· 2.2.1.5
action less needle retention ········ 9.2.8.2
activating blood flow and removing
　　stagnated blood ············· 8.1.10.20
activating the blood and dispelling wind
　　····································· 8.1.10.32
activating the blood and moving stagnation
　　····································· 8.1.10.24
activating the blood and nourishing the blood
　　····································· 8.1.10.27
activating the blood and relaxing the sinews
　　····································· 8.1.10.31
activating the blood and relieving pain
　　····································· 8.1.10.33
activating the blood and resolving
　　accumulation ················· 8.1.10.25
active needle retention ············· 9.2.8.3
acupoint blocking therapy ········ 9.2.13.5.3
acupoint drug therapy ············· 9.2.13.5.2
acupoint fluid pressure therapy ··· 9.2.13.5.1

acupoint immunity therapy ······· 9.2.13.5.4
acupoint magnetic therapy ········· 9.3.3.11
acupuncture and moxibustion ········ 9.1.1
acupuncture anesthesia ············· 8.3.13
acupuncture therapy ················· 9.1.8
acute conjunctivitis due to liver heat ······ 6.4.3
acute eclampsia ····················· 6.11.7.1
acute enteritis ······················· 6.3.14.1
acute pharyngolaryngitis ············ 6.10.5.3
acute suppurative disease ··········· 6.6.30
adjacent point selection ············· 9.1.18.2
adverse qi pattern ····················· 7.13.2
aerosol therapy ······················· 8.2.68
air acupuncture therapy ············ 9.2.13.7
air injection method ··············· 9.2.13.7.3
alcohol cupping method ·········· 9.3.3.7.1.4
alcohol fire-separated method ····· 9.3.3.7.1.5
alopecia ································· 6.8.6
alternate preponderance among five elements
　　······································· 1.3.18
alternative predominance of yin and yang
　　····································· 4.2.2.3.4
anemia of blood and body fluids pattern ······ 7.2.4.3
anger as liver emotion ············· 2.1.4.11.8
anger causing the qi to rise ········· 4.1.5.2
angle of needle insertion ············ 9.2.6.2
anterior-posterior points combination ··· 9.1.19.9
anuria ································· 6.5.25
anuria due to damage ·············· 6.5.25.1
anuria due to heat ·················· 6.5.25.2
anuria due to stasis or abortive obstruction
　　····································· 6.5.25.3
anus dampness heat pattern ········ 7.7.55

apoplexy	6.1.7
appendicitis	6.3.15
arresting seminal emission	8.1.10.126
arrow needle	9.2.2.4
arrowhead-like needle	9.2.1.1
arthralgia	6.11.13.3
arthralgia syndrome	6.11.13
arthritis	6.11.13.2
ascending of stomach qi pattern	7.7.25
ascites	6.3.22
aspiration or deglutition pneumonia	6.2.14
asthenic dyspnea	6.2.3.2
asthma	6.2.3
astringing for arresting lactation	8.1.10.130
attack of lung by liver fire	4.2.4.8.22
attack of lung by water-cold	4.2.4.8.9
attack of spleen by liver qi	4.2.4.8.23
attack of stomach by liver qi	4.2.4.8.24
auriculo-acupuncture therapy	8.3.14
awaiting qi	9.1.15

B

back inflammation due to hot evil	6.6.17
bandage-fixing therapy	8.2.70
basic theory of Chinese veterinary medicine	1.1.1.1
bedsore	6.6.22
bending of needle	9.2.10.3
bi jing deficiency in channels of water ring pattern	7.12.7.12
bi syndrome due to heat pathogen heat arthralgia	6.11.13.4
bile	2.1.5.4.3
bile heat and sputum disturbance pattern	7.8.35
bite wound of insect	6.8.19
black nail like boil	6.6.24
bladder	2.1.8
bladder blood amassment pattern	7.9.16
bladder cold deficiency pattern	7.9.17
bladder dominating fluid storage	2.1.8.1
bladder heat amassment pattern	7.9.14
bladder meridian of hindlimbs-taiyang	3.1.4.3.3
bladder pathogenesis	4.2.4.7.2
bladder qi blockage	4.2.4.7.2.4
bladder water amassment pattern	7.9.15
blazing of both qi and blood levels	4.2.6.7
blazing of both qi and nutrient levels	4.2.6.6
blazing of heart-liver fire	4.2.4.8.3
blazing of stomach fire	4.2.4.5.2.6
bleeding at the six points	5.1.3
bleeding due to blood heat	4.2.6.4.2
bleeding due to collateral injury pattern	7.13.2.13
bleeding or hemorrhagic syndrome	6.1.2
blood	2.2.2
blood and sweat share the same source	2.2.13
blood cold	4.2.3.1.2.3
blood cold stasis pattern	7.3.2.8
blood collapse	4.2.3.1.2.5
blood conveying qi	2.2.6.1
blood counterflow with qi	4.2.3.1.3.6
blood deficiency	4.2.3.1.2.1
blood deficiency alveolar dysplasia pattern	7.12.11.11
blood deficiency and clod coagulation pattern	7.4.2.3
blood deficiency and intestine dryness pattern	7.7.45
blood deficiency and lip dryness pattern	7.12.12.5
blood deficiency and loss of nose pattern	7.12.9.11
blood deficiency and wind dryness pattern	7.2.4.2
blood deficiency carrying sputum pattern	7.4.2.2
blood deficiency causing evil toxin detaining pattern	7.4.6.7
blood deficiency dryness pattern	7.12.8.13
blood deficiency gingival meat loss pattern	7.12.11.10
blood deficiency in flesh ring pattern	7.12.5.7
blood deficiency mixed excess pattern	7.4.2

blood deficiency mixed stasis pattern	7.4.2.1
blood deficiency of ear pattern	7.12.8.12
blood deficiency of heart	6.1.5.2
blood deficiency of the heart and liver pattern	7.10.22
blood deficiency pattern	7.2.4
blood deficiency stirring wind pattern	7.2.4.1
blood flowing outside channels	4.2.3.1.2.6
blood governs nourishing and moistening	2.2.2.1
blood heat	4.2.3.1.2.4
blood heat and intestine dryness pattern	7.7.47
blood heat disturbing mind pattern	7.5.23
blood heat injuring yin pattern	7.4.9.1
blood injection therapy	9.2.13.2
blood is the mother of qi	2.2.6
blood nourishes qi	2.2.6.2
blood obstruction water ring pattern	7.12.7.11
blood phase pattern	7.3.6.3
blood ring pattern	7.12.4
blood stasis	4.2.3.1.2.2
blood stasis and head offense pattern	7.12.1.7
blood stasis and qi stagnation pattern	7.3.13.1
blood stasis and sperm blocking essence chamber pattern	7.9.29
blood stasis and waist offense pattern	7.13.2.10
blood stasis blocking body fluids pattern	7.3.13.4
blood stasis blocking lung channel pattern	7.6.33
blood stasis blocking stomach and intestine pattern	7.7.72
blood stasis blocking stomach channel pattern	7.7.38
blood stasis causing wind dryness pattern	7.4.11
blood stasis due to qi deficiency	4.2.3.1.3.3
blood stasis due to qi stagnation	4.2.3.1.3.2
blood stasis gingival pattern	7.12.11.7
blood stasis in chest and stress pattern	7.15.1.8
blood stasis in ear pattern	7.12.8.8
blood stasis in qi ring pattern	7.12.3.3
blood stasis moving blood pattern	7.3.13.2
blood stasis pattern	7.3.13
blood stasis pattern	7.8.13
blood stasis pattern flesh ring pattern	7.12.5.1
blood stasis transforming heat pattern	7.3.13.3
bloodletting for chronic blood stasis	9.1.17.7
blood-letting therapy	8.2.57
bloody stool	6.3.21
bloody stranguria	6.5.7.4
body deficiency cold	6.2.1.3
body deficiency cold	6.2.2.4
body fluids	2.2.3
body fluids and blood share the same source	2.2.11
body fluids carries qi	2.2.10
body fluids deficiency pattern	7.2.14
body fluids deficiency pattern	7.2.14.1
body fluids exhausting pattern	7.2.14.2
body fluids loss and heat accumulating pattern	7.4.5
bone	2.1.10.3
bone impotence due to kidney asthenia	6.5.5
bone spavin	6.7.18
both body fluids and qi deficiency pattern	7.2.15
both exhausting of yin and yang pattern	7.2.13.3
both heart and kidney deficiency of qi and yin pattern	7.10.6
both heart and kidney deficiency of yin and yang pattern	7.10.7
both heart and lung deficiency of yin and qi pattern	7.10.10
both qi and blood deficiency in heart pattern	7.5.3
both qi and blood deficiency moving wind pattern	7.2.10.3
both qi and yin deficiency in heart pattern	7.5.4
both qi and yin deficiency in lung pattern	7.6.2
both qi and yin fluids deficiency pattern	7.2.11
both stomach qi and yin deficiency pattern	7.7.23
both vital essence and blood deficiency pattern	7.2.17
bovine acute conjunctivitis	6.10.6
brain	2.1.10.1
brain as sea of marrow	2.1.10.1.1

breaking the blood and moving stasis ······ 8.1.10.21
burning warships ···································· 8.2.72

C

calculus ·· 4.1.8.3
callus like disease of the lips ······················ 6.8.20
carbuncle of neck ··· 6.6.34
carbuncle of pus ·· 6.6.38
carbuncle of shoulder ··································· 6.6.35
cardialgia ··· 6.1.11
catching therapy ·· 8.2.67
catgut embedding therapy ·························· 8.3.12
catgut-embedding with lumbar puncture
　　needle ··· 9.3.3.9.1
catgut-embedding with suture needle ··· 9.3.3.9.2
cattle and sheep bloating ···························· 6.3.18
causative factors ·· 4.1.1.2
cauterization therapy ···································· 9.3.2.3
cellulitis in joints and bones ······················· 6.6.33
cellulitis in skin and muscle ······················· 6.6.32
cellulitis or phlegmon ································· 6.6.31
channel hit by wind pattern ························· 7.13.1
checking profuse sweating and securing
　　the exterior ······································· 8.1.10.122
cheek swelling due to hot evil ···················· 6.6.14
chest and diaphragm disturbed by heat
　　pattern ·· 7.15.1.1
chest binding pattern with cold fluid pattern
　　··· 7.15.1.3
chest binding pattern with heat sputum
　　pattern ·· 7.15.1.4
chest diaphragm evil disturbance pattern ··· 7.15.1
chest pain ··· 6.2.6
chest swelling due to hot evil ····················· 6.6.18
child element ·· 1.3.11.2
Chinese veterinary medicine（CVM）········ 1.1.1
chong meridian ··· 3.1.8.3
chronic eclampsia ·· 6.11.7.2
chronic edema ··· 6.5.17
chronic enteritis ·· 6.3.14.2

chronic laminitis due to stasis ······················ 6.7.26
chuan huang needle·· 9.2.2.7
circling moxibustion ································· 9.3.2.1.2.3
classification of five elements······················ 1.3.2.1
clearing and diffusing stagnant heat ········ 8.1.6.5
clearing and disinhibiting the triple energizer
　　·· 8.1.7.20
clearing and purging deficiency-heat········ 8.1.6.6
clearing and purging diaphragm heat ······ 8.1.6.28
clearing and purging gallbladder heat······· 8.1.6.19
clearing and purging internal heat ············ 8.1.6.7
clearing and purging intestinal heat ······· 8.1.6.29
clearing and purging ministerial fire ······ 8.1.6.30
clearing and purging the lung and stomach
　　·· 8.1.6.27
clearing away heart-heat and dispelling
　　summer ··· 8.1.6.40
clearing away heart-heat for resuscitation
　　·· 8.1.10.95
clearing away heat and resolving dampness
　　·· 8.1.7.18
clearing dryness and moistening the nose ··· 8.2.28
clearing heat and cooling the blood ········ 8.1.6.3
clearing heat and dissipating phlegm ······ 8.1.10.76
clearing heat and dissolving distension ··· 8.1.7.39
clearing heat and drying dampness to
　　remove toxin ······································· 8.1.7.21
clearing heat and eliminating dampness ··· 8.1.7.16
clearing heat and eliminating steaming ··· 8.1.6.34
clearing heat and evacuating pus ··············· 8.2.5
clearing heat and improving vision ········· 8.2.14
clearing heat and inhibiting wind ·········· 8.1.10.64
clearing heat and moistening dryness ····· 8.1.10.50
clearing heat and preventing abortion ······ 8.1.6.32
clearing heat and producing fluid ············ 8.1.6.33
clearing heat and promoting diuresis ······ 8.1.7.17
clearing heat and purgation ······················· 8.1.3.1
clearing heat and purging dampness ······ 8.1.7.19
clearing heat and purging fire ··················· 8.1.6.1
clearing heat and purging the lung ········· 8.1.6.15

clearing heat and removing food stagnation 8.1.7.37	
clearing heat and removing toxin 8.1.6.2	
clearing heat for resuscitation 8.1.10.94	
clearing heat to relieve stranguria 8.1.6.31	
clearing heat with pungent-cool 8.1.1.3	
clearing heat, inducing expectoration and removing stagnated blood 8.1.10.90	
clearing liver-heat and inhibiting wind ... 8.1.10.66	
clearing lung and moistening dryness 8.1.10.51	
clearing summer and promoting diuresis ... 8.1.6.38	
clearing summer and removing toxin 8.1.6.37	
clearing summer and replenishing qi 8.1.6.39	
clearing summer heat 8.1.6.35	
clearing summer to ventilate the stagnated lung qi 8.1.6.41	
clearing the gallbladder and promoting diuresis 8.1.7.11	
clearing the heart and cooling the blood ... 8.1.6.12	
clearing the heart and cooling the nutrient 8.1.6.11	
clearing the heart and downbearing heat ... 8.1.6.9	
clearing the heart and nourishing yin 8.1.6.13	
clearing the heart and promoting diuresis ... 8.1.7.10	
clearing the heart and purging fire 8.1.6.8	
clearing the heart and purging the kidney ... 8.1.6.24	
clearing the heart and purging the liver ... 8.1.6.22	
clearing the heart and purging the lung ... 8.1.6.23	
clearing the heart and purging the spleen ... 8.1.6.21	
clearing the heart and removing toxin 8.1.6.10	
clearing the heart and tranquilizing the spirit 8.1.6.14	
clearing the liver and gallbladder 8.1.6.20	
clearing the liver and purging fire 8.1.6.18	
clearing the nutrient and cooling the blood ... 8.1.6.4	
clearing the spleen and purging heat 8.1.6.17	
clearing the stomach and purging heat ... 8.1.6.16	
coagulated cold in womb pattern 7.9.18	
coagulated sputum in womb pattern 7.9.19	
cold 6.2.1	

cold accumulated in liver meridian 4.2.4.6.1.13	
cold accumulation in large intestine ... 4.2.4.4.2.2	
cold and dampness attacking waist pattern 7.13.2.7	
cold and heat in complexity 4.2.2.4.7	
cold attributing to congealing 4.1.2.2.2	
cold attributing to contraction 4.1.2.2.3	
cold beats pain and bi pattern 7.13.2.2	
cold coagulation and qi stagnation pattern ... 7.3.2.2	
cold coagulation astringent skin pattern 7.11.23	
cold coagulation pattern 7.3.2	
cold constipation 6.3.20.4	
cold constipation 6.11.1	
cold damaging spleen and stomach 4.2.4.8.16	
cold damp diarrhea 6.3.19.4	
cold damp dysentery 6.3.13.3	
cold dampness affecting the spleen pattern ... 7.7.20	
cold dampness blocking channels pattern ... 7.13.1.7	
cold dampness causing stagnation pattern ... 7.3.2.1	
cold dampness causing stasis and stagnation pattern 7.3.2.5	
cold dampness due to kidney deficiency pattern 7.9.10	
cold dampness smolder at skin pattern 7.11.15	
cold dampness transforming heat pattern ... 7.3.2.4	
cold dryness pattern 7.3.5.2	
cold evil attacking stomach and intestine pattern 7.7.73	
cold evil attacking stomach pattern 7.7.36	
cold evil stagnated in heart pattern 7.5.14	
cold fluids accumulating in lung pattern 7.6.21	
cold heat mixed pattern 7.3.2.7	
cold in both exterior and interior 4.2.2.5.3.3	
cold in both superficies and interior pattern 7.15.12	
cold or heat of exterior or interior 4.2.2.5.1	
cold pattern with pseudo heat pattern 7.3.2.6	
cold sputum blocking lung pattern 7.6.24	
cold sputum pattern 7.3.7.2	
cold stagnated intestine pattern 7.7.39	

cold stagnating in the liver vessel pattern ··· 7.8.31
cold stagnation and blood deficiency pattern
　··· 7.4.7.2
cold stagnation and blood stasis pattern ······ 7.3.2.3
cold stagnation causing yang deficiency
　pattern·· 7.4.7.1
cold stagnation in channels pattern ········· 7.13.1.3
cold syndrome transforming to heat syndrome
　··· 4.2.2.4.9.1
cold tending to injure yang ················ 4.1.2.2.1
cold water accumulating in stomach pattern ··· 7.7.37
cold-dampness in liver and gallbladder ···4.2.4.8.29
cold-dampness in taiyin meridian ·········· 4.2.5.4.1
cold-phlegm ································· 4.1.8.1.1.1
collapse of yin and yang ················· 4.2.2.3.7
collaterals ···································· 3.1.1.1.2
coloclysis therapy ···························· 8.2.69
combined pathogens ···························· 4.1.2.7
combined zang-fu visceral pathogenesis ··· 4.2.4.8
common cold of wind cold ···················· 6.2.1.1
common cold of wind cold ···················· 6.2.2.2
common cold of wind heat ···················· 6.2.1.2
common cold of wind heat ···················· 6.2.2.3
compound reinforcing and reducing method
　·· 9.2.11.10
cone ·· 9.3.2.1.1.1
constipation ······································ 6.3.20
constipation of young stock ···················· 6.11.9
constitutional insufficiency···················· 4.1.14
constraining lung to relieve asthma ··· 8.1.10.124
constraining lung to stop cough ······ 8.1.10.123
constraining yin fluids and inducing
　astringency ······························ 8.1.10.134
consumption of body fluid ················ 4.2.3.2.2
consumptive disease with deficiency of both
　kidney yin and kidney yang pattern ········7.9.9
consumptive diseases ·························· 6.11.3
contaminated food ···························· 4.1.6.2
continuous lifting-thrusting method ········ 9.2.7.5
contraction of flexor tendons ················ 6.7.22

contralateral meridian point selection······ 9.1.18.11
contusion of the elbow joint ·····················6.7.9
contusion of the fetlock joint ················ 6.7.12
cool dryness attacking lung pattern············ 7.6.27
cooling the blood and inhibiting wind ··· 8.1.10.65
cooling the blood and resolving stasis ··· 8.1.10.23
coordinating the heart and kidney ······ 8.1.10.143
coordination between heart and kidney
　··2.1.4.8.15
corresponding master points combination
　·· 9.1.19.13
corrosion therapy ···························· 8.2.54
cotton-adhibiting method ·················· 9.3.3.7.1.3
cough ···6.2.2
cough with internal injury ···················· 6.2.2.5
counterflow rise of lung qi ················ 4.2.4.4.1.6
counterflow rise of stomach qi ········ 4.2.4.5.2.10
crossing meridian points combination ··· 9.1.19.6
crossing point selection ···················· 9.1.18.12
cup removing method ························ 9.3.3.7.7
cupping therapy ···························· 9.3.3.7
cupping therapy ······························8.3.8
cupping with medicinal liquid inside ······ 9.3.3.7.5
cutaneous needle therapy ·················· 9.3.3.3.1
cysticercus attacking brain pattern ············ 7.5.31

D

dai meridian ·································· 3.1.8.4
damage to paired meridians ················ 4.2.1.4.7
damp heat affecting the liver channel pattern
　··· 7.8.25
damp heat affecting the spleen pattern ······ 7.7.18
damp heat attacking mouth and lips pattern
　·· 7.12.12.3
damp heat attacking the waist pattern······ 7.13.2.9
damp heat brewing in the kidney pattern ··· 7.9.11
damp heat diarrhea ························ 6.3.19.3
damp heat dysentery ························ 6.3.13.1
damp heat in lower jiao pattern ·············· 7.15.5
damp heat in middle jiao pattern············ 7.15.4

damp heat in the spleen and stomach pattern 7.7.63
damp heat in the upper jiao pattern ············ 7.15.3
damp heat in the urinary bladder and blood stasis pattern ·································· 7.9.13.2
damp heat in the urinary bladder and qi stasis pattern ·································· 7.9.13.1
damp heat in the urinary bladder pattern ··· 7.9.13
damp heat smolder at skin pattern ············ 7.11.16
dampness and cold in liver and gallbladder pattern ·································· 7.8.40
dampness and poison accumulate to the skin pattern ································· 7.11.12
dampness attacking nose pattern ············ 7.12.9.3
dampness attributing to heaviness and turbidity ································· 4.1.2.4.3
dampness attributing to viscosity and stagnation ································· 4.1.2.4.4
dampness blocking causing qi stagnation pattern ·································· 7.3.4.1
dampness blocking intestine pattern ·········· 7.7.48
dampness blocking qi pattern ·················· 7.3.4
dampness blocking spleen and stomach pattern ·································· 7.7.64
dampness characterized by downward going ·································· 4.1.2.4.5
dampness conquers bi pattern ·················· 7.13.2.3
dampness hampering qi movement ········ 4.1.2.4.1
dampness heat and toxin accumulation pattern ·································· 7.3.4.6
dampness heat attacking body pattern ······ 7.3.4.5
dampness heat attacking ear pattern ······ 7.12.8.4
dampness heat blocking bi pattern ·········· 7.13.2.5
dampness heat blocking channel pattern··· 7.13.1.6
dampness heat blocking channels pattern··· 7.13.1.5
dampness heat blocking qi pattern ·········· 7.3.4.4
dampness heat blocks essence chambe pattern ·································· 7.9.27
dampness heat causing congestion pattern ·································· 7.3.4.7

dampness heat diffusing downward pattern ·································· 7.15.5.1
dampness heat in qi ring pattern ············ 7.12.3.2
dampness heat in uterus pattern ············ 7.9.22
dampness heat in wind ring pattern ······ 7.12.6.2
dampness heat pattern in flesh ring pattern ·································· 7.12.5.4
dampness heat steam throat pattern ······ 7.12.10.6
dampness heat steaming nose pattern ······ 7.12.9.6
dampness heat steaming tongue pattern ···· 7.12.12.6
dampness heat steaming tooth pattern ···· 7.12.11.4
dampness heat tangling pattern ················ 7.3.4.2
dampness sputum pattern ····················· 7.3.7.3
dampness sputum smolder at skin pattern ···7.11.17
dampness stasis the lower jiao pattern ··· 7.15.5.3
dampness tending to injure yang············ 4.1.2.4.2
dampness toxin stagnation pattern ·········· 7.3.14.7
dampness-heat in bladder ··················· 4.2.4.7.2.1
dampness-heat in large intestine ······ 4.2.4.4.2.3
dampness-heat in liver and gallbladder ···4.2.4.8.27
dampness-heat in liver meridian ······ 4.2.4.6.1.10
dampness-heat in lower energizer ··········· 4.2.7.3
dampness-heat in middle energizer ·········· 4.2.7.2
dampness-heat in spleen and stomach ···4.2.4.8.18
dampness-heat in triple energizer ·········· 4.2.7.4
dampness-heat in upper energizer ·········· 4.2.7.1
damp-phlegm ··································· 4.1.8.1.1.3
decline of yang ································· 4.2.2.3.2.1
decline of yang and safekeeping of yin ··· 1.2.1.12
decline of yin ································· 4.2.2.3.2.3
deep needling ································· 9.2.6.4.2
defense qi coming from lower jiao ·········· 2.2.1.9.3
defense qi coming from middle jiao ······ 2.2.1.9.2
defense qi coming from upper jiao ········ 2.2.1.9.1
defensive qi deficiency pattern ··············· 7.2.18
deficiencies of ying and qi pattern ·········· 7.2.19
deficiency complicated with excess ······ 4.2.2.2.3
deficiency complicated with excess ··· 4.2.2.2.3.1
deficiency fire attacking mouth and lips pattern ································· 7.12.12.4

deficiency in both exterior and interior	4.2.2.5.3.8
deficiency of both qi and blood pattern	7.2.10
deficiency of both qi and yin	4.2.3.2.11
deficiency of both spleen and kidney pattern	7.10.57
deficiency of both spleen and lung pattern	7.10.53
deficiency of heart yin induces fire hyperactivity pattern	7.5.9
deficiency of kidney essence pattern	7.10.27
deficiency of liver qi pattern	7.8.3
deficiency of liver yang pattern	7.8.4
deficiency of liver yin and hyperactivity of liver yang pattern	7.8.7
deficiency of qi and yin in spleen and stomach pattern	7.7.61
deficiency of the heart and spleen pattern	7.10.15
deficiency of vital essence pattern	7.2.16
deficiency of yang affecting yin	4.2.2.3.3.1
deficiency of yin and blood pattern	7.2.12
deficiency of yin and body fluids pattern	7.2.5.5
deficiency of yin and yang pattern	7.2.13
deficiency of yin induces fire hyperactivity pattern	7.4.3.2
deficiency or excess of exterior or interior	4.2.2.5.2
deficiency transforming into excess	4.2.2.2.5.2
deficiency yang floating upward pattern	7.2.9
deficiency-cold	4.2.2.4.2
deficiency-cold in large intestine	4.2.4.4.2.5
deficiency-cold in small intestine	4.2.4.3.2.1
deficiency-cold of bladder	4.2.4.7.2.3
deficiency-cold of spleen and stomach	4.2.4.8.17
deficiency-cold of taiyin meridian	4.2.5.4.2
deficiency-cold of yangming meridian	4.2.5.2.4
deficiency-excess transformation	4.2.2.2.5
deficiency-fire	4.2.2.4.6
deficiency-heat	4.2.2.4.5
deficient cold in uterus pattern	7.9.21
deficient constipation	6.3.20.2
deficient fire flaring gum pattern	7.12.11.9
delayed qi movement	4.2.3.1.1.5
depressed gallbladder with harassing sputum pattern	7.8.34
dermal needle	9.3.3.3
destroying parasites and dispersing accumulation	8.1.10.112
deterioration of the lung	6.2.8
diabetes insipidus	6.5.12
diarrhea	6.3.19
diarrhea due to deficiency of kidney	6.3.19.5
diarrhea due to improper feeding	6.3.19.1
diarrhea due to spleen asthenia	6.3.19.2
diarrhoea due to dampness and heat evil	6.11.8.2
diarrhoea due to milk	6.11.8.1
diarrhoea due to spleen deficiency	6.11.8.3
diarrhoea in foal	6.11.8
dietotherapy	8.5.1
different treatments for the same disease	5.2.6
differentiate pattern to reveal an etiology	4.1.1.1
diffusing and dissipating pathogenic dampness	8.1.7.3
diffusing and moistening dryness	8.1.10.49
diffusing and unblocking the triple energizer	8.1.10.142
diffusing the lung and calming panting	8.1.10.7
diffusing the lung and disinhibiting water	8.1.7.26
diffusing the lung and downbearing qi	8.1.10.6
diffusing the lung and unblocking qi	8.1.10.5
diffusive dampness heat in sanjiao pattern	7.15.2
direct attack	4.2.1.4.8
direct moxibustion	9.3.2.1.1.3
direct penetrating	9.2.6.3.1
directing fire back to its origin	8.1.10.140
directional reinforcing and reducing method	9.2.11.9.1
disease involving both defense and nutrient levels	4.2.6.8
disease involving both defense and qi levels	4.2.6.5
disease involving two or more meridians	4.2.1.4.9

disease of both exterior and interior 4.2.2.5.3
disease of one meridian involving another
　　meridian 4.2.1.4.10
disharmony between liver and spleen 4.2.4.8.26
disharmony between the heart and kidney
　　pattern................................... 7.10.2
disharmony between yin and yang 4.2.2.3
disharmony of nutrient-defensive qi 4.2.5.1.1
disharmony of the thoroughfare and controlling
　　vessels pattern 7.9.24
disinhibiting the pharynx 8.2.31
disinhibiting water and alleviating edema... 8.1.7.22
dislocation 6.7.4
disorder of blood 4.2.3.1.2
disorder of movement of qi............... 4.2.3.1.1.2
disorder of qi 4.2.3.1.1
disorder of qi and blood 4.2.3.1.3
disorder of stomach qi.................... 4.2.4.5.2.8
dispelling cold and resolving fluid retention
　　.. 8.1.10.81
dispelling cold for resuscitation 8.1.10.101
dispelling heat from the ying fen system for
　　resuscitation 8.1.10.97
dispelling pathogenic wind from muscles
　　.. 8.1.10.62
dispelling phlegm and disinhibiting the pharynx
　　... 8.2.34
dispelling stasis and expediting childbirth
　　.. 8.1.10.39
dispelling stasis and promoting regeneration
　　.. 8.1.10.26
dispelling stasis and unblocking the collaterals
　　.. 8.1.10.28
dispelling stasis and unblocking the nose ... 8.2.29
dispelling summer and inducing resuscitation
　　....................................... 8.1.10.106
dispelling summer and reliving the exterior
　　.. 8.1.6.36
dispelling summer for resuscitation 8.1.10.98
dispelling wind and drying dampness...... 8.1.7.14

dispelling wind and expelling parasites ... 8.1.10.113
dispelling wind and improving vision.......... 8.2.13
dispelling wind and promoting eruption 8.1.1.5
dispelling wind and unblock the nose 8.2.26
dispersing dampness for resuscitation ... 8.1.10.102
dispersing mass and expelling retention 8.1.3.2
dispersing pathogenic factors and releasing
　　the external 8.1.1.4
dispersing the external and unblocking the
　　meridians 8.1.1.6
dispersing the stagnated liver qi 8.1.10.3
dispersing wind and diffusing the ears 8.2.19
dispersing wind and disinhibiting the pharynx
　　... 8.2.32
dispersing wind and disinhibiting water ... 8.1.7.27
dissipating cold and disinhibiting the pharynx
　　... 8.2.33
dissipating cold and disinhibiting water ... 8.1.7.29
dissipating cold and eliminating dampness
　　.. 8.1.7.15
dissipating cold and unblocking the nose ... 8.2.27
dissipating phlegm for eliminating goiter... 8.1.10.88
dissipating phlegm for resuscitation ... 8.1.10.103
dissipating wind-phlegm 8.1.10.79
distant point selection 9.1.18.3
distant-local points combination 9.1.19.12
disturbance of body fluid 4.2.3.2
downbearing counterflow and checking
　　vomiting 8.1.10.14
downbearing counterflow and relieving
　　panting 8.1.10.9
downbearing fire to strengthen the teeth 8.2.43
drainage therapy 8.2.56
draining dampness and disinhibiting water
　　.. 8.1.7.23
drug allergy 6.8.17
drying dampness and dissipating phlegm
　　.. 8.1.10.75
drying dampness and harmonizing the
　　stomach 8.1.7.13

drying dampness and moving qi ·········· 8.1.7.12
dryness affecting the clear orifices pattern
　·· 7.12.13.1
dryness attacking nose pattern ············ 7.12.9.7
dryness characterized by aridity and astringency
　·· 4.1.2.5.1
dryness evil attacking lung pattern ··········· 7.6.26
dryness sputum evil blocking lung pattern ··· 7.6.29
dryness sputum stagnation pattern ········· 7.3.7.5
dryness tending to injure lung ············· 4.1.2.5.2
dryness toxin pattern ························ 7.3.14.15
dryness-heat in yangming meridian ······ 4.2.5.2.1
dryness-heat injuring lung ················ 4.2.4.4.1.9
dryness-phlegm ······························ 4.1.8.1.1.4
du meridian ·· 3.1.8.1
dual deficiency of qi and blood ········· 4.2.3.1.3.1
dual deficiency of yin and yang ········· 4.2.2.3.3.3
dynamic waxing and waning between yin
　and yang ······································· 1.2.1.2.3
dysentery ··· 6.3.13
dysfunction of spleen in transportation ··· 4.2.4.5.1.6
dyspepsia pattern ····································· 7.3.15
dystocia ·· 6.9.6
dysuria due to hot and wet ····················· 6.5.23

E

ear acupuncture therapy ························ 9.3.3.6
ear drop therapy ······································ 8.2.62
ear pattern ··· 7.12.8
ear swelling due to hot and poison ········· 6.6.15
ear washing therapy ································ 8.2.63
ear-insufflation therapy ··························· 8.2.61
early death of the embryo and abortion ······ 6.9.13
earth counter-restricting wood ·············· 1.3.14.5
earth generating metal ··························· 1.3.11.5
earth over-restricting water ··················· 1.3.13.2
earth restricting water ··························· 1.3.12.4
eclampsia of young stock ·························· 6.11.7
eczema due to wet poison ·························· 6.8.3
edema ··· 6.11.2

edema due to spleen deficiency ·················· 6.3.9
edema of posterior limbs due to kidney asthenia
　·· 6.5.3
effulgent liver and weak spleen pattern ····· 7.10.34
eight extraordinary meridians ··················· 3.1.8
electroacupuncture apparatus ················· 9.2.3.1
electroacupuncture therapy ···················· 9.2.13.4
electro-acupuncture therapy ······················ 8.3.9
electrothermal needle apparatus ············· 9.2.3.2
electrothermal needle therapy ················· 9.3.3.2
elevating and inducing astringency ······· 8.1.10.136
eliminating cold for stopping pain ············ 8.1.5.4
eliminating dampness and astringing sores ······ 8.2.8
eliminating dampness and unblocking the
　collaterals ······································· 8.1.7.31
eliminating pathogen to support vital qi ······ 5.2.5.2
eliminating pathogens from lung for
　expelling fluid retention ················ 8.1.10.82
eliminating slough and promoting granulation ··· 8.2.6
elimination after reinforcement ··············· 5.2.5.3
elimination containing reinforcement ········· 5.2.5.7
empyema of frontal sinus ························ 6.2.4.3
endogenous dampness ····························· 4.1.4.3
endogenous dryness ································ 4.1.4.4
endogenous etiological factors ··············· 4.1.1.2.2
endogenous fire ······································· 4.1.4.5
enriching and reinforcing heart yin ········· 8.1.8.21
enriching and reinforcing kidney yin ······· 8.1.8.29
enriching and reinforcing liver yin ·········· 8.1.8.25
enriching and reinforcing lung yin ·········· 8.1.8.22
enriching and reinforcing the heart and lung
　·· 8.1.8.31
enriching and reinforcing the spleen and
　stomach ·· 8.1.8.24
enriching the kidney to receive qi ········· 8.1.8.30
enriching the liver and improving vision ··· 8.2.16
enriching the liver and kidney to moisten the
　ears ·· 8.2.24
enriching yin ··· 8.1.8.20
enriching yin and calming the liver ······· 8.1.8.28

enriching yin and checking sweating	8.1.8.40
enriching yin and clearing heat	8.1.8.32
enriching yin and cooling the blood	8.1.8.34
enriching yin and cooling the nutrient ...	8.1.8.33
enriching yin and downbearing fire	8.1.8.37
enriching yin and producing fluid	8.1.8.35
enriching yin and reinforcing blood	8.1.8.66
enriching yin and reinforcing stomach ...	8.1.8.23
enriching yin and reinforcing yang	8.1.8.62
enriching yin and relaxing bowels	8.1.8.41
enriching yin and softening the liver	8.1.8.26
enriching yin and soothing the liver	8.1.8.27
enriching yin and subduing yang	8.1.8.38
enriching yin and suppressing cough	8.1.8.39
enriching yin to moisten the ears	8.2.23
enriching yin to moisten the teeth	8.2.41
enriching yin to quench thirst	8.1.8.36
entering of heat into blood level	4.2.6.4.1
enteritis	6.3.14
epidemic pestilence	4.1.3
epidemic toxic dysentery	6.3.13.2
epidemic toxin attacking pattern	7.3.14.11
epidemic toxin blocking pattern	7.3.14.12
epilepsy	6.1.9
epistaxis	6.2.9
equilibrium between yin and yang	1.2.1.4
equine colic	6.3.10
equine intestinal tympany	6.3.17
erysipelas	6.6.29
essence	2.2.4
essence and blood share the same source ...	2.2.12
essence prostration	4.2.4.7.1.8
essential qi	2.2.4.3
etiological theory	4.1.1
evil attack of wei–defence surface pattern	
....	7.11.1
evil crime clearing the orifice pattern	7.12.13
evil entering shaoyang pattern	7.15.7
evil heat injuring yin pattern	7.4.9
evil stop in ear pattern	7.12.8.6
evil toxin attacking while healthy qi exhausting pattern	7.4.13
excess complicated with deficiency ...	4.2.2.2.3.2
excess evil attacking the eyes pattern	7.12.2
excess evil attacking the head pattern	7.12.1
excess evil blocking and healthy qi exhausting pattern	7.4.14
excess feeding	4.1.6.1.1
excess heat in the small intestine pattern ...	7.7.41.1
excess heat in water ring pattern	7.12.7.1
excess heat pattern in blood ring pattern ...	7.12.4.1
excess in both exterior and interior	4.2.2.5.3.7
excess pattern carrying deficiency pattern ...	7.4.12
excess pattern with pseudo deficiency pattern	7.3.18
excess transformation into deficiency ...	4.2.2.2.5.1
excess yin or yang	4.2.2.3.1
excess–cold	4.2.2.4.1
excess–fire	4.2.2.4.4
excess–heat	4.2.2.4.3
excess–heat in small intestine	4.2.4.3.2.2
excessive anger damaging the liver	4.1.5.1
excessive body fluids stagnated in pericardium pattern	7.5.16
excessive fetal moovement	6.9.1
excessive fluid collecting internally pattern ...	7.3.8
excessive heat in liver and gallbladder ...	4.2.4.8.28
excessive heat in middle jiao pattern	7.15.4.1
excessive yang repelling yin	4.2.2.3.6.3
excessive yin rejecting yang	4.2.2.3.6.1
excessiveness of yangming fu viscera ...	4.2.5.2.2
exogenous cough	6.2.2.1
exogenous etiological factors	4.1.1.2.1
exogenous injury attacking eye channel pattern	7.12.2.3
expelling and killing ascarid	8.1.10.110
expelling parasites and tranquilizing mind	8.1.10.114
expelling parasites, removing toxin and ventilating lung	8.1.10.115

expelling retained toxin	8.2.9
expelling wind	8.1.10.61
exterior attacked by summer heat dampness pattern	7.11.6
exterior cold	4.2.2.5.1.1
exterior cold and interior heat	4.2.2.5.3.1
exterior cold and lung heat pattern	7.6.25
exterior deficiency	4.2.2.5.2.2
exterior deficiency and interior excess	4.2.2.5.3.6
exterior excess	4.2.2.5.2.1
exterior excess and interior deficiency	4.2.2.5.3.5
exterior heat	4.2.2.5.1.2
exterior heat and interior cold	4.2.2.5.3.2
exterior-interior meridian points combination	9.1.19.2
exterior-interior meridian points selection	9.1.18.8
external dryness attacking the exterior pattern	7.11.7
external dryness pattern	7.3.5
external injury	4.1.9
external oculopathy	6.11.17
external wind evil pattern	7.3.1
extinguishing wind and relieving spasm	8.1.10.73
extraordinary fu organs	2.1.10
extreme yang transforming into yin	1.2.1.2.4.2
extreme yin transforming into yang	1.2.1.2.4.1
exuberance of pathogenic qi leading to excessiveness	4.2.2.2.1
exuberance of pathogenic qi with decrease of vital qi	4.2.2.2.8
exuberance of yang	4.2.2.3.1.1
exuberance of yin	4.2.2.3.1.2
exuberant heat in qi level	4.2.6.2.1
eye drop therapy	8.2.60

F

facial paralysis	6.11.18
facial wandering wind	6.8.16
failure in qi transformation	4.2.3.2.6
failure of blood to nourish tendons	4.2.3.1.2.7
failure of kidney to receive qi	4.2.4.7.1.3
failure of spleen to control blood	4.2.4.5.1.7
failure of stomach qi to descend	4.2.4.5.2.9
failure of water to nourish wood	4.2.4.8.30
failure of water to transform into qi	4.2.3.2.7
fainting during acupuncture treatment	9.2.10.1
fear as kidney emotion	2.1.4.12.14
fear causing the qi to descend	4.1.5.10
fear/fright damaging the kidney	4.1.5.9
fei huang	6.2.12
fetal poisoning pattern	7.3.14.16
fetal restlessness due to blood heat	6.9.1.2
fetal restlessness due to deficiency	6.9.1.1
fetal restlessness due to trauma	6.9.1.3
fetal toxin	4.1.11
fifteen collaterals	3.1.9
filiform needle	9.2.1.7 or 9.2.2.1
filiform needle therapy	9.2.13.1.1
fingernail-pressing method	9.2.4.2
fire cauterizing therapy	8.2.58
fire characterized by flaming	4.1.2.6.3
fire counter-restricting water	1.3.14.3
fire cupping method	9.3.3.7.1
fire evil injuring channels of water ring pattern	7.12.7.10
fire generating earth	1.3.11.4
fire heat injuring yin pattern	7.4.9.2
fire hyperactivity in liver and gallbladder pattern	7.8.38
fire hyperactivity in liver channel pattern	7.8.23
fire hyperactivity of the heart and liver pattern	7.10.20
fire needle	9.2.2.6
fire needling therapy	9.2.13.3
fire needling therapy	8.3.1
fire over-restricting metal	1.3.13.4
fire pathogen tending to cause carbuncle	4.1.2.6.8
fire poison entering channels pattern	7.13.1.4
fire restricting metal	1.3.12.6

fire tending to cause bleeding	4.1.2.6.5
fire tending to cause wind	4.1.2.6.4
fire tending to consume qi and fluid	4.1.2.6.6
fire tending to disturb heart	4.1.2.6.7
fire throwing cupping method	9.3.3.7.1.1
fire toxin attacking ying pattern	7.3.14.5
fire toxin moving randomly pattern	7.3.14.4
fistula	6.6.41
fistula of inner and outer orifice	6.6.6
fistula of outer orifice	6.6.7
five body constituents	2.1.4.4
five changes	1.3.5
five colors	1.3.6
five directions	1.3.10
five elements	1.3.2
five elements mutually counter-restrain one another	1.3.14
five elements mutually generate	1.3.11
five elements mutually over-restrain one another	1.3.13
five elements mutually restrain	1.3.12
five endogenous pathogenic factors	4.1.4
five flavors	1.3.7
five humors	2.1.4.2
five notes	1.3.8
five periods of year	1.3.3
five qi	1.3.4
five sense organs	1.3.9
five signal apertures on the head	2.1.4.5
five zang deficiency	4.2.2.2.2.2
five zang excess	4.2.2.2.1.2
five zang organs	2.1.4
five zang viscera pathogenesis	4.2.4.1
five-element theory	1.3.1
flaccidity syndrome	6.11.14
flaming of deficiency-fire	4.2.2.3.2.3.4
flash cupping method	9.3.3.7.1.6
flash-fire cupping method	9.3.3.7.1.2
flatulence caused by qi deficiency	4.2.3.1.1.1.1
flesh ring pattern	7.12.5
flesh ring qi deficiency pattern	7.12.5.6
flexor tendinitis	6.7.10
floating collaterals	3.1.11
floating upward of deficiency-yang	4.2.2.3.6.2
fluid depletion	4.2.3.2.3
fluid depletion causing blood stasis	4.2.3.2.10
fluid exhaustion causing blood dryness	4.2.3.2.9
fluid of the heart is sweat	2.1.4.8.11
fluid of the kidney is thick saliva	2.1.4.12.15
fluid of the liver are tears	2.1.4.11.16
fluid of the lung is nasal discharge	2.1.4.9.17
fluid of the spleen is thin saliva	2.1.4.10.14
fluid retained in chest and hypochondrium pattern	7.15.1.7
flushing and washing therapy	8.2.52
food preference	4.1.6.3
food stagnating in the stomach venter	4.2.4.5.2.12
food stagnation in stomach and intestine pattern	7.7.70
food toxin pattern	7.3.14.14
foot rot	6.7.27
fortifying the spleen and disinhibiting water	8.1.7.24
fortifying the spleen and resolving dampness	8.1.7.7
fourteen meridians	3.1.1.1.3
fracture	6.6.8
fright and fear damage the kidney pattern	7.9.31
fright causing the qi to become chaotic	4.1.5.11
fumigating and washing therapy	8.2.51
fumigation therapy	8.6.3
furuncle	6.6.20
furuncle or pyogenic infection on body surface	6.6.10
fuviscera in transportation of transformed products	2.1.5.3

G

gallbladder	2.1.5.4

gallbladder being essence of fu viscera ··· 2.1.5.4.1
gallbladder dominating decision ············ 2.1.5.4.4
gallbladder enlargement ························6.4.2
gallbladder meridian of hindlimbs-shaoyang
  ···································· 3.1.4.3.2
gallbladder pathogenesis ················ 4.2.4.6.2
gallbladder qi ···························· 2.1.5.4.2
gallbladder qi deficiency pattern ············ 7.8.33
gallbladder stagnation with phlegm
  disturbance ···················· 4.2.4.6.2.2
garlic interposed moxibustion ············ 9.3.2.1.1.8
gastric dilatation ···························· 6.3.26
generation and restriction among five elements
  ···································· 1.3.15
gentle moxibustion ···················· 9.3.2.1.2.2
genuine qi ································ 2.2.1.6
gill swelling due to hot evil ············ 6.6.16
ginger interposed moxibustion ········· 9.3.2.1.1.7
gingival fire poison pattern ············ 7.12.11.6
glaucoma ··································6.11.16
gout ··6.7.3
gradual onset ······················· 4.2.1.4.4
grief as lung emotion ·················2.1.4.9.16
growth of yang and generation of yin ······ 1.2.1.11
gum pattern ···························7.12.11

H

hand replace needle hammer holding method
  ······································ 9.2.5.8
handle-flicking method ···················· 9.2.7.7
handle-scraping method ···················· 9.2.7.8
handle-waggling method ···················· 9.2.7.9
hard skin nodule ························ 6.8.25
harmful hyperactivity checked for harmony
  ···································· 1.3.19
harmonizing and releasing shaoyang ········ 8.1.4.1
harmonizing the blood and preventing abortion
  ···································· 8.1.10.38
harmonizing the collaterals ············ 8.1.10.36
harmonizing the liver and spleen ········ 8.1.4.2

harmonizing the middle and relaxing
  convulsion ·························· 8.1.4.9
harmonizing the nutrient to promote
  regeneration ·················· 8.1.10.141
harmonizing the stomach and downbearing
  counterflow ···················· 8.1.10.13
healing sore and relieving pain ·············· 8.2.10
healthy qi ·································· 4.2.1.2
heart ···································· 2.1.4.8
heart and kidney fiery pattern ···············7.10.14
heart and kidney yin deficiency pattern ······ 7.10.1
heart and liver blood deficiency and stasis
  pattern ································7.10.24
heart and liver blood stasis pattern ···········7.10.21
heart and liver both qi deficiency and blood
  stasis pattern ·························7.10.26
heart and liver deficiency of qi and blood
  pattern ································7.10.25
heart and lung exuberant heat pattern ········7.10.13
heart and lung yang deficiency pattern ······7.10.12
heart and lung yin deficiency ang blood stasis
  pattern ································7.10.11
heart and lung yin deficiency pattern ········ 7.10.9
heart and small intestine pathogenesis ······ 4.2.4.3
heart and spleen heat accumulation pattern ···7.10.19
heart and spleen qi deficiency pattern ········7.10.16
heart and spleen qi deficiency pattern ········7.10.18
heart and spleen yang deficiency pattern ···7.10.17
heart blood ···························· 2.1.4.8.2
heart blood deficiency ···················· 4.2.4.3.1.2
heart blood deficiency pattern ···············7.5.7
heart blood stagnation and obstruction ··· 4.2.4.3.1.9
heart blood stasis pattern ···················· 7.5.12
heart deficiency ····························6.1.5
heart deficiency and mind timidity pattern ··· 7.5.30
heart dominating blood ·················· 2.1.4.8.8.1
heart dominating vessel ·················· 2.1.4.8.8.2
heart fire blazing pattern ···················· 7.5.17
heart governs the blood and vessels ······ 2.1.4.8.8
heart heat and yin deficiency pattern ········ 7.5.19

heart heat stagnation and blood stasis pattern
 ·· 7.5.12.2
heart housing mind ·················· 2.1.4.8.6
heart intolerating heat ···············2.1.4.8.12
heart kidney qi deficiency pattern ········· 7.10.5
heart liver yin deficiency pattern ·······7.10.23
heart manifesting in complexion············ 2.1.4.8.9
heart meridian of forelimb-shaoyin ······ 3.1.4.1.3
heart opens into the tongue ···············2.1.4.8.10
heart pathogenesis ······················· 4.2.4.3.1
heart qi ································ 2.1.4.8.1
heart qi deficiency ···················· 4.2.4.3.1.1
heart qi deficiency causing blood stasis pattern
 ·· 7.5.2
heart qi deficiency pattern ················7.5.1
heart qi stagnation and blood stasis pattern··· 7.5.12.1
heart qi stagnation pattern ················ 7.5.15
heart system ······················· 2.1.4.8.5
heart yang ························· 2.1.4.8.4
heart yang deficiency ·················· 4.2.4.3.1.3
heart yang deficiency causing blood stasis
 pattern··7.5.6
heart yang deficiency pattern ··········· 7.5.5.2
heart yin ···························· 2.1.4.8.3
heart yin and blood deficiency pattern ········7.5.8
heart yin and yang deficiency pattern ········ 7.5.11
heart yin deficiency··················· 4.2.4.3.1.5
heart yin deficiency causing blood stasis
 pattern······································ 7.5.10
heart-gallbladder qi deficiency ·········4.2.4.8.11
heart-kidney yang deficiency ············· 4.2.4.8.6
heart-kidney yin deficiency ············· 4.2.4.8.5
heart-liver blood deficiency ············· 4.2.4.8.4
heart-lung qi deficiency ················· 4.2.4.8.1
heat accumulation in bladder ··········· 4.2.4.7.2.2
heat accumulation in large intestine ··· 4.2.4.4.2.1
heat attacking heart ying pattern ············· 7.5.22
heat attacking ying blood pattern ········· 7.3.6.4
heat blazing in both qi and blood phases
 pattern····································· 7.3.6.7

heat blazing in both qi and ying phases
 pattern······································ 7.3.6.6
heat blocking pericardium pattern ··········· 7.5.20
heat constipation ····················· 6.3.20.3
heat deficiency of blood ring pattern ······ 7.12.4.2
heat distressing large intestine ········ 4.2.4.4.2.4
heat disturbing mind ················ 4.2.4.3.1.8
heat disturbing mind pattern ············· 7.5.21
heat excess chest bind pattern ············· 7.15.1.2
heat in both exterior and interior········· 4.2.2.5.3.4
heat in both exterior and interior pattern······7.15.13
heat induced obstruction of bi pattern······ 7.13.2.4
heat injuring ying yin pattern ················· 7.4.9.4
heat pathogen ························· 4.1.2.6.1
heat sputum pattern ···················· 7.3.7.4
heat stroke ·······························6.1.8
heat syncope pattern ·················· 7.3.6.11
heat syndrome transforming to cold syndrome
 ·· 4.2.2.4.9.2
heat tangling in the large intestine pattern ··· 7.7.42
heat toxicity and liver stasis pattern ·········· 7.8.27
heat toxicity attacking throat pattern ······ 7.12.10.7
heat toxicity attacking tongue pattern ······ 7.12.12.7
heat toxicity in the skin pattern ···············7.11.13
heat toxicity in wind ring pattern ··········· 7.12.6.3
heat toxicity stagnated in head and face
 pattern···································· 7.12.1.9
heat toxin accumulating in intestine pattern··· 7.7.43
heat toxin at anus pattern ················ 7.7.54
heat toxin blocking lung pattern ··············· 7.6.34
heat toxin in blood level ················ 4.2.6.4.4
heat toxin in flesh ring pattern ············· 7.12.5.3
heat toxin in qi ring pattern················· 7.12.3.4
heat toxin in spleen channel pattern ········ 7.7.19
heat toxin inward penetration pattern······ 7.3.14.6
heat wind in flesh ring pattern ············· 7.12.5.2
heat wind in qi ring pattern·················· 7.12.3.1
heat-phlegm ························ 4.1.8.1.1.2
heatstroke pattern ······················ 7.3.3.2
heavy reduction ························ 9.2.11.8

hematuria ································· 6.5.26	impaction of omasum ················6.3.6
hematuria due to blood type incompatibility	impaction of rumen ················ 6.3.23
······································ 6.11.10.1	impairment of yin affecting yang ······ 4.2.2.3.3.2
hematuria of foal ···················6.11.10	impotence ·························· 6.5.21
hematuria of foal ··············· 6.11.10.2	improper feeding ·····················4.1.6
hepatobiliary wind ·····················6.4.7	improving vision ···················· 8.2.11
hernia ································ 6.11.5	inappetence due to spleen deficiency ············6.3.5
herpetic dermatosis due to hot evil ············6.8.4	incarceration or strangulation ··············· 6.3.32
holding needle like using pen ··············· 9.2.5.2	incision therapy ······················ 8.2.55
holding needle with································ 9.2.5.6	increasing fluid and moistening the lung ··· 8.1.10.55
holding needle with both hands ··············· 9.2.5.4	indigestion due to improper feeding ············6.3.3
holding needle with full grip ··············· 9.2.5.5	indigestion due to improper watering ············6.3.4
holding needle with two fingers ··············· 9.2.5.3	indigestion due to spleen deficiency ············6.3.8
holistic concept ················ 1.1.1.1.1	indirect moxibustion ··············· 9.3.2.1.1.6
hoof injure due to stampede ··············· 6.7.19.1	indirect moxibustion on navel············ 9.3.2.1.1.10
hoof injure due to trim foot················ 6.7.19.2	inducing astringency to restrain teardrop
horn fracture ························6.6.9	····································· 8.1.10.131
hot compress therapy ··············· 9.3.2.2	inducing astringency to stop bleeding··· 8.1.10.127
hot compress therapy ··············· 8.2.46	inducing expectoration and dispersing
hot wax therapy ·······················8.6.1	dampness for resuscitation ············ 8.1.10.104
house of essence qi without leakage ········ 2.1.4.7	inducing expectoration and dredging channel
house of original mentality ···············2.1.10.1.2	blockage ····················· 8.1.10.89
hydro-acupuncture therapy ··············· 8.3.11	inducing expectoration and expelling
hydroptysis spleen and heart deficiency ·········6.1.6	parasite ······················ 8.1.10.91
hydrothorax ························ 6.2.13	inducing expectoration, regulating the
hygropyretic bloody stool ··············· 6.3.21.1	stagnated qi and removing toxin ······ 8.1.10.92
hygropyretic vomiting···················· 6.3.1.2	inducing resuscitation and opening closed
hyperactivity of liver yang ··············· 4.2.4.6.1.11	resistance ···················· 8.1.10.93
hyperactivity of the liver yang pattern ·········7.8.5	inducing resuscitation with aromatics··· 8.1.10.100
hypoabdominal blood stasis pattern ······ 7.15.6.1	inducing vomit of phlegm and drool ········ 8.1.2.1
hypoabdominal dampness heat block pattern	inducing vomit of phlegm and retained food
······································ 7.15.6.4	······································ 8.1.2.3
hypoabdominal heat stagnation pattern ····· 7.15.6.3	inducing vomit of retained food ············ 8.1.2.4
hypoabdominal of evil offenders pattern ··· 7.15.6	inducing vomit of wind-phlegm ············ 8.1.2.2
hypoabdominal qi stagnation pattern ······ 7.15.6.2	infectious vesicular stomatitis of rabbits······ 6.10.7
hypogalactia ························ 6.9.10	inflammation due to wind cold or hot evil ··· 6.6.37
	influenza ···························· 6.10.3
I	infrared moxibustion ················ 9.3.3.13
imbalance between cold and heat ············ 4.2.2.4	injuries to the hoof ···················· 6.7.19
impaction of intestines ··············· 6.3.11	injury by animal and insect···················· 4.1.12

injury of yang	4.2.2.3.7.4
injury of yin	4.2.2.3.7.2
innate essence	2.2.4.1
innate qi	2.2.1.4
inner invasion of exterior syndrome	4.2.2.5.3.10
insecurity of defensive qi	4.2.5.1.2
insecurity of kidney qi	4.2.4.7.1.2
insecurity of the thoroughfare and controlling vessels pattern	7.9.25
insertion of needle with tube	9.2.6.8
insufficiency of both heart and spleen	4.2.4.8.2
insufficiency of both lung and spleen	4.2.4.8.12
insufficient body fluid	4.2.3.2.1
insufficient feeding	4.1.6.1.2
insult of water to heart	4.2.4.8.8
intense lung stomach fire pattern	7.10.62
interaction of yin and yang	1.2.1.2
interior cold	4.2.2.5.1.3
interior deficiency	4.2.2.5.2.4
interior disease involving superficies	4.2.2.5.3.11
interior excess	4.2.2.5.2.3
interior heat	4.2.2.5.1.4
internal cold	4.1.4.2
internal stirring of liver wind	4.2.4.6.1.12
internal stirring of liver wind pattern	7.8.32
internal wind	4.1.4.1
intestinal sabulous	6.3.12
intestinal stasis pattern	7.7.75
intestinal wind bleeding	6.3.16
intestine cold dampness pattern	7.7.49
intestine dampness heat pattern	7.7.40
intestine dryness and fluids exhausting pattern	7.7.44
intestine excess heat pattern	7.7.41
intestine heat and qi stagnation pattern	7.7.41.3
intestine heat and yin deficiency pattern	7.7.41.2
invasion of pericardium by heat	4.2.6.3.1
invigorating water to control yang	8.1.10.138
involving both wei-defence and qi phases pattern	7.3.6.5
ironing and cauterization technique	9.3.2.3.2
irregular eating	4.1.6.1
irregular excess and insufficient feeding	4.1.6.1.3

J

jaundice	6.4.1
jaundice due to cold damp	6.4.1.2
jaundice due to heat damp	6.4.1.1
jia qi needle	9.2.2.8
jia qi needle acupuncture	9.2.13.7.1
joy as heart emotion	2.1.4.8.7
joy causing the qi to slack	4.1.5.4
jueyin cold reversal pattern	7.14.9.1
jueyin disease pattern	7.14.9
jueyin heat reversal pattern	7.14.9.2

K

kidney	2.1.4.12
kidney and bladder pathogenesis	4.2.4.7
kidney cold	6.5.2
kidney controls two lower orifices	2.1.4.12.13
kidney deficiency and blood stasis pattern	7.9.10.3
kidney deficiency and intestinal withdrawal pattern	7.10.67
kidney deficiency and marrow depletion pattern	7.9.8
kidney deficiency cold sputum pattern	7.9.10.2
kidney deficiency cold stasis pattern	7.9.10.1
kidney dominating bone	2.1.4.12.9
kidney dominating innateness	2.1.4.12.5.1
kidney dominating reproduction	2.1.4.12.5.2
kidney dominating storage	2.1.4.12.8
kidney essence	2.1.4.12.1
kidney essence depletion pattern	7.9.7
kidney essence insufficiency	4.2.4.7.1.7
kidney governs qi reception	2.1.4.12.6
kidney governs water	2.1.4.12.7
kidney insufficiency with water diffusion	4.2.4.7.1.5
kidney intolerating dryness	2.1.4.12.16

kidney manifesting in teeth	2.1.4.12.11
kidney meridian of hindlimbs-shaoyin	3.1.4.4.3
kidney opens into the ears	2.1.4.12.12
kidney pathogenesis	4.2.4.7.1
kidney produces marrow	2.1.4.12.10
kidney qi	2.1.4.12.2
kidney qi deficiency	4.2.4.7.1.1
kidney qi deficiency and water retention pattern	7.9.3.1
kidney qi deficiency pattern	7.9.1
kidney qi insecurity pattern	7.9.2
kidney stores essence	2.1.4.12.5
kidney yang	2.1.4.12.4
kidney yang deficiency	4.2.4.7.1.4
kidney yang deficiency pattern	7.9.4
kidney yin	2.1.4.12.3
kidney yin deficiency	4.2.4.7.1.6
kidney yin deficiency and fire hyperactivity pattern	7.9.6
kidney yin deficiency pattern	7.9.5
knock needling	9.3.3.3.1.1
knock needling along the meridian	9.3.3.3.1.3

L

laceration of the muzzle	6.6.2
laminitis	6.7.25
lance needle	9.2.1.4
large intestine	2.1.7
large intestine dominating conveyance	2.1.7.1
large intestine governs the thin body fluids	2.1.7.2
large intestine meridian of forelimb-yangming	3.1.4.2.1
large intestine pathogenesis	4.2.4.4.2
large needle	9.2.1.9
large splenic collateral channel	3.1.9.1
laryngeal swelling	6.10.5.1
laser acupuncture apparatus	9.2.3.3
laser irradiation therapy of point	9.3.3.10
laser-acupuncture therapy	8.3.10
latent pathogenic qi	4.2.1.4.2
left-right points combination	9.1.19.10
leucorrhea due to kidney asthenia	6.5.4
life gate	2.1.4.12.18
lift skin air acupuncture therapy	9.2.13.7.2
lifting method	9.2.7.2
lifting-thrusting reinforcing and reducing method	9.2.11.9.3
lingering pathogenic qi due to deficient vital qi	4.2.2.2.9
liquid depletion	4.2.3.2.4
liquid injection therapy	9.2.13.5
liquid insufficiency of large intestine	4.2.4.4.2.6
liquor moxibustion	9.3.2.2.3
lithiasis blocking pattern	7.3.17
lithiasis blocking qi stagnation pattern	7.3.17.2
lithiasis blocking qiji pattern	7.3.17.1
liver	2.1.4.11
liver and gallbladder pathogenesis	4.2.4.6
liver and kidney tron same source	2.1.4.11.19
liver and kidney yin deficiency and hyperactive yang pattern	7.10.29
liver and kidney yin deficiency pattern	7.10.28
liver and lung heat flourishing pattern	7.10.51
liver and spleen blood stasis pattern	7.10.37
liver and spleen dampness heat pattern	7.10.35
liver and spleen deficiency of qi and blood pattern	7.10.31
liver and spleen deficiency of qi and yin pattern	7.10.32
liver and spleen qi stagnation pattern	7.10.36
liver and stomach cold deficiency pattern	7.10.48
liver and stomach disharmony pattern	7.10.39
liver and stomach heat filled pattern	7.10.40
liver and stomach qi deficiency and blood stasis pattern	7.10.44
liver and stomach qi stagnation and blood stasis pattern	7.10.43
liver and stomach qi stagnation and yin deficiency pattern	7.10.45
liver and stomach qi stagnation pattern	7.10.42

liver and stomach yin deficiency and blood stasis pattern ……7.10.47
liver being firm-characierized zang viseus …… 2.1.4.11.11
liver being yin in substance and yang in function …… 2.1.4.11.12
liver blood ……2.1.4.11.2
liver blood deficiency …… 4.2.4.6.1.2
liver blood deficiency and stirring wind pattern…… 7.8.32.4
liver blood pattern ……7.8.2
liver deficiency and blood stasis pattern …… 7.8.14
liver depression transforming into fire pattern …… 7.8.17
liver dominating design of strategy………2.1.4.11.7
liver dominating rise and dispersion … 2.1.4.11.10
liver dominating tendon …… 2.1.4.11.13
liver enlargement……6.4.6
liver fire attacking ear pattern …… 7.12.8.1
liver fire attacking the head pattern …… 7.12.1.10
liver fire attacking the lung pattern……7.10.49
liver fire attacking the stomach pattern ……7.10.41
liver fire blazing pattern …… 7.8.21
liver fire flaming …… 4.2.4.6.1.7
liver gallbladder dampness heat pattern …… 7.8.37
liver gallbladder qi stagnation pattern……… 7.8.42
liver gallbladder stasis pattern …… 7.8.41
liver governs the free flow of qi ……2.1.4.11.6
liver heat and blood stasis pattern …… 7.8.29
liver heat and qi stasis pattern……… 7.8.28
liver heat and spleen deficiency pattern ……7.10.38
liver heat and yin deficiency pattern …… 7.8.30
liver heat epistaxis …… 6.2.9.2
liver heat stirring wind pattern …… 7.8.32.2
liver huang ……6.4.8
liver intestinal qi stagnation pattern ……7.10.68
liver intolerating wind……… 2.1.4.11.17
liver lung wind heat pattern ……7.10.50
liver manifesting in nail …… 2.1.4.11.14
liver meridian of hindlimbs-jueyin …… 3.1.4.4.2

liver opens into the eyes …… 2.1.4.11.15
liver pathogenesis …… 4.2.4.6.1
liver qi ……2.1.4.11.1
liver qi deficiency …… 4.2.4.6.1.1
liver qi deficiency and blood stasis pattern… 7.8.16
liver qi stagnant and sputum stasis pattern … 7.8.20
liver qi stagnation …… 4.2.4.6.1.5
liver qi stagnation and blood heat pattern … 7.8.18
liver qi stagnation and sputum heat pattern … 7.8.19
liver spleen deficiency pattern ……7.10.30
liver stagnant and kidney deficiency pattern ……7.10.52
liver stagnant and spleen deficiency pattern ……7.10.33
liver stagnation with spleen insufficiency ……4.2.4.8.25
liver stasis transforming heat pattern …… 7.8.12
liver stores blood……2.1.4.11.9
liver system ……2.1.4.11.5
liver yang ……2.1.4.11.4
liver yang deficiency …… 4.2.4.6.1.4
liver yang transforming into wind pattern… 7.8.32.1
liver yin ……2.1.4.11.3
liver yin deficiency …… 4.2.4.6.1.3
liver yin deficiency and blood stasis pattern …… 7.8.15
liver yin deficiency and stirring wind pattern …… 7.8.32.3
liver yin pattern ……7.8.1
local knock needling …… 9.3.3.3.1.2
local point selection …… 9.1.18.1
locked throat …… 6.2.15
long needle …… 9.2.1.8
long summer …… 1.3.3.1
loosening the chest and disinhibiting the diaphragm …… 8.1.10.146
loosening the middle and promoting diuresis …… 8.1.7.9
loss of essential qi leading to deficiency …… 4.2.2.2.2

loss of moistening of throat due to yin
　　deficiency pattern ·················· 7.12.10.12
low abdominal stasis and heat pattern ··· 7.15.6.6
low abdominal stasis pattern ············· 7.15.6.5
lower jiao ··································· 2.1.9.3
lower jiao dominates discharge ············ 2.1.9.3.2
lower jiao resembles a sluice ············· 2.1.9.3.1
lubricating the orifices and canals ······ 8.1.10.145
lumbago and paralysis of the posterior limbs
　　······································· 6.5.22.2
lumbago pains due to wind cold and wet evil
　　·· 6.7.13
lumbar swelling ···························· 6.5.20
lung ··· 2.1.4.9
lung and large intestine pathogenesis ········ 4.2.4.4
lung and stomach wind heat pattern ········7.10.61
lung as canopy································2.1.4.9.11
lung as container of phlegm ············· 2.1.4.9.8.2
lung as delicate zang viscus ·············2.1.4.9.12
lung as upper source of water ············ 2.1.4.9.8.1
lung channel fire stagnation pattern ········ 7.6.13
lung deficiency and intestinal withdrawal
　　pattern································7.10.65
lung dominating management and regulation
　　·····································2.1.4.9.10
lung dominating voice··················· 2.1.4.9.5.2
lung dryness and heat stagnation pattern ··· 7.6.30
lung dryness and intestine blocking pattern ··· 7.6.32
lung dryness and intestine heat pattern ······ 7.6.31
lung governs breathing ················ 2.1.4.9.5.1
lung governs descent and purification ··· 2.1.4.9.7
lung governs qi ························· 2.1.4.9.5
lung governs skin and body hair·········2.1.4.9.13
lung governs upward and outward diffusion
　　·· 2.1.4.9.6
lung heat and blood stasis pattern ········· 7.6.14
lung heat and body fluids stagnation pattern··· 7.6.23
lung heat and yin deficiency pattern ·······7.6.8
lung heat caused intestinal dysfunction pattern
　　··7.6.9

lung heat disturbing nose pattern ·········· 7.12.9.9
lung heat epistaxis ·························· 6.2.9.1
lung heat over vigorous pattern ··················7.6.7
lung intolerating cold ·····················2.1.4.9.18
lung manifesting in hair ···················2.1.4.9.14
lung meridian of forelimb-taiyin ········ 3.1.4.1.1
lung opens into the nose ··················2.1.4.9.15
lung pathogenesis ························ 4.2.4.4.1
lung presides over the hundred vessels ··· 2.1.4.9.9
lung protective qi deficiency pattern ··········7.6.5
lung qi ···································· 2.1.4.9.1
lung qi deficiency ····················· 4.2.4.4.1.1
lung qi deficiency cough················· 6.2.2.6
lung qi deficiency pattern ···················7.6.1
lung qi failing in dispersion ··········· 4.2.4.4.1.4
lung qi failing in purification and descent
　　······································ 4.2.4.4.1.5
lung qi stagnation and body fluids
　　stagnation pattern ···················· 7.6.22
lung regulates waterways ················· 2.1.4.9.8
lung system ······························ 2.1.4.9.4
lung yang···································· 2.1.4.9.3
lung yang deficiency ··················· 4.2.4.4.1.2
lung yang deficiency pattern ·················7.6.3
lung yin ···································· 2.1.4.9.2
lung yin deficiency ···················· 4.2.4.4.1.3
lung yin deficiency cough ················ 6.2.2.7
lung yin deficiency pattern ··················7.6.4
lung-kidney qi deficiency ··············4.2.4.8.14
lung-kidney yin deficiency················4.2.4.8.13

M

manifestations of five zang viscera············ 2.1.4.3
manipulating needle to await qi ············ 9.1.15.2
marrow ··································· 2.1.10.2
massage therapy ·····························8.4.1
mastitis ·······································6.9.8
medial-lateral points combination ········ 9.1.19.11
medicated ironing therapy ···················· 8.2.45
medicated liquid nose drops therapy ········ 8.2.66

medicated liquor therapy	8.5.2
medicinal bath therapy	8.6.2
mei dao needle	9.2.2.9
melancholy injuring lung	4.1.5.7
meningo encephalitis	6.1.12
mental exhaustion	4.1.7.1.2
meridian passage	3.1.3
meridian theory	3.1.1
meridians	3.1.1.1.1
meridians and collaterals	3.1.1.1
metacarpal pains	6.7.11
metal counter-restricting fire	1.3.14.2
metal generating water	1.3.11.6
metal over-restricting wood	1.3.13.5
metal restricting wood	1.3.12.7
method of lifting-thrusting needle	9.2.7.4
microwave acupuncture apparatus	9.2.3.4
microwave acupuncture therapy	9.3.3.14
middle jiao	2.1.9.2
middle jiao resembles foam	2.1.9.2.1
middle jiao resembles transformation	2.1.9.2.2
migratory arthralgia	6.11.13.1
mild heat stroke	6.1.8.1
mild stimulation	9.1.12.3
mildly regulating cold and heat	8.1.4.12
mind confusion by phlegm	4.2.4.3.1.11
moderate cold parasite disturbance pattern	7.15.4.2
moderate stimulation	9.1.12.2
moistening dryness and quenching thirst	8.1.10.57
moistening dryness and removing toxin	8.1.10.59
moistening dryness and suppressing cough	8.1.10.58
moistening dryness for relaxing bowels	8.1.3.4
moistening dryness with sweet and cold drugs	8.1.10.60
moistening lung and dissipating phlegm	8.1.10.77
moistening the intestines and purging heat	8.1.3.6
monkshood cake interposed moxibustion	9.3.2.1.1.9
moon blindness	6.4.9
mother element	1.3.11.1
mother-son points combination	9.1.19.5
mother-son reinforcing and reducing method	9.2.11.9.10
mouth and lips pattern	7.12.12
mouth and tongue ulcers	6.1.1
moving qi and breaking the blood	8.1.10.18
moving qi and downbearing counterflow	8.1.10.12
moving qi and harmonizing the stomach	8.1.10.10
moving qi and resolving dampness	8.1.7.4
moxa burner	9.3.1.4
moxa burner moxibustion	9.3.2.1.3
moxa cone	9.3.1.2
moxa cone moxibustion	9.3.2.1.1
moxa floss	9.3.1.1
moxa floss moxibustion therapy	9.3.2.1
moxa stick	9.3.1.3
moxa stick moxibustion therapy	9.3.2.1.2
moxibustion	9.1.10
moxibustion for sunken pulse condition	9.1.17.6
moxibustion on garlic therapy	8.3.6
moxibustion on ginger therapy	8.3.5
moxibustion therapy	8.3.7
moxibustion with moxa tube	9.3.2.1.3.1
muscle and bone injury pattern	7.13.2.11
mutual impairment between yin and yang	4.2.2.3.3
mutual promotion between lung and kidney	2.1.4.9.20

N

nail like boil	6.6.23
nail like boil of exposed membrane	6.6.25
nail like boil of exudate	6.6.27
nail like boil of foamy pus	6.6.26
nail like boil of pus and blood	6.6.28
nasal discharge	6.2.4
nasal discharge due to lung fire	6.2.4.1
nasal discharge due to over exertion	6.2.4.2
nasal discharge of pus and blood	6.6.36

nasal fire poison pattern	7.12.9.8
nasal pattern	7.12.9
nasal swelling due to hot evil	6.6.13
natural harmony of yin and yang	1.2.1.5
nebula over the eyes	6.4.4
neck rheumatism	6.11.4
needle breakage	9.2.10.4
needle hammer holding method	9.2.5.7
needle manipulation	9.2.7.1
needle retention	9.2.8.1
needle withdrawal	9.2.9.1
needle-flying insertion	9.2.6.7
needle-holding hand	9.2.5.1
needle-medicated cupping method	9.3.3.7.6
needle-twisting method	9.2.7.10
needle-withdrawing method	9.2.9.4
needling associated with cupping	9.3.3.7.4
needling depth	9.2.6.4
needling direction	9.2.6.1
needling technique	9.1.9
nephralgia	6.5.1
nervous pattern	7.5.33
neutral reinforcing and reducing method	9.2.11.9.9
new contraction	4.2.1.4.1
nine kinds of classical needles	9.2.1
ninereinforcing and six-reducing	9.2.11.9.11
non-acclimatization	4.1.13
non-interaction of heart and kidney	4.2.4.8.7
non-scarring moxibustion	9.3.2.1.1.5
nose insertion therapy	8.2.64
nose insufflation therapy	8.2.65
nourishing blood and moistening dryness	8.1.10.56
nourishing kidney and soothing the fetus	8.1.10.128
nourishing kidney to induce astringency	8.1.10.137
nourishing liver and inhibiting wind	8.1.10.69
nourishing the blood and activating the blood	8.1.8.16
nourishing the blood and improving vision	8.2.17
nourishing the blood and preventing abortion	8.1.8.17
nourishing the blood and stopping bleeding	8.1.8.18
nourishing the blood to moisten the intestines	8.1.8.15
nourishing yin and inhibiting wind	8.1.10.70
nourishing yin and moistening dryness	8.1.10.52
nutrient qi derived from middle jiao	2.2.1.8.1
nutrient qi moving in vessels	2.2.1.8.2
nutrient yin injury	4.2.6.3.4
nyctalopia	6.4.10

O

oat-vinegar moxibustion	9.3.2.2.1
oblique insertion	9.2.6.2.2
obstruction of colon	6.3.30
obstruction parasite dampness pattern	7.13.1.11
obstructive dysuria	6.5.10
obtaining qi	9.1.13
occurrence of secondary disease	4.2.1.4.5
oesophagus obstruction	6.3.2
ointment application therapy	8.2.49
onset of disease	4.2.1.4
open-closed reinforcing and reducing method	9.2.11.9.8
opening the locked jaw for resuscitation	8.1.10.109
opposition between yin and yang	1.2.1.2.1
ordinary needling therapy	9.2.13.1
organ of drainnge	2.1.9.4
origin of life	1.2.1.13
osteomalacia	6.3.31
other meridian point selection	9.1.18.6
outburst of heart and stomach fire	4.2.4.8.10
over consumption of heart nutrient	4.2.6.3.3
over discharge in vagina	6.9.2.1

over discharge in vagina due to spleen
　　defcience ·················· 6.3.34
over exertion cough················ 6.2.2.8
over-restriction and counter-restriction among
　　five elements ················ 1.3.16
overwhelming joy impairing heart ·········· 4.1.5.3
overwork ···················· 4.1.7.1
overwork leading to qi consumption ··· 4.2.3.1.1.1.2
oxone point-injection therapy ··········· 9.3.3.4

P

pain in shoulder and up arm ············· 6.7.1
pain in the head of the hoof············ 6.7.24
pain of the stifle joint ············· 6.7.16
pain on chest and foreleg ············· 6.2.11
pains of the lumbus and hip ············ 6.7.14
pains of the lumbus and hip due to cold evil
　　·························· 6.7.14.1
pains of the lumbus and hip due to sudden
　　sprain ····················· 6.7.14.2
pains of the shoulder joint ············· 6.7.2
pairing between the heart and small intestine
　　························ 2.1.4.8.13
pairing between the kidney and urinary
　　bladder ··················· 2.1.4.12.17
pairing between the liver and gallbladder
　　························ 2.1.4.11.18
pairing between the lung and large intestine
　　························· 2.1.4.9.19
pairing between the spleen and stomach
　　························· 2.1.4.10.16
paradoxical treatment ················ 5.2.3
paralysis due to wind evil ············· 6.11.12
paralysis of the radial nerve ············ 6.7.6
parasite ······················ 4.1.10
parasite accumulating in intestine pattern ···· 7.7.51
parasite accumulation and chancre pattern ··· 7.12.2.4
parasite clumping pattern ············· 7.3.16
parasite disturbing biliary and diaphragmatic
　　pattern····················· 7.8.36

parasite gathering at anus pattern ········· 7.7.52
parasite poison attacking skin pattern········ 7.11.19
parasite poison smolder at skin pattern ······ 7.11.14
parasite toxin accumulating pattern ······ 7.3.14.10
parasite toxin attacking lung pattern ········ 7.6.35
paste application therapy ············· 8.2.47
pathogenesis ··················· 4.2.1.1
pathogenesis of blood level············· 4.2.6.4
pathogenesis of defense level ············ 4.2.6.1
pathogenesis of defense qi nutrient blood
　　syndromes ··················· 4.2.6
pathogenesis of deficiency syndrome ··· 4.2.2.2.2.1
pathogenesis of eight principal syndromes ······ 4.2.2
pathogenesis of excess syndrome ······ 4.2.2.2.1.1
pathogenesis of exterior and interior ········ 4.2.2.5
pathogenesis of half-exterior and half-interior
　　······················· 4.2.5.3.1
pathogenesis of jueyin meridian ··········· 4.2.5.6
pathogenesis of nutrient level ············ 4.2.6.3
pathogenesis of qi level ··············· 4.2.6.2
pathogenesis of qi, blood and body fluid
　　syndromes ··················· 4.2.3
pathogenesis of shaoyang meridian ········ 4.2.5.3
pathogenesis of shaoyin meridian ········· 4.2.5.5
pathogenesis of six meridians syndromes ······ 4.2.5
pathogenesis of taiyang meridian ·········· 4.2.5.1
pathogenesis of taiyin meridian ··········· 4.2.5.4
pathogenesis of triple energizer syndromes ··· 4.2.7
pathogenesis of yangming meridian ········ 4.2.5.2
pathogenesis of zang-fu viscera syndromes ··· 4.2.4
pathogenesis theory················· 4.2.1
pathogenic cold ·················· 4.1.2.2
pathogenic dampness ················ 4.1.2.4
pathogenic dryness ················· 4.1.2.5
pathogenic fire ··················· 4.1.2.6
pathogenic qi ···················· 4.2.1.3
pathogenic qi retreating with deficient vital qi
　　························· 4.2.2.2.7
pathogenic summer heat ··············· 4.1.2.3
pathogenic wind ·················· 4.1.2.1

pathological product that can cause disease ··· 4.1.8
pattern clinical manifestation ·············· 1.1.1.1.3
pattern differentiation by the eight principles
　　······································· 1.1.1.1.5
pattern differentiation of qi, blood and body
　　fluids ································· 1.1.1.1.7
pattern differentiation of six meridians ··· 1.1.1.1.8
pattern differentiation of triple warmer ···1.1.1.1.10
pattern differentiation of wei-defence, qi,
　　ying nutrients and blood ············· 1.1.1.1.9
pattern differentiation of zang-fu organs ··· 1.1.1.1.6
pattern qi deficiency with blood stasis of
　　water ring pattern ··················· 7.12.7.5
pattern qi deficiency with blood stasis of
　　water ring pattern ··················· 7.12.7.6
patterns of disharmony between ying and
　　wei-defence pattern ···················· 7.15.9
patterns of exterior cold and interior heat
　　pattern ································7.15.10
penetration needling ························ 9.2.6.3
penial prolapse ····························· 6.5.22
pensiveness as spleen emotion ········· 2.1.4.10.13
pensiveness injuring spleen ··················· 4.1.5.5
pensiveness leading to qi knotting ············ 4.1.5.6
pericardium ······························2.1.4.8.14
pericardium meridian of forelimb-jueyin ··· 3.1.4.1.2
perpendicular insertion ···················· 9.2.6.2.1
pharyngeal pattern ···························7.12.10
phlegm··································· 4.1.8.1.1
phlegm retention of diaphragm ············· 6.3.35
phlegm-fire disturbing heart ········ 4.2.4.3.1.10
phlegm-fluid retention ······················ 4.1.8.1
phlegm-heat accumulated in lung ······ 4.2.4.4.1.11
phthisis of bones and joints ··················· 6.6.39
physical exhaustion ························· 4.1.7.1.1
physical inactivity ····························· 4.1.7.2
pinching pressing method ···················· 9.2.4.4
plaster application therapy ···················· 8.2.48
plaster therapy ······························ 9.3.3.15
point catgut-embedding therapy ············· 9.3.3.9

point selection ······························ 9.1.18
point selection according to corresponding
　　portion ······························ 9.1.18.9
point selection according to corresponding
　　portion of same name meridian ········ 9.1.18.5
point selection according to symptom ··· 9.1.18.7
point selection according to syndrome
　　differentiation ························ 9.1.18.10
point selection along the meridian ········ 9.1.18.4
points combination ··························· 9.1.19
poison fire attack lip pattern ············· 7.12.12.2
poor appetite and digestion ············· 4.2.4.5.2.11
postpartum abdominal pain······················6.9.5
postpartum fever ······························6.9.9
postpartum paralysis ······················· 6.9.11
post-recovery prevention ················· 5.1.1.3
predominance of yang making yin suffer
　　································4.2.2.3.1.1.1
predominance of yin making yang suffer
　　································4.2.2.3.1.2.1
pregnant edema ···························· 6.9.12
premature ejaculation or spermatorrhea ··· 6.5.22.1
preponderance of cold or heat ············· 4.2.5.6.1
pressing hand ······························· 9.2.4.1
pressing moxibustion ····················· 9.3.2.1.2.5
preventing a disease before it arises ········ 5.1.1.1
prevention of progress of disease ············ 5.1.1.2
preventive treatment ·························5.1.1
pricking therapy ·······························8.3.4
pricking-cupping bloodletting method ··· 9.3.3.7.2
producing fluid and disinhibiting the pharynx
　　······································· 8.2.36
producing fluid and moistening dryness ··· 8.1.10.53
producing fluid and quenching thirst ······ 8.1.10.54
production of phlegm by spleen deficiency
　　····································4.2.4.5.1.12
production of wind by spleen deficiency
　　····································4.2.4.5.1.11
progressive edema ························· 6.5.18
prolapse of vagina ·····························6.9.7

prolonged lochia ·················· 6.9.3
promoting blood circulation and detoxication ··· 8.2.2
promoting diuresis and resolving turbidity ··· 8.1.7.2
promoting diuresis with bland drugs ······· 8.1.7.8
proteinuric edema ··················· 6.5.16
protrusion of the swollen third eyelid ······· 6.11.15
pruritus ························· 6.8.5
pruritus and alopecia ··············· 6.2.7
pruritus and alopecia ··············· 6.8.2
psoriasis ························ 6.8.23
puffiness by wind cold wet and hot ········ 6.5.15
pulmonary abscess ················· 6.2.5
puncturing drills on a paper pad ·········· 9.2.12.1
puncturing drills with a cotton ball ········ 9.2.12.2
puncturing drills with copper coin ········· 9.2.12.4
puncturing drills with floating fruit········· 9.2.12.3
purging the liver and clearing the lung ··· 8.1.6.25
purging the liver and clearing the stomach ··· 8.1.6.26
purging the lung and calming panting······ 8.1.10.8
purpura ·························· 6.8.21
pushing and thrusting needle ··········· 9.2.11.4
pyrography ······················· 9.3.2.3.1

Q

qi ····························· 2.2.1
qi and blood stagnating at anus pattern ······ 7.7.56
qi blockage ······················ 4.2.3.1.1.2.1
qi blocked pattern ·················· 7.3.12
qi circulates blood ················· 2.2.5.2
qi circulates body fluids ············· 2.2.8
qi collapse ······················ 4.2.3.1.1.4
qi collapse following heavy blood loss ··· 4.2.3.1.3.5
qi controls blood ·················· 2.2.5.3
qi controls body fluids ············· 2.2.9
qi counterflow ···················· 4.2.3.1.1.2.2
qi deficiency ····················· 4.2.3.1.1.1
qi deficiency and throat insufficiency
    pattern························ 7.12.10.11
qi deficiency and tooth movement pattern
    ···························· 7.12.11.12
qi deficiency bloody stool ············· 6.3.21.2
qi deficiency causing blood stasis pattern ··· 7.4.1.7
qi deficiency causing evil detaining pattern
    ······························ 7.4.6.1
qi deficiency causing evil toxin detaining
    pattern························ 7.4.6.2
qi deficiency causing heat detaining pattern··· 7.4.6.3
qi deficiency causing inner heat pattern ······ 7.4.2.4
qi deficiency causing qi stagnation pattern··· 7.4.1.6
qi deficiency ear insufficiency pattern ··· 7.12.8.11
qi deficiency mixed body fluids stagnation
    pattern························ 7.4.1.4
qi deficiency mixed cold coagulation pattern
    ······························ 7.4.1.5
qi deficiency mixed dampness pattern ····· 7.4.1.3
qi deficiency mixed excess pattern ············7.4.1
qi deficiency mixed heat pattern ········· 7.4.1.1
qi deficiency mixed sputum stagnation pattern
    ······························ 7.4.1.2
qi deficiency nose insufficiency pattern ··· 7.12.9.10
qi deficiency of heart ················ 6.1.5.1
qi deficiency of the heart and lung pattern ··· 7.10.8
qi deficiency of the lung and kidney pattern
    ·····························7.10.59
qi deficiency pattern ················ 7.2.1
qi deficiency water ring pattern ··········· 7.12.7.4
qi desertion pattern ···················7.2.3
qi engenders blood ·················· 2.2.5.1
qi engenders body fluids···············2.2.7
qi failing control blood pattern ········· 7.2.10.2
qi failing to contain blood ············ 4.2.3.1.3.4
qi fen dampness heat pattern ·············· 7.3.4.3
qi governs warming ·················· 2.2.1.12
qi impediment ····················· 4.2.3.1.1.8
qi is the commander of the blood ···········2.2.5
qi loss due to blood depletion pattern ······ 7.2.10.1
qi monism ······················· 2.2.1.1
qi movement ····················· 2.2.1.3
qi of meridians ·····················3.1.2
qi of the middle jiao ················· 2.2.1.11

qi of the zang-fu organs	2.2.1.10
qi phase pattern	7.3.6.1
qi prostration following liquid depletion	4.2.3.2.5
qi ring pattern	7.12.3
qi ring yin deficiency pattern	7.12.3.5
qi sinking	4.2.3.1.1.3
qi sinking pattern	7.2.2
qi stagnation	4.2.3.1.1.6
qi stagnation and blood stasis of water ring pattern	7.12.7.7
qi stagnation and blood stasis pattern	7.3.10.1
qi stagnation and coagulated sputum in throat pattern	7.12.10.4
qi stagnation blocking body fluid pattern	7.3.10.5
qi stagnation caused lung injuring pattern	7.6.36
qi stagnation causing dampness stagnation pattern	7.3.10.2
qi stagnation causing mind syncope pattern	7.5.29
qi stagnation causing severe heat pattern	7.3.10.3
qi stagnation in ear pattern	7.12.8.7
qi stagnation in intestine pattern	7.7.50
qi stagnation of vocal cords pattern	7.12.10.8
qi stagnation pattern	7.3.10
qi stagnation transforming fire pattern	7.3.10.4
qi transformation	2.2.1.2
qi transformation of the urinary bladder	2.1.8.2
qi upwards reverse pattern	7.3.11
qi-blood disharmony	4.2.3.1
quadriplegia	6.7.7
quick needling for heat	9.1.17.4
quick-slow reinforcing and reducing method	9.2.11.9.6
quittor	6.7.23

R

rampancy of heart fire	4.2.4.3.1.6
recurrence	4.2.1.4.6
reducing child-element in excess condition	5.2.10.4
reducing child-element in excess condition	9.1.17.13
reducing needle by one inserting and three lifting	9.2.11.6
reducing needle manipulation	9.2.11.5
reducing phlegm and resolving masses	8.1.10.87
reducing urination for preventing enuresis	8.1.10.132
reducing with ignited moxa	9.3.2.1.4.2
refreshing the head and eyes	8.1.10.144
regulate and balance yin and yang	5.2.9
regulate and harmonize qi and blood	5.2.10
regulating and harmonizing qi and blood	8.1.4.10
regulating blood and inhibiting wind	8.1.10.71
regulating nutrient and defensing	8.1.1.8
regulating qi	9.1.14
regulating qi according to changes	9.1.17.8
regulating qi and disinhibiting the pharynx	8.2.35
regulating qi and dissolving distension	8.1.10.15
regulating qi and fortifying the spleen	8.1.10.11
regulating qi and harmonizing the nutrient	8.1.4.11
regulating qi and moving stagnation	8.1.10.1
regulating qi and preventing abortion	8.1.10.19
regulating qi and releasing depression	8.1.10.2
regulating qi and releasing superficies	8.1.1.10
regulating qi and relieving pain	8.1.10.16
regulating qi and resolving stasis	8.1.10.17
regulating qi-flowing for eliminating phlegm	8.1.10.80
regulating the intestines and stomach	8.1.4.7
regulating the spleen and stomach	8.1.4.8
regulating the stomach and dissolving distension	8.1.7.40
regulating the thoroughfare vessel and conception vessel	8.1.4.16
regulating yin and yang	5.2.3.5
reinforce healthy qi and remove pathogenic factors simultaneously	5.2.5.5

reinforce healthy qi to eliminate pathogenic factors 5.2.5
reinforce healthy qi to strengthen the body... 5.2.5.1
reinforcement after elimination 5.2.5.4
reinforcement containing elimination 5.2.5.6
reinforcing and reducing manipulations of needle 9.2.11.1
reinforcing and reducing with moxibustion 9.3.2.1.4
reinforcing and replenishing essence and marrow 8.1.8.64
reinforcing and replenishing middle qi 8.1.8.4
reinforcing and replenishing the heart and kidney..............8.1.8.65
reinforcing blood..............8.1.8.11
reinforcing heart and lung 8.1.8.7
reinforcing manipulation..............9.2.11.2
reinforcing mother-element in deficiency condition..............5.2.10.3
reinforcing mother-element in deficiency condition..............9.1.17.14
reinforcing needle by three inserting and one lifting 9.2.11.3
reinforcing qi 8.1.8.1
reinforcing qi and securing collapse 8.1.8.10
reinforcing spleen and kidney..............8.1.8.9
reinforcing spleen and lung..............8.1.8.8
reinforcing the blood and nourishing the heart 8.1.8.12
reinforcing the blood and nourishing the liver 8.1.8.13
reinforcing the blood and nourishing the spirit 8.1.8.19
reinforcing the blood and securing collapse 8.1.8.14
reinforcing the heart qi 8.1.8.2
reinforcing the kidney qi..............8.1.8.6
reinforcing the kidney to check diarrhea ...8.1.8.53
reinforcing the liver qi 8.1.8.5
reinforcing the lung qi..............8.1.8.3
reinforcing with ignited moxa..............9.3.2.1.4.1
reinforcing yang 8.1.8.43
reinforcing yang and replenishing qi 8.1.8.42
relapse due to drugs 4.2.1.4.6.3
relapse due to improper diet 4.2.1.4.6.1
relapse due to overwork 4.2.1.4.6.2
releasing superficies with pungent-cool......8.1.1.2
releasing superficies with pungent-warm ...8.1.1.1
releasing the external and clearing the internal 8.1.9.3
releasing the external and disinhibiting water 8.1.7.28
releasing the external and purging the internal 8.1.9.2
releasing the external and warming the internal 8.1.9.4
relieving oppression and masses..............8.1.7.38
relieving phlegm and food 8.1.10.85
removing dampness and releasing superficies 8.1.1.9
removing food stagnation and regulating the stomach 8.1.7.33
removing food stagnation and relieving pain 8.1.7.35
removing food stagnation and removing toxin 8.1.7.36
removing food stagnation and stopping vomiting 8.1.7.34
removing nebula and improving vision 8.2.12
removing phlegm and removing blood stasis 8.1.10.86
removing the wind and inducing expectoration for resuscitation 8.1.10.105
removing toxin and disinhibiting the ears ...8.2.20
removing toxin and disinhibiting the gums ...8.2.44
removing toxin and inducing resuscitation 8.1.10.107
removing toxin and inhibiting wind 8.1.10.67
removing toxin and reducing swelling 8.2.4
removing toxin for eliminating carbuncles......8.2.1

removing toxin for protecting yin	8.2.3
removing toxin for resuscitation	8.1.10.96
removing toxin to ventilate lung and inducing resuscitation	8.1.10.108
removing water retention by purgation	8.1.3.8
ren meridian	3.1.8.2
renal syncope	6.5.6
replenishing fire to disperse yin	8.1.10.139
replenishing qi and enriching yin	8.1.8.63
replenishing qi and unblocking the ears	8.2.22
replenishing qi to induce astringency	8.1.10.133
replenishing qi to relax the bowels	8.1.3.5
replenishing qi to strengthen the teeth	8.2.39
reservoir of five zang and six fu viscera	2.1.5.5.1
resolving carbuncle and expulsing boil	8.2.7
resolving dampness and harmonizing the middle	8.1.7.6
resolving dampness and harmonizing the nutrient	8.1.7.5
resolving dampness with aromatics	8.1.7.1
resolving food stagnation	8.1.7.32
resolving stasis and clearing heat	8.1.10.22
resolving stasis and diffusing the lung	8.1.10.41
resolving stasis and disinhibiting water	8.1.10.29
resolving stasis and dispersing swelling	8.1.10.30
resolving stasis and harmonizing the stomach	8.1.10.44
resolving stasis and loosening the chest	8.1.10.42
resolving stasis and loosening the heart	8.1.10.43
resolving stasis and nourishing the liver	8.1.10.47
resolving stasis and nourishing the stomach	8.1.10.45
resolving stasis and regulating the spleen	8.1.10.48
resolving stasis and soothing the liver	8.1.10.46
resolving stasis and stopping bleeding	8.1.10.34
resolving stasis and unblocking the brain	8.1.10.40
respiratory reinforcing and reducing method	9.2.11.9.7
restricted object	1.3.12.1
restriction and generation among five elements	1.3.17
resuscitation through vomiting	8.1.2.5
retained cupping method	9.3.3.7.1.7
retained fluid	4.1.8.1.2
retained needling for cold	9.1.17.5
retaining needle to await qi	9.1.15.1
retention of cold-dampness in spleen	4.2.4.5.1.10
retention of placenta	6.9.4
retum prolapse	6.11.6
reviving yang for resuscitation	8.1.5.1
ringworm	6.8.10
round-pointed needle	9.2.1.2
round-sharp needle	9.2.1.6 or 9.2.2.2
round-sharp needle therapy	9.2.13.1.2
routine treatment	5.2.2

S

sadness consuming the qi	4.1.5.8
salivation due to pulmonary cold evil	6.2.10
same meridian points combination	9.1.19.1
same name meridian points combination	9.1.19.4
same treatment for different diseases	5.2.7
san jiao	2.1.9
sandcrack	6.7.21
sanjiao meridian of forelimb-shaoyang	3.1.4.2.2
scabies	6.8.12
scalds and burns	6.6.3
scalp acupuncture therapy	9.3.3.5
scare caused mental disorder pattern	7.5.32
scarring moxibustion	9.3.2.1.1.4
science of acupuncture and moxibustion	9.1.2
science of veterinary acupuncture and moxibustion	9.1.3
scourge toxin pouring downward pattern	7.15.5.2
scraping therapy	9.3.3.8
scraping therapy	8.2.59
scrofula	6.6.40
scrotal hernia	6.3.28

scrotal hernia	6.11.5.1
seasonal formula	5.1.2
seasonal pestilence	6.10.2
seeking yang from yin	5.2.9.4
seeking yin from yang	5.2.9.3
selection of source point for zang-viscus disease	9.1.17.15
self-blood injection therapy	9.2.13.6
separating elimination from the external and internal	8.1.4.15
separating elimination from the upper and lower	8.1.4.14
separating elimination through urination and defecation	8.1.4.13
separation between yin and yang	4.2.2.3.8
separation of the refined from residue	2.1.6.4
sepsis toxin accumulating pattern	7.3.14.13
septic kidney pattern	7.9.12
seven emotions	4.1.5
seven-star needling therapy	8.3.2
severe evil toxin pattern	7.3.14
severe heat causing qi stagnation pattern	7.3.6.10
severe heat causing suppuration pattern	7.3.6.12
severe heat injuring body fluids pattern	7.4.9.3
severe heat moving blood pattern	7.3.6.9
severe heat moving wind pattern	7.3.6.8
severe heat pattern	7.3.6
severe heat stroke	6.1.8.2
severe heat toxin pattern	7.3.14.3
severe wind toxin pattern	7.3.14.2
sexual exhaustion	4.1.7.1.3
shallow needling	9.2.6.4.1
shaoyang disease pattern	7.14.6
shaoyin cold transformation pattern	7.14.8.1
shaoyin disease pattern	7.14.8
shaoyin heat transformation pattern	7.14.8.2
sheath tube hernia	6.11.5.3
shoulder and up arm pain due to cold evil	6.7.1.1
shoulder and up arm pain due to sudden sprain	6.7.1.2
shoulder dislocation	6.7.5
side-by-side finger pressing method	9.2.4.5
sideward counterflow of liver qi	4.2.4.6.1.6
simple reinforcing and reducing method	9.2.11.9
simultaneous treatment of both the exterior and the interior	8.1.9.1
sinking of spleen qi	4.2.4.5.1.3
sinking of spleen qi pattern	7.7.2
six fu organs	2.1.5
six fu viscera descended in function	2.1.5.2
six fu viscera pathogenesis	4.2.4.2
six fu viscera unobistructed in function	2.1.5.1
six pathogenic factors	4.1.2
skeleton	2.1.10.3.1
skin dysplasia pattern	7.11.24
skin is closed and the water stasis pattern	7.11.21
skin spreading pressing method	9.2.4.3
skin stasis pattern	7.11.22
small intestine	2.1.6
small intestine dominating digestion	2.1.6.2
small intestine dominating reception of residue as container	2.1.6.1
small intestine governs thick body fluids	2.1.6.3
small intestine meridian of forelimb-taiyang	3.1.4.2.3
small intestine pathogenesis	4.2.4.3.2
arrow needle therapy	9.2.13.1.3
snake venom attacking pattern	7.3.14.18
soft burning method	9.3.2.2.4
softening and resolving hard mass	8.1.7.41
softening hardness and moistening dryness	8.1.3.7
solitary yang failing to rise	1.2.1.6.1
solitary yin failing to increase	1.2.1.6.2
soothing the liver and disinhibiting the gallbladder	8.1.10.4
soothing the liver and fortifying the spleen	8.1.4.4
soothing the liver and harmonizing the stomach	8.1.4.6

soothing the liver and regulating the spleen 8.1.4.3	spleen deficiency and dampness heat stagnation pattern 7.7.14
soothing the liver and relieving yang 8.1.10.72	spleen deficiency and intestinal withdrawal pattern 7.10.66
sores 6.6.21	
sores due to unacclimatization 6.8.14	spleen deficiency and parasite stagnation pattern 7.7.17
sores due to wet evil 6.8.15	
sores of pus 6.8.9	spleen deficiency and sputum dampness stagnation pattern 7.7.15
sores of wet 6.8.13	
sound yin and firm yang 1.2.1.3	spleen deficiency caused blood dryness pattern 7.7.9
source of granary supply 2.1.5.10	
source-connecting points combination ... 9.1.19.3	spleen deficiency caused food stagnating pattern 7.7.16
sparrow pecking-like moxibustion 9.3.2.1.2.4	
spasm of diaphragm 6.11.11	spleen deficiency caused qi stagnation pattern 7.7.10
spasm of diaphragm due to cold and deficiency 6.11.11.1	spleen deficiency caused water retention pattern 7.7.11
spasm of diaphragm due to heat evil 6.11.11.2	spleen deficiency with dampness pattern ... 7.7.13
spasmodic colic 6.3.27	spleen deficiency with dampness retention 4.2.4.5.1.8
spirit adhering to vessel 2.1.10.4.2	
spittoon obstructing diaphragm pattern ... 7.15.1.5	spleen dominating acquirement 2.1.4.10.5.1
spleen 2.1.4.10	spleen dominating limbs 2.1.4.10.9
spleen and kidney qi deficiency and water stasis pattern 7.10.57.1	spleen dominating muscle 2.1.4.10.10
	spleen earth obstructed and hepatic qi stagnated 4.2.4.8.21
spleen and kidney qi deficiency pattern 7.10.56	
spleen and lung both qi and yin deficiency pattern 7.10.54	spleen failing to control the blood pattern 7.7.4
	spleen governs transportation and transformation 2.1.4.10.5
spleen and stomach excess heat pattern 7.7.62	
spleen and stomach pathogenesis 4.2.4.5	spleen housing nutrient 2.1.4.10.5.4
spleen and stomach qi deficiency pattern ... 7.7.57	spleen intolerating dampness 2.1.4.10.15
spleen and stomach qi disharmonyp pattern 7.7.66	spleen kidney yang deficiency and water stasis pattern 7.10.57.2
spleen and stomach yang deficiency causing qi stagnation pattern 7.7.60	spleen manifesting in lips 2.1.4.10.11
	spleen meridian of hindlimbs-taiyin 3.1.4.4.1
spleen and stomach yang deficiency pattern 7.7.59	spleen opens into the mouth 2.1.4.10.12
	spleen pathogenesis 4.2.4.5.1
spleen as source of phlegm 2.1.4.10.5.3	spleen qi 2.1.4.10.1
spleen as source of qi-blood formation 2.1.4.10.5.2	spleen qi deficiency 4.2.4.5.1.1
	spleen qi deficiency pattern 7.7.1
spleen ascends the nutrients 2.1.4.10.7	spleen qi failing to ascend 4.2.4.5.1.2
spleen contains blood 2.1.4.10.6	spleen qi stagnation pattern 7.7.3
spleen deficiency and blood depletion pattern ... 7.7.8	

spleen system	2.1.4.10.4
spleen yang	2.1.4.10.3
spleen yang deficiency	4.2.4.5.1.5
spleen yang deficiency caused water retention pattern	7.7.12
spleen yang deficiency pattern	7.7.5
spleen yin	2.1.4.10.2
spleen yin deficiency	4.2.4.5.1.4
spleen yin deficiency pattern	7.7.6
spleen ying deficiency pattern	7.7.7
spleen-kidney yang deficiency	4.2.4.8.20
spleen-stomach deficiency	4.2.4.8.15
spleen-stomach yin deficiency	4.2.4.8.19
splint-fixing therapy	8.2.71
splint-fixing therapy	8.6.4
spoon-like needle	9.2.1.3
sprain and contusion	6.6.5
sprain of the hip joint	6.7.15
sprain of triceps	6.7.8
spreading skin ulcer	6.8.24
sputum and static blood tangling at water ring pattern	7.12.7.9
sputum and toxin stagnation pattern	7.3.7.13
sputum blocked heart mind pattern	7.5.26
sputum blocking heart channel pattern	7.5.13
sputum damp obstruction pattern	7.13.2.15
sputum dampness accumulating in the lung pattern	7.6.16
sputum dampness accumulating in the lung pattern	7.6.18
sputum dampness attacking ear pattern	7.12.8.5
sputum dampness blocking channels pattern	7.13.1.8
sputum dampness blocking essence chamber pattern	7.9.28
sputum dampness blocking zhongjiao pattern	7.7.74
sputum dampness causes lumbar pattern	7.13.2.8
sputum dampness coagulation blocking throat pattern	7.12.10.3
sputum dampness flesh ring pattern	7.12.5.5
sputum dampness stagnation pattern	7.3.7.16
sputum fire in water ring pattern	7.12.7.2
sputum food tangling pattern	7.3.7.14
sputum forming nuclear lump pattern	7.3.7.17
sputum gathering nasal orifice	7.12.9.4
sputum heat attacking nose pattern	7.12.9.5
sputum heat blocking lung pattern	7.6.17
sputum heat blocking mind pattern	7.3.7.11
sputum heat causing qi stagnation pattern	7.3.7.9
sputum heat causing yin deficiency pattern	7.4.10
sputum heat disturbing mind pattern	7.5.25
sputum heat interference pattern	7.3.7.10
sputum heat moving wind pattern	7.3.7.12
sputum parasite tangling pattern	7.3.7.15
sputum pattern	7.3.7
sputum poison obstructing throat pattern	7.12.10.5
sputum qi tangling pattern	7.3.7.6
sputum stasis blocking diaphragm pattern	7.15.1.6
sputum stasis blocking lung pattern	7.6.19
sputum stasis tangling pattern	7.3.7.7
sputum stasis transforming heat pattern	7.3.7.8
stagnant fire	4.2.2.5.1.5
stagnant heat attacking channels pattern	7.13.1.9
stagnant heat of gallbladder channel pattern	7.8.39
stagnated heat in liver meridian	4.2.4.6.1.9
stagnation of heat in blood level	4.2.6.4.3
stagnation of heat in gallbladder meridian	4.2.4.6.2.1
stagnation of liver qi and blood deficiency pattern	7.8.9
stagnation of liver qi and blood stasis pattern	7.8.10
stagnation of liver qi and dampness heat pattern	7.8.26
stagnation of liver qi and yin deficiency pattern	7.8.11
stagnation of liver qi pattern	7.8.8
stagnation of nutrient qi	4.2.5.1.3

stagnation of pathogen in lung defense level
.... 4.2.6.1.1
stagnation of pathogen in shaoyang meridian
.... 4.2.5.3.2
stasis and swelling 6.6.11
stasis blocking clear orifice pattern 7.12.13.2
stasis heat attacking the head pattern 7.12.1.8
stasis of muscles and bones pattern 7.13.2.12
stasis of vocal cords pattern 7.12.10.9
stasis turbidity blocking essence chambe
　　pattern 7.9.30
static blood 4.1.8.2
static blood blocking channels pattern 7.13.1.10
static blood blocking in womb pattern 7.9.20
static blood obstructing the network vessels
　　pattern 7.15.1.9
static blood stagnated in throat pattern 7.12.10.10
steeping and washing therapy 8.2.53
sthenia constipation 6.3.20.1
sthenic dyspnea 6.2.3.1
stiletto needle 9.2.1.5
stimulating quantity of acupuncture and
　　moxibustion 9.1.11
stimulus intensity of acupuncture and
　　moxibustion 9.1.12
stomach 2.1.5.5
stomach and intestine dampness heat pattern
.... 7.7.67
stomach and intestine excess heat pattern 7.7.68
stomach and intestine food stagnation pattern
.... 7.7.69
stomach and intestine qi stagnation pattern 7.7.71
stomach cavity 2.1.5.5.2
stomach cold 4.2.4.5.2.4
stomach cold 6.3.24
stomach cold and qi upward inversion pattern
.... 7.7.25.2
stomach dominating descent 2.1.5.8
stomach dominating residue descent 2.1.5.8.1
stomach dryness injuring body fluids pattern 7.7.35

stomach fire burnt tooth pattern 7.12.11.3
stomach fire flaring gum pattern 7.12.11.2
stomach fluid 2.1.5.5.5.1
stomach governs decomposition 2.1.5.7
stomach governs receiving and holding 2.1.5.6
stomach heat 4.2.4.5.2.5
stomach heat 6.3.25
stomach heat and qi stagnation pattern 7.7.32
stomach heat and qi upward inversion
　　pattern 7.7.25.1
stomach heat and spleen deficiency pattern
.... 7.7.65
stomach heat and yin deficiency pattern 7.7.34
stomach heat epistaxis 6.2.9.3
stomach heat injuring body fluids pattern 7.7.33
stomach heat pattern 7.7.31
stomach meridian of hindlimbs-yangming
.... 3.1.4.3.1
stomach pathogenesis 4.2.4.5.2
stomach preferring softness and moisture 2.1.5.9
stomach qi 2.1.5.5.3
stomach qi deficiency 4.2.4.5.2.1
stomach qi deficiency and blood stasis pattern
.... 7.7.24
stomach qi deficiency pattern 7.7.22
stomach qi stagnation and blood stasis pattern
.... 7.7.27
stomach stagnation and qi upward inversion
　　pattern 7.7.25.3
stomach yang 2.1.5.5.4
stomach yang deficiency 4.2.4.5.2.2
stomach yang deficiency and blood stasis
　　pattern 7.7.26.2
stomach yang deficiency and qi stagnation
　　pattern 7.7.26.1
stomach yang deficiency pattern 7.7.26
stomach yin 2.1.5.5.5
stomach yin deficiency 4.2.4.5.2.3
stomach yin deficiency and blood stasis
　　pattern 7.7.30

stomach yin deficiency and qi stagnation
　　pattern ·································· 7.7.29
stomach yin deficiency pattern ············· 7.7.28
stomach-intestinal excessiveness ········ 4.2.5.2.3
stomatitis due to hot and poison ············· 6.6.12
stony stranguria ····························· 6.5.7.3
stopping bleeding and soothing the fetus··· 8.1.10.129
strangles ···································· 6.10.5.2
stranguria due to chyluria ················· 6.5.7.5
stranguria due to deficiency of qi and blood　6.5.24
stranguria due to heat ····················· 6.5.7.1
stranguria due to overstrain················ 6.5.7.2
stranguria syndrome ··························6.5.7
strengthening resistance and relieving exterior
　　syndrome ····························· 8.1.1.11
strengthening spleen and expelling intestinal
　　parasites ····························· 8.1.10.111
strengthening the teeth ····················· 8.2.37
string halt ····································· 6.7.17
stroke ··· 6.1.13
strong defensive qi and weak nutrient qi ··· 4.2.5.1.1.1
strong spleen being pathogen resistant ···2.1.4.10.8
strong stimulation ·························· 9.1.12.1
struggle between anti-pathogenic qi and
　　pathogenic factors ················· 4.2.2.1
stuck needle································· 9.2.10.2
su shui guan································· 9.2.2.10
subduing liver and inhibiting wind ········ 8.1.10.63
subject of acupuncture and moxibustion
　　technique ······························9.1.6
subject of acupuncture and moxibustion therapy
　　···9.1.7
subject of acupuncture points ················9.1.5
subject of meridian and collateral ··········· 9.1.4
sublingual blood stasis pattern ············ 7.12.12.8
suction cupping method ··················· 9.3.3.7.3
sudden collapse of heart yang ·········· 4.2.4.3.1.4
sudden collapse of heart yang pattern ········ 7.5.5.1
sudden hyperactivity of the liver yang pattern ···7.8.6
sudden onset ································ 4.2.1.4.3

sudden sprain ································6.6.4
summer dampness pattern ················ 7.3.3.3
summer heat blocking mind pattern ········ 7.5.24
summer heat blocking qi pattern ··········· 7.3.3.5
summer heat hurt pattern ················· 7.3.3.1
summer heat injuring body fluids and qi pattern
　　···7.4.8
summer heat moving wind pattern ········· 7.3.3.4
summer heat pattern ·························7.3.3
summer-heat attributing to burning heat ··· 4.1.2.3.1
summer-heat attributing to rise and dispersion
　　·· 4.1.2.3.2
summer-heat injured lung channel pattern··· 7.6.15
summer-heat tending to be mixed with
　　dampness ··························· 4.1.2.3.3
summer-heat tending to consume fluid
　　and qi ································ 4.1.2.3.5
summer-heat tending to disturb heart······ 4.1.2.3.4
sunburn ······································· 6.8.18
superficies heat and interior cold pattern ··· 7.15.11
suppressing the liver and reinforcing the
　　spleen ··································· 8.1.4.5
suppressing yang and inhibiting wind······ 8.1.10.68
suspended moxibustion ················· 9.3.2.1.2.1
sweating syndrome ···························6.1.3
swelling of the scrotum ····················· 6.5.19
swift insertion and slow lifting ············9.2.11.9.5
swift lifting and slow insertion ············ 9.2.11.9.4
swollen and rigid tongue······················6.1.4
swollen of leg due to deficiency of kidney··· 6.5.27
symptom, root-cause, non-urgency and
　　urgency ···································5.2.4

T

tai yi moxa stick moxibustion············ 9.3.2.1.2.7
taiyang channel pattern ··················· 7.14.1.1
taiyang cold attack pattern ················7.14.1.1.2
taiyang disease pattern ····················· 7.14.1
taiyang disease with stagnated blood pattern
　　··· 7.14.2.2

taiyang water amassment pattern ……… 7.14.2.1
taiyang wind attack pattern …………7.14.1.1.1
taiyang-fu organ pattern …………… 7.14.2
taiyin disease pattern ……………… 7.14.7
teeth are the extension of the bone ……… 2.1.10.3.2
teeth dryness due to yin deficiency pattern
　……………………………… 7.12.11.8
tender point selection …………… 9.1.18.13
tendon rupture …………………… 6.7.20
tertiary collaterals ……………… 3.1.10
testitis due to yang evil ………… 6.5.19.2
testitis due to yin evil …………… 6.5.19.1
tetanus …………………………… 6.10.4
the deeper the heat, the severer the limb
　coldness ………………………… 4.2.5.6.2
the milder the heat, the milder the limb
　coldness ………………………… 4.2.5.6.3
the order of perfusion in the twelve meridians
　…………………………………… 3.1.14
theory of visceral state ……………… 2.1.3
theory, principle, prescription and medicinal
　………………………………… 1.1.1.1.4
therapy of medicinal application on navel … 9.3.3.16
thick fluids ……………………… 2.2.3.2
thin fluids ……………………… 2.2.3.1
thoroughfare and controlling vessel stasis
　obstruction pattern ……………… 7.9.26
those dominated by five zang viscera ……… 2.1.4.1
those intolerated by five zang viscera ……… 2.1.4.6
three-edged needle ……………… 9.2.2.5
three kinds of laryngo-pharyngeal diseases … 6.10.5
three times needle of one-inserting and
　three-lifting ……………………… 9.2.11.7
three-edged needle therapy …………9.2.13.2.2
three-edged needling therapy …………8.3.3
three-yang meridians of forelimb ……… 3.1.4.2
three-yang meridians of posterior limb … 3.1.4.3
three-yin meridians of forelimb ……… 3.1.4.1
three-yin meridians of posterior limb……… 3.1.4.4
thrusting insertion ……………… 9.2.6.5

thrusting method ………………… 9.2.7.3
thunder-fire moxibustion …………… 9.3.2.1.2.6
tinea of pudendum ……………… 6.8.11
tonifying qi and blood …………… 8.1.8.61
tonifying the blood and fortifying the teeth … 8.2.40
tonifying the blood and nourishing the ears … 8.2.21
tonifying the kidney and improving vision … 8.2.15
tonifying the kidney to strengthen the teeth … 8.2.38
tooth fire poison offender pattern ……… 7.12.11.5
torsion incarceration and intussusception … 6.3.29
torsion of the urinary bladder …………… 6.5.9
toxic fire attacking ear pattern ………… 7.12.8.2
toxin attacking heart and liver pattern ……7.10.64
toxin attacking ying blood pattern ……… 7.3.14.9
toxin blocked heart apertures pattern ……… 7.5.28
toxin moving randomly pattern ………… 7.3.14.1
tranquilizing mind by enriching blood … 8.1.10.117
tranquilizing mind by nourishing kidney
　………………………………… 8.1.10.118
tranquilizing mind by nourishing the heart
　………………………………… 8.1.10.116
tranquilizing mind by nourishing yin … 8.1.10.120
tranquilizing mind by replenishing qi … 8.1.10.119
tranquilizing mind by suppressing heart … 8.1.10.121
tranquilizing mind for resuscitation …… 8.1.10.99
transformation between cold and heat … 4.2.2.4.9
transformation between exterior and interior
　………………………………… 4.2.2.5.3.9
transformation between yin and yang …… 4.2.2.3.5
transformation of pathogen in shaoyin
　into cold ………………………… 4.2.5.5.2
transformation of pathogen in shaoyin
　into heat ………………………… 4.2.5.5.1
transformation of qi depression into fire … 4.2.3.1.1.7
transforming qi and disinhibiting water … 8.1.7.30
transportation of transformed products without
　storing them …………………… 2.1.5.3.1
transverse insertion ……………… 9.2.6.2.3
transverse penetration …………… 9.2.6.3.2
traumatic epistaxis ……………… 6.2.9.5

traumatic reticuloperitonitis	6.3.7
traumatic stasis pattern	7.13.2.14
treat according to place	5.2.8.2
treat according to time	5.2.8.1
treat both the tip and root	5.2.4.3
treat cold with heat	5.2.2.1
treat deficiency with reinforcement	4.2.2.3
treat excess with purgation/reduction	5.2.2.4
treat false cold with cold	5.2.3.1
treat false heat with heat	5.2.3.2
treat heat with cold	5.2.2.2
treat the root in remissive stages	5.2.4.2
treat the tip first in acute conditions	5.2.4.1
treat uncontrolled discharge by unblocking	5.2.3.4
treating blood for qi disorder	5.2.10.1
treating deficiency with reinforcement	9.1.17.2
treating diarrhea with astringents	8.1.10.125
treating disease under the root	5.2.1.1
treating excess with expelling	9.1.17.1
treating fu for zang viscus disease	5.2.10.5
treating inapparent asthenia and sthenia with point on the meridian	9.1.17.3
treating left disease with right point	9.1.17.12
treating lower disorder with upper point	9.1.17.10
treating obstructive syndrome with tonifying methods	5.2.3.3
treating qi for blood disorder	5.2.10.2
treating right disease with left point	9.1.17.11
treating upper disorder with lower point	9.1.17.9
treating zang for fu viscus disease	5.2.10.6
treatment based on pattern identification	1.1.1.1.2
treatment in accordance with animals individuality	5.2.8.3
treatment in accordance with three factors	5.2.8
treatment of yang for yin disease	5.2.9.2
treatment of yin for yang disease	5.2.9.1
treatment principles	5.2.1
true cold with false heat	4.2.2.4.8.2
true deficiency with false excess	4.2.2.2.4.1
true excess with false deficiency	4.2.2.2.4.2
true heat with false cold	4.2.2.4.8.1
true-false of cold and heat	4.2.2.4.8
true-false of deficiency and excess	4.2.2.2.4
turbid phlegm obstructing lung	4.2.4.4.1.10
turbid urine	6.5.8
turbidness of sputum attacking the head pattern	7.12.1.5
twelve cutaneous regions	3.1.7
twelve divergent meridians	3.1.5
twelve meridians	3.1.4
twelve muscle regions	3.1.6
twirling insertion	9.2.6.6
twirling method	9.2.7.6
twirling needle withdrawal method	9.2.9.2
twirling reinforcing and reducing method	9.2.11.9.2

U

ultraviolet irradiation therapy of point	9.3.3.12
umbilical hernia	6.11.5.2
unblocking the collaterals and promoting lactation	8.1.10.35
unblocking the ears	8.2.18
unblocking the meridian and collateral	9.1.16
unblocking the meridians and stopping itching	8.1.10.37
unblocking the nose	8.2.25
unblocking the nose with aroma	8.2.30
un-restricted object	1.3.12.2
upper body exuberance and lower body deficience pattern	7.15.14
upper cold and lower heat	4.2.2.4.7.1
upper cold and lower heat pattern	7.15.16
upper heat and lower cold	4.2.2.4.7.2
upper heat and lower cold pattern	7.15.15
upper jiao	2.1.9.1
upper jiao dominates reception	2.1.9.1.2
upper jiao resembles mist	2.1.9.1.1
upper-lower points combination	9.1.19.8

upward flaming of heart fire ············ 4.2.4.3.1.7
upward flaming of heart fire pattern ········ 7.5.18
upward flaming of liver heat pattern ········ 7.8.22
urinary incontinence ················ 6.5.11
urticaria ························ 6.8.1
uterine blood heat pattern ············ 7.9.23
uterus with its appendages ············ 2.1.10.5

V

ventilating lung and dissipating phlegm ··· 8.1.10.74
ventilating lung and releasing superficies ··· 8.1.1.7
vesicular disease due to hot evil ············ 6.8.22
vessel ···················· 2.1.10.4
vessel as house of blood ··············· 2.1.10.4.1
vinegar-liquor moxibustion ·············· 9.3.2.2.2
visceral manifestation ··················· 2.1.1
vital qi deficiency causing evil detaining
　　pattern ···························· 7.4.6
vital qi deficiency causing evil toxin attacking
　　pattern ························· 7.4.6.5
vital qi deficiency causing evil toxin detaining
　　pattern ························· 7.4.6.4
vital qi deficiency causing sepsis detaining
　　pattern ························ 7.4.6.10
vital qi increasing with pathogenic qi
　　decreasing ···················· 4.2.2.2.6
vomiting ··························· 6.3.1
vomiting due to cold of insufficiency type ··· 6.3.1.3
vomiting due to improper diet ············ 6.3.1.1

W

waning of yin or yang ················ 4.2.2.3.2
warm and toxic attacking skin pattern ······ 7.11.10
warm disease ····················· 6.10.1
warm dryness attacking lung pattern ········ 7.6.28
warm dryness pattern ················ 7.3.5.1
warm pathogen ···················· 4.1.2.6.2
warming and dredging small intestine ····· 8.1.5.10
warming and purging accumulated cold ····· 8.1.3.3
warming and reinforcing heart yang ······· 8.1.8.44

warming and reinforcing kidney yang ··· 8.1.8.49
warming and reinforcing life fire ········ 8.1.8.50
warming and reinforcing liver yang ······ 8.1.8.48
warming and reinforcing lung yang ······ 8.1.8.45
warming and reinforcing receive qi ······ 8.1.8.51
warming and reinforcing spleen yang ······ 8.1.8.46
warming and reinforcing stomach yang ··· 8.1.8.47
warming and reinforcing the heart and kidney
　　···························· 8.1.8.57
warming and reinforcing the heart and lung
　　···························· 8.1.8.54
warming and reinforcing the lower
　　primordium ···················· 8.1.8.52
warming and reinforcing the spleen and kidney
　　···························· 8.1.8.56
warming and reinforcing the spleen and
　　stomach ······················ 8.1.8.55
warming channel and activating blood
　　circulation ····················· 8.1.5.11
warming channel for stopping bleeding
　　···························· 8.1.5.12
warming for resolving cold-phlegm ······ 8.1.10.78
warming jingmai for dispelling cold ········ 8.1.5.3
warming kidney for dispelling cold ········· 8.1.5.7
warming liver for dispelling cold ··········· 8.1.5.8
warming lung for dispelling cold ·········· 8.1.5.6
warming needling ·················· 9.3.3.1
warming phlegm and fluiding retention
　　··························· 8.1.10.84
warming stomach for dispelling cold ········ 8.1.5.9
warming the kidney and disinhibiting water
　　···························· 8.1.7.25
warming the middle and astringing the
　　intestines ····················· 8.1.8.58
warming the middle and stopping bleeding
　　···························· 8.1.8.59
warming yang and inducing astringency
　　··························· 8.1.10.135
warming yang and moving qi ············ 8.1.8.60
warming yang for dispelling cold ·········· 8.1.5.5

warming yang to dissipate fluid retention ·················· 8.1.10.83
warming yang to fortify the teeth ············ 8.2.42
warming zhongjiao for dispelling cold ······ 8.1.5.2
warts ··6.8.7
warts of perineum ·····································6.8.8
water counter-restricting earth ············· 1.3.14.4
water dampness stagnate at water ring pattern ····················· 7.12.7.8
water dampness stagnation pattern ···············7.3.9
water generating wood ························· 1.3.11.7
water over-restricting fire ······················ 1.3.13.3
water restricting fire ····························· 1.3.12.5
water retention affecting the heart pattern ··· 7.10.4
water retention causing qi stagnation ······· 4.2.3.2.8
water retention due to deficiency of kidney yang pattern ····························· 7.9.3.2
water retention due to kidney deficiency pattern·································7.9.3
water ring pattern ······························· 7.12.7
waxing and waning of anti-pathogenic qi and pathogenic factors ···················· 4.2.2.2
weak defensive qi and strong nutrient qi ····································· 4.2.5.1.1.2
wei-defensive qi··············· 2.2.1.9
wet compress therapy ························· 8.2.50
wheat-grain size cone moxibustion ···· 9.3.2.1.1.2
wheezing and pulse ···························2.1.10.4.3
wide needle ·· 9.2.2.3
wide needle blood therapy ··················9.2.13.2.1
wind and poison attacking the exterior pattern ··· 7.11.8
wind and water retention with each other pattern································7.11.20
wind attributing to mobility ··············· 4.1.2.1.3
wind attributing to opening and releasing ··· 4.1.2.1.1
wind being leading cause of diseases ······ 4.1.2.1.5
wind being mobile and changeable········· 4.1.2.1.4
wind cold and dampness stagnation of muscles and bones pattern ············ 7.13.2.6
wind cold assailing the pharyngeal pattern ······································ 7.12.10.1
wind cold attacking head pattern ·········· 7.12.1.1
wind cold attacking nose pattern············· 7.12.9.1
wind cold evil attacking lung pattern ········ 7.6.20
wind cold fettering the exterior pattern ······ 7.11.3
wind cold transforming heat pattern ········· 7.3.1.8
wind dampness attacking exterior pattern ··· 7.3.1.2
wind dampness attacking exterior pattern ··· 7.11.5
wind dampness attacking eye pattern ······ 7.12.2.2
wind dampness attacking head pattern ··· 7.12.1.3
wind dampness carrying toxin pattern ······ 7.3.1.5
wind dampness transform to fire pattern ··· 7.3.1.7
wind edema ·· 6.5.14
wind evil attacking exterior pattern ·········· 7.3.1.1
wind evil attacking exterior pattern ·········· 7.11.2
wind evil attacking mouth and lips pattern ·································· 7.12.12.1
wind fire attacking teeth pattern ············· 7.12.11.1
wind heat accumulated in the liver channel ···6.4.5
wind heat affecting the liver channel pattern ······································· 7.8.24
wind heat and sputum toxin pattern ········· 7.3.1.4
wind heat assailing the outer body pattern ··· 7.3.1.3
wind heat assailing the pharyngeal pattern ······································ 7.12.10.2
wind heat attacking ear pattern ············· 7.12.8.3
wind heat attacking exterior pattern ·········· 7.11.4
wind heat attacking head pattern············· 7.12.1.2
wind heat attacking nose pattern············· 7.12.9.2
wind heat attacking the eyes pattern ······· 7.12.2.1
wind heat attacking the lung pattern ········· 7.6.10
wind heat blocking lung pattern ················ 7.6.11
wind heat carrying dampness pattern ······ 7.3.1.6
wind heat causing blood dryness pattern ······7.4.7
wind heat in channels pattern ················ 7.13.1.2
wind heat in lung channel pattern ············ 7.6.12
wind heat in wind ring pattern ·············· 7.12.6.1
wind heat stagnation skin pattern ···············7.11.18
wind heat toxin pattern ························· 7.3.14.8

wind injuring intestine channel pattern 7.7.53
wind poison attacking skin pattern 7.11.11
wind poison attacking the exterior pattern ... 7.11.9
wind prevails over the marching bi pattern
.... 7.13.2.1
wind produced by blood deficiency 4.1.4.1.3
wind produced by blood dryness 4.1.4.1.4
wind produced by extreme heat 4.1.4.1.5
wind produced by liver yang 4.1.4.1.1
wind produced by yin deficiency 4.1.4.1.2
wind ring pattern 7.12.6
wind sputum attacking channel pattern ... 7.13.1.1
wind sputum attacking head pattern 7.12.1.4
wind sputum blocked mind pattern........... 7.5.27
wind sputum pattern 7.3.7.1
wind tending to attack yang portion of body
.... 4.1.2.1.2
wind yang harassing the upper body pattern
.... 7.12.1.6
wind-cold attacking lung 4.2.4.4.1.8
wind-heat in liver meridian 4.2.4.6.1.8
wind-heat invading lung................ 4.2.4.4.1.7
wind-phlegm 4.1.8.1.1.5
withdraw needle quickly 9.2.9.3
wood counter-restricting metal 1.3.14.1
wood generating fire 1.3.11.3
wood over-restricting earth 1.3.13.1
wood restricting earth 1.3.12.3
work-rest imbalance 4.1.7
worry injuring spleen qi pattern 7.7.21
wounds 6.6.1

X

xin huang 6.1.10

Y

yang 1.2.1.1.2
yang collapse 4.2.2.3.7.3
yang deficiency and external affections
pattern 7.4.4.8
yang deficiency causing blood stasis pattern
.... 7.4.4.5
yang deficiency causing dampness blocking
pattern 7.4.4.2
yang deficiency causing evil toxin detaining
pattern 7.4.6.8
yang deficiency causing water retention
pattern 7.4.4.4
yang deficiency ear loss of heat pattern ... 7.12.8.10
yang deficiency leading to cold 4.2.2.3.2.2
yang deficiency leading to yin fluids
deficiency pattern 7.2.13.2
yang deficiency mixed body fluids
stagnation pattern 7.4.4.3
yang deficiency mixed excess pattern........... 7.4.4
yang deficiency mixed qi stagnation pattern
.... 7.4.4.1
yang deficiency nose loss of warm pattern
.... 7.12.9.13
yang deficiency of heart pattern 7.5.5
yang deficiency of lung and kidney pattern
.... 7.10.60
yang deficiency of the heart and kidney
pattern 7.10.3
yang deficiency of the spleen and kidney
pattern 7.10.55
yang deficiency pattern 7.2.7
yang deficiency with cold coagulation
pattern 7.4.4.7
yang deficiency with sputum coagulation
pattern 7.4.4.6
yang edema 6.11.2.2
yang excess leading to heat............ 4.2.2.3.1.1.2
yang forming qi 1.2.1.1.5
yang meridian 3.1.12
yang qi 1.2.1.1.4
yang qi exhausting pattern 7.2.8
yang syndrome transforming to yin
syndrome 4.2.2.3.5.1
yang within yang 1.2.1.10

yang within yin	1.2.1.8	yin deficiency mixed severe heat pattern	7.4.3.13
yangming channel pattern	7.14.4	yin deficiency mixed sputum dampness blocking pattern	7.4.3.12
yangming disease pattern	7.14.3	yin deficiency mixed sputum heat pattern	7.4.3.11
yangming-fu organ pattern	7.14.5	yin deficiency moving blood pattern	7.2.5.3
yangqiao meridian	3.1.8.8	yin deficiency moving wind pattern	7.2.5.4
yangwei meridian	3.1.8.6	yin deficiency nose loss pattern	7.12.9.12
yellow urine due to renal meridian fever	6.5.13	yin deficiency of liver and stomach pattern	7.10.46
yin	1.2.1.1.1	yin deficiency of lung and stomach pattern	7.10.63
yin and yang mutually transform	1.2.1.2.4	yin deficiency of the lung and kidney pattern	7.10.58
yin and yang reciprocally root	1.2.1.2.2	yin deficiency of the spleen and stomach pattern	7.7.58
yin collapse	4.2.2.3.7.1	yin deficiency of water ring pattern	7.12.7.3
yin deficiency and blood dryness pattern	7.2.5.2	yin deficiency of wind ring pattern	7.12.6.4
yin deficiency and blood heat pattern	7.4.3.4	yin deficiency pattern	7.2.5
yin deficiency and external affections pattern	7.4.3.5	yin deficiency with blood congestion pattern	7.4.3.8
yin deficiency and inner heat pattern	7.4.3.6	yin deficiency with qi stagnation pattern	7.4.3.7
yin deficiency and intestine dryness pattern	7.7.46	yin edema	6.11.2.1
yin deficiency and loss of osmosis from ear pattern	7.12.8.9	yin excess leading to cold	4.2.2.3.1.2.2
yin deficiency and lung dryness pattern	7.6.6	yin exhaustion and yang collapse	4.2.2.3.7.5
yin deficiency and yang floating pattern	7.2.5.1	yin fluids deficiency leading to yang deficiency pattern	7.2.13.1
yin deficiency and yang hyperactivity pattern	7.4.3.3	yin fluids exhausting pattern	7.2.6
yin deficiency causing evil toxin detaining pattern	7.4.6.6	yin forming substance	1.2.1.1.6
yin deficiency causing heat detaining pattern	7.4.6.9	yin injury by blazing heat	4.2.6.2.2
yin deficiency causing inner heat pattern	7.4.3.1	yin meridian	3.1.13
yin deficiency epistaxis	6.2.9.4	yin qi	1.2.1.1.3
yin deficiency leading to fire hyperactivity	4.2.2.3.2.3.3	yin syndrome transforming to yang syndrome	4.2.2.3.5.2
yin deficiency leading to heat	4.2.2.3.2.3.1	yin toxin pattern	7.3.14.17
yin deficiency leading to yang hyperactivity	4.2.2.3.2.3.2	yin within yang	1.2.1.9
yin deficiency mixed body fluids stagnation pattern	7.4.3.9	yin within yin	1.2.1.7
yin deficiency mixed dampness heat pattern	7.4.3.10	yin yang rejection	4.2.2.3.6
yin deficiency mixed excess pattern	7.4.3	ying phase pattern	7.3.6.2
		ying-nutrient qi	2.2.1.8

yinqiao meridian	3.1.8.7
yinwei meridian	3.1.8.5
yin-yang	1.2.1.1
yin-yang points combination	9.1.19.7
yin-yang separation	1.2.1.6
yin-yang theory	1.2.1

Z

zang-fu	2.1.2
zhen bang（zhang）	9.2.2.12
zhen chui	9.2.2.11
zong-pectoral qi	2.2.1.7
zou suo zi	9.2.2.13